Florian Wiedmann

Post-oil Urbanism
in the Gulf

Florian Wiedmann

Post-oil Urbanism
in the Gulf

New Evolutions in Governance and the Impact on
Urban Morphologies

Südwestdeutscher Verlag für Hochschulschriften

Impressum/Imprint (nur für Deutschland/only for Germany)
Bibliografische Information der Deutschen Nationalbibliothek: Die Deutsche Nationalbibliothek verzeichnet diese Publikation in der Deutschen Nationalbibliografie; detaillierte bibliografische Daten sind im Internet über http://dnb.d-nb.de abrufbar.
Alle in diesem Buch genannten Marken und Produktnamen unterliegen warenzeichen-, marken- oder patentrechtlichem Schutz bzw. sind Warenzeichen oder eingetragene Warenzeichen der jeweiligen Inhaber. Die Wiedergabe von Marken, Produktnamen, Gebrauchsnamen, Handelsnamen, Warenbezeichnungen u.s.w. in diesem Werk berechtigt auch ohne besondere Kennzeichnung nicht zu der Annahme, dass solche Namen im Sinne der Warenzeichen- und Markenschutzgesetzgebung als frei zu betrachten wären und daher von jedermann benutzt werden dürften.

Coverbild: www.ingimage.com

Verlag: Südwestdeutscher Verlag für Hochschulschriften GmbH & Co. KG
Heinrich-Böcking-Str. 6-8, 66121 Saarbrücken, Deutschland
Telefon +49 681 37 20 271-1, Telefax +49 681 37 20 271-0
Email: info@svh-verlag.de

Approved by: Stuttgart, University, Diss., 2010

Herstellung in Deutschland (siehe letzte Seite)
ISBN: 978-3-8381-3365-2

Imprint (only for USA, GB)
Bibliographic information published by the Deutsche Nationalbibliothek: The Deutsche Nationalbibliothek lists this publication in the Deutsche Nationalbibliografie; detailed bibliographic data are available in the Internet at http://dnb.d-nb.de.
Any brand names and product names mentioned in this book are subject to trademark, brand or patent protection and are trademarks or registered trademarks of their respective holders. The use of brand names, product names, common names, trade names, product descriptions etc. even without a particular marking in this works is in no way to be construed to mean that such names may be regarded as unrestricted in respect of trademark and brand protection legislation and could thus be used by anyone.

Cover image: www.ingimage.com

Publisher: Südwestdeutscher Verlag für Hochschulschriften GmbH & Co. KG
Heinrich-Böcking-Str. 6-8, 66121 Saarbrücken, Germany
Phone +49 681 37 20 271-1, Fax +49 681 37 20 271-0
Email: info@svh-verlag.de

Printed in the U.S.A.
Printed in the U.K. by (see last page)
ISBN: 978-3-8381-3365-2

Preface

Since the end of the 20th century the governments of various Gulf States have been attempting to diversify their oil-dependent economies. This has resulted in a new type of urbanism, often referred to as post-oil urbanism. The first model of post-oil urbanism was the Emirate of Dubai due to its pioneering efforts during the 1990s when it initiated its economic transformation into a regional service hub by introducing open market policies. This liberalisation included the local real-estate market, which opened up for regional and international investment. This new strategy made Dubai a role model of exponential urban growth in the region. Hence, there has been widespread imitation of its development strategy by other GCC countries such as the Kingdom of Bahrain.

In the case of Bahrain the liberalised real-estate market has led to the transformation of its urban morphologies due to the construction of several mega-projects and hundreds of high-rises along its coasts. Subsequently, the population has increased from around just 660,000 inhabitants in 2001 to more than 1 million in 2008. In order to attract investment and sustain the speed of urban growth, restrictions have become more and more relaxed, leading to a new form of urban governance in which semi-public developers have become the main driving force. Thus, a growing number of developers have gained the right to design the individual master plans of their projects, leading to an urban development that is not guided by any overall plan but is instead the accumulation of individual case-by-case decisions. The speed and size of recent projects in combination with shrinking restrictions have resulted in decreasing urban qualities in certain areas, particularly due to a deficit of technical and social infrastructure. The recent development of Bahrain's capital Manama and its three main expansion areas is an example of the current urbanisation in the Gulf, where speculation-driven development has led to fast urban growth without sufficiently integrating the needs of liveable cities.

Today, the introduction of an integrated development strategy in order to encourage balanced and consolidated urban growth has become the biggest challenge of urbanism in the Gulf. While comprehensive regulations are needed, the pressure to attract investors forces a continuation of de-centralisation and liberalisation. Subsequently, urban governance in the Gulf is challenged to elaborate holistic plans that integrate various development strategies in addition to a system of implementation providing flexibility and effectiveness. Thus, urban planning in the Gulf has reached a turning point in which the outdated preparation of local master plans is being replaced by holistic visions based on economic diversification strategies. The recent outbreak of the global financial crisis has put additional pressure on current post-oil urbanism to define its way between the establishment of 'event hubs' and 'sustainable cities'.

1

Contents:

I
Background: From Pre-oil to Post-oil Urbanism

1. The Urbanisation of the Arabian Peninsula

1.1. The Islamic Oasis Town as an Indigenous Type of Settlement

1.1.1 The Socio-economic Conditions before the Oil Production
The Arabian Peninsula, with a size of more than three million square kilometres, is the world's biggest peninsula and geologically, it belongs to the old African continent, from which it is divided by the Red Sea. Tectonically, however, the peninsula is located separately on the Arabian Plate. In addition to its coastal borders in the west, south and east, the edge of the so-called 'fertile crescent' marks its geographical border in the north. Along the coast of the Red Sea, the so-called Tihana, a 20- to 40-kilometre wide area of lowland, ascends toward the west to a highland, which reaches up to 3,000 metres in height[1]. Another mountain area exists in the south-eastern part of the peninsula where the Arabian and Eurasian plates join not far from the coast. The centre of the peninsula is occupied by a plateau called Nejd as well as large deserts such as the Nefud in the north and the Rub Al Khali in the south. The peninsula descends from the higher regions of the west down to the lower regions of the east into the Persian Gulf. The annual rainfalls are small and amount to over 170 mm a year in certain regions only, such as mountain areas. Furthermore, temperatures can reach up to 50° C in the summer months and along the coasts the humidity, at over 80%, is very high[2].

These environmental circumstances have made the Arabian Peninsula one of the least populated regions in the world. Water, which is essential for any human settlement, was rare and in most parts of the peninsula it could only be found at oases, where millennia-old water reservoirs either provide natural springs or are close enough to the surface to enable the creation of man-made springs. In spite of the fact that there is no annual water-bearing river, the so-called wadis[3] have often been annual sources of water in the highlands because of the accumulation of underground water at certain spots. In some regions such as Oman, these river oases were the starting points of complex water management systems, the so-called falaj, in which small channels distributed water to widespread fields in the mountains[4]. Thus, the biggest areas for agricultural utilisation were found in highlands, where the annual rainfalls provided enough water, and at certain inland or coastal locations, where natural oases made the agricultural use of land possible. The most favoured agricultural crop was dates because of their durability, making them an important factor of survival in the desert. Apart from the settled farmers, a large part of the population lived a nomadic life as bedouins and moved from place to place with their herds and flocks. It was the camel and its suitability as a desert vehicle that enabled man to cross desert areas and find new fertile oases for further settlements.

In addition to nomadic herds and date plantations, fishing and trading were further sources of livelihood. Due to its geopolitical location between India, central Asia, Africa and the Mediterranean Sea, many ancient trading routes crossed the peninsula and went along its coasts. Consequently, many harbours were built and developed into important trading centres in the region, from where caravans carried goods through the deserts stopping at oases for rest and trade. At certain crossing points of trading routes, regional centres started to prosper and reached scales of more than ten thousand inhabitants. Not only merchants but also various kinds of craftsmen settled down and founded new markets. These market places generally developed centrally in areas of agricultural use or at harbours along the coasts. Since the beginning of Islam and its successful spread all over the Middle East and beyond, Mecca and Medina developed into the most important centres of Islamic culture. As traditional oasis settlements and market places, these two cities started to develop further because of their religious and thus political dimensions. Every year, thousands of pilgrims from various regions visited these cities, consequently developing their economies on the basis of these high numbers of visitors.

The family and its structure as a tribe was at all times the most important key factor of survival in a desert area. Its social network helped weak members to survive and its clear hierarchy with a tribal leader as the sheikh made for an effective organisation that defended common interests. The size and wealth of a tribe determined the amount of land that it had under its control. As result of the constant struggle to survive, many tribal conflicts and wars have occurred during the course of the history of the Arabian Peninsula and have led to the strong identity and segregation of tribes. While cross marriages between tribes were a common tool to unite and to create allies, a strong force of unity between the tribes has been Islam, which replaced the pre-Islamic religion of Arabic polytheism, the characteristics of which differed from tribe to tribe. The Quran was not only the foundation of a united belief, it was also a source of reform for society through its introduction of new laws and principles. The political consequence of the new religion was the installation of the caliphate as the government of all Muslims in the defence of the interests of the community and the spread of Islam. Since the beginning of its existence the caliphate was under the control of two major dynasties – the tribe of the Umayyads and the tribe of the Abbasids – until the Mongolian assault put an end to the Arab caliphate and the Ottoman empire with its Turkish hegemony occupied the peninsula in the 16th century[5].

1.1.2 The Traditional Form of Urban Governance

Through their struggle to survive and improve their living conditions in the harsh environment of the Arabian Peninsula, the people who settled there passed on from generation to generation the knowledge they gained regarding the best way of building settlements that were well suited to the area. This millennia-old attempt to survive in this part of the world led to the accumulation of a broad base of indigenous knowledge as to how to use the few local building materials to create shelter that not only provided protection from the climate but also a better standard of living. In some regions certain knowledge was exchanged and adapted. An example of this is the import of the concept of wind towers by Persian merchants, who were settling at harbours along the Gulf coast. Apart from the desert climate, the religion of Islam dictated clear rules and principles of building. The Islamic rule that non-related women and men are to be separated from each other and the introduction of the hijab[6] led to a strong sense of privacy. A stranger's view inside someone's house was made impossible through certain building regulations such as the heights of the houses[7]. Furthermore, the Friday mosque became the most important public arena for the community beside its function as religious centre. It was often used as a court or school, particularly in smaller settlements[8]. Its simple cubic form with an additional square, often enclosed by walls to form a courtyard, could be easily expanded according to the growth of the settlement. Thus, the size of the Friday mosque was often an expression of the number of inhabitants of an oasis town.

While residential houses were generally built individually by each family and with the support of related neighbours, the whole local community helped collectively to build public buildings such as the mosque. It was only in bigger towns that it was common for people to practice the profession of craftsmen as builders, the so-called mualims, who constructed public facilities and even private homes[9]. The process of town planning was a collective effort based on indigenous knowledge and culture instead of an idea or plan from a central authority. Land was usually distributed between tribes, clans and ethnic minorities[10]. In general, land that was not built-up was not owned by any individual and was free to use as long as the utilisation was not seen as harmful for the tribal community. Furthermore, the leading sheikh had to reconcile any kind of issue and therefore he was in charge of solving arguments about buildings and coordinating the allocation of land for public use, for example, mosques, cemeteries or markets[11]. His judgement was unchallengeable law and normally based on Islam. Therefore, traditional building laws are often found as part of fiqh literature[12]. These laws helped to clarify issues about streets and the locational restriction of uses that might cause harm such as noise or pollution. Furthermore, regulations regarding overlooking elements, walls between neighbours and their rights of ownership and use and even the drainage of rain and waste water were part of these laws[13].

1.1.3 The Typical Urban Structure of Oasis Towns

The oasis settlement was normally placed close to the oasis on land which had no agricultural value itself and often on higher ground in order to defend it more easily. A ring of walls made of mud, stones or rocks helped to protect the bigger settlements and towns from hostile tribes and sand storms. Urban growth was therefore often restricted to within the borders of this protective barrier and expansion beyond was only possible in directions where it did not harm fertile cropland. Large gates on various sides of the

Fig. 1: The historic Souq in Riyadh.

Fig. 2: Adobe architecture.

Fig. 3: A gated neighbourhood.

Fig. 4: A traditional sabat.

settlement were often the only entrances and exits of the settlement. The main streets, made crooked by the haphazard positions of irregularly placed individual houses and buildings, ran from these gates and normally met each other in the centre of the settlement, which was often formed by the ensemble of Friday mosque, court and palace. Along these roads, which had to be the width of at least two packed camels[14], the traditional market, the souq, extended in a linear fashion, often sheltered by roofs. The souq was strictly divided into various sections such as the gold souq, cloth souq or spice souq[15]. Animal trading was usually conducted outside of the settlement in front of the gates, where bedouins settled temporarily during certain months of the year. The whole public space of oasis towns was restricted to the market and the mosques with their multiple use as the centre of the community. In bigger towns, bath houses and tea houses were common public meeting points for men and the only public leisure space. Parks and representative public squares have never been part of the Islamic oasis culture on the Arabian Peninsula[16].

Furthermore the Islamic oasis town was generally dominated by a strong segregation of public and private life, wherein the area of private housing occupied the most land. Smaller alleys led from the main roads to the private homes of the oasis population[17]. These narrow labyrinthine streets usually had to be at least the width of one packed camel while their height also defined how low the so-called finas[18] could be constructed[19]. The built narrowness had two major purposes – on the one hand, to save land within the settlement and on the other, the shadows of the buildings were used as natural protection from the sun for the streets and the walls of neighbouring houses. Apart from these functional purposes, the network of narrow side roads and culs-de-sac emphasised the private character of the neighbourhoods, known as ferej[20]. These urban cells were developed by a system of branching side streets, which ended in an arrangement of houses of related clans[21]. In some cases, these neighbourhoods were even protected by gates from the rest of the town. Thus, traditional oasis towns were strongly segregated according to tribal, ethnic or religious differences among groups. Beside the fact that most land was used for housing and was owned privately, small public centres with mosques and markets were often built as the core of neighbourhood areas. The so-called majlis[22] were used by related families to meet for religious purposes but also to discuss issues concerning the community.

The architecture of oasis towns was generally vernacular and uniform because of the use of the same building rules, materials and techniques in the construction of the buildings, resulting in the typical typology of this type of settlement[23]. In addition to the traditional courtyard house, which formed the most common housing typology, simple cubic buildings were often built in rural settlements. The height of houses was standardised and often restricted to two floors in order to prevent overlooking into private homes. Bedouins constructed temporary one-storey houses on plots at the outskirts of towns, which were surrounded by fences or walls. Courtyard houses provided not only a protected open space for private use but also a better supply of air ventilation and light in the narrow built settlements. Furthermore, the flat rooftops were important as open-air areas that the family could use for cooking or sleeping[24]. The ground floor, which normally had very few window openings because of statical reasons and privacy issues, was often used as a storage space and private majlis, where guests were entertained. Only in the hot summer months were the ground floors favoured as housing space for the family, which usually used the first floors as a private living area in addition to the rooftops. In some towns it was common for first floors to be extended over the street to the opposite neighbouring house. Such a room bridging a street was called a sabat and it could greatly increase the private living space[25].

The architecture was generally determined by the building materials that were available in the local region. For instance, in the case of settlements along the coast, where apart from sun-dried adobe, coral stone and gypsum were often used for constructing walls. Furthermore, poor families often lived in barasti huts, which were simply constructed using palm fronds[26]. More inland, adobe, deposits of which could be found along the wadis, was used as basic building material for walls and ceilings, which were carried by beams made of palm trunks[27]. Adobe was not only in plentiful supply as a local building material, it also improved the indoor climate because of its good natural insulation properties and ability to absorb air moisture[28]. Apart from the use of adobe as a building material, constructive elements such as openings helped to cool down the indoor temperature by increasing natural ventilation. In the wall just above the ground and just below the ceiling, small openings helped to maintain constant air movement and exchange[29]. This system of natural ventilation through constructive elements was brought to perfection by the invention of wind towers, which were up to 15 metres in height and had at least two separate chambers, one for catching the wind and one for releasing the air[30]. Although the

11

architectural design was mainly dominated by such adjustments to outer circumstances, there was a widespread use of elements such as wooden screens, geometrical block-outs or crenellated roofs to decorate the façades. These decorative elements could differ from region to region but were uniform within one town. The public expression of the individual through façade architecture was not part of the traditional Islamic community, which had a strong commitment to its unity.

For over 5,000 years, harbours were built for trade along the coast of the Persian Gulf due to its geopolitical location[31]. In the north, the emirate of Kuwait developed itself over centuries to become one of the most important harbours of the region, which attracted colonial interests as did many other coastal settlements. During the 16th and 17th centuries, the Portuguese built many forts along the coast and occupied important harbours to protect their colonial interests and trading routes, particularly toward India. From the 18th century onward, the British, represented by the East India Company, increased their political influence in the Gulf and opposed the Ottoman empire by offering contracts to the Arab tribes[32]. Like the Portuguese, the British wanted to protect their trading routes and did not intend to settle in the peninsula itself. Thus, there was no remarkable change in the traditional built environment due to colonial influence in the Gulf as there was, for example, in North Africa or elsewhere in the Arab world.

On the island of Bahrain, such influence was initially Persian, which had been growing since the defeat of the Portuguese at the beginning of the 17th century, but only about 100 years later, the Persian hegemony was put to an end by the Arab tribe Al Khalifa, which was able to sustain its power against the Persian empire because of the protectorate of Great Britain[33]. Apart from agriculture and fishing it was trade, particularly of pearls, that made Manama and Muharraq important regional harbours. Apart from smaller market places such as Doha on the Qatari peninsula, there were several coastal villages and harbours such as, for example, Abu Dhabi, Dubai or Sharjah, along the southern coast of the Persian Gulf. These emirates were famous as the so-called 'Trucial States' since Great Britain had concluded contracts with all seven emirs in 1820[34]. At the end of the 19th century, Dubai was growing into a big harbour town when Persian merchants emigrated to the small emirate and helped build up the trade of pearls, which became a lucrative regional product for the global market. The Omani capital Muscat, located beyond the Strait of Hormuz in the Gulf of Oman, was a centre of trade for thousands of years because of its fortunate geopolitical location in relation to India and East Africa. Furthermore many oasis settlements were founded inland in the Hajar mountains, for example, Al Buraimi, Al Ain or Al Nizwa. Along the southern coast of the Peninsula, several small agricultural settlements or fishing villages began to develop. In the south-west, the town of Sanaa, which had about 20,000 inhabitants at the beginning of the 20th century, was the biggest oasis town in the region[35]. Surrounded by several smaller settlements in the middle of the Sarawat Mountains, it was the urban centre of the most fertile agricultural region in the entire Arabian Peninsula.

Beside smaller villages along the coast of the Red Sea, Jeddah developed into the central

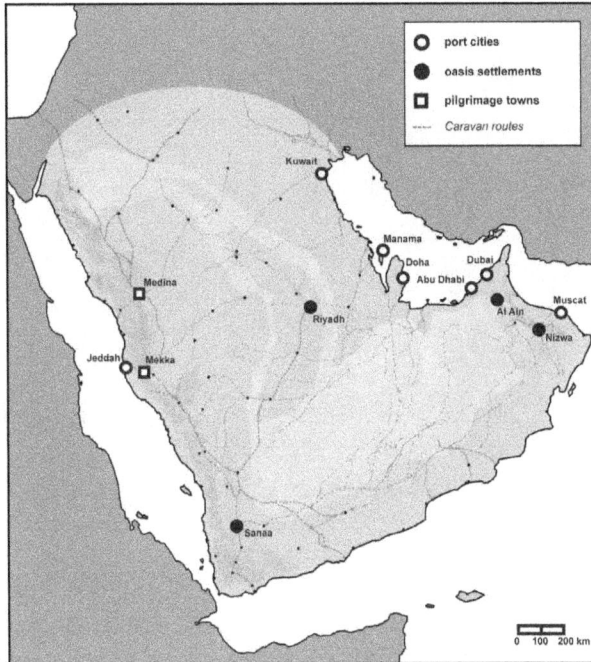

Fig. 5: The major pre-oil settlements.

harbour in the west as it was the arriving point for many Muslim pilgrims heading for Mecca. Although the western part of the peninsula from Jerusalem to Sanaa was under the control and administration of the Ottoman caliphate from the 16th to the beginning of the 20th century, the influence of the Turkish rulers on the built environment was rather minor apart from the import of certain building materials and techniques such as dome architecture[36]. The most important cities in the western part of the Arabian Peninsula were Mecca and Medina because of their religious and therefore decisive political role. Many smaller settlements were founded in the western highlands and the central plateau, where Riyadh evolved as a flourishing oasis town at the joining point of two caravan routes, one of which connected the coasts of the Persian Gulf and the Red Sea[37]. Like many other towns, it developed over centuries and changed size, expansion and location. At the beginning of the 20th century, about 20,000 inhabitants lived in the town, which was home of the abundant tribe Al Saud[38]. Located close to two wadis, Riyadh had a plentiful supply of water for agricultural use, namely, large date-tree plantations. The settlement was surrounded by walls that were eight metres tall with nine gates and occupied an area of about one square kilometre[39]. Furthermore, the centre was formed by the Al Masmak palace[40], the Friday mosque and the courthouse Qasr al Hokum and the local adobe architecture featured the typical regional characteristics of small triangular block-outs in the walls and crenellated roofs[41].

The economic basis of pre-oil settlements was oasis agriculture, fishing or trade and the

13

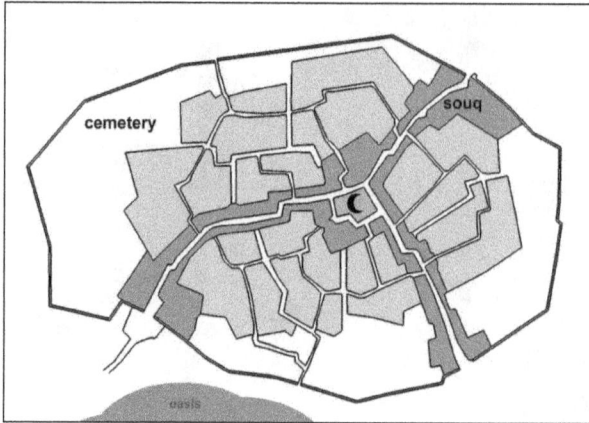

Fig. 6: The typical structure of an oasis settlement.

basic key to survival was strong tribal organisation with one leading sheikh. All knowledge of building was based on the experience of previous generations, who tried to survive and improve their living conditions. Thus the rules of building have been dictated mainly by the desert climate and Islamic culture. In cases where these rules were violated, the sheikh had the authority of arbitration. Furthermore, it was his decision as ruler to decide the general distribution of land on a macro-level[42] and to allocate land for public areas and buildings such as cemeteries, markets and mosques. It was also the ruler's initiative to start revitalisation activities during eras of security and prosperity[43]. The common rule that unbuilt land was free to use on a micro-level based on the individual initiative of an inhabitant led to a collective building process with strong participation on the part of every family. Intervention in this process by the ruling authority only occurred if such building involved some sort of harm for other inhabitants. The oasis settlements were generally built on infertile land and often on higher ground with a surrounding protective wall, which largely determined the outer form. Moreover, the centre was dominated by the mosque, which functioned as a public arena for religious and political meetings, as well as the representative palace of the sheikh. Apart from the mosque, the souq was the only other public space while private housing occupied most of the urban area.

The main roads lined with souq shops did not follow any plan but they were defined by a minimum width while the direction and pattern of their lines developed over time in accordance with as and how buildings were erected. Because of the collective manner in which building took place without the planning of a street grid, the way buildings were organised in order to provide protection from sandstorms and a wish to create closeness and intimacy against the wide surrounding desert landscape, long linear roads were rare. Thus, the side streets, which led to private homes, were generally narrow and labyrinthine in order to protect and preserve the privacy of families. Every family or clan settled separately and therefore tribal, ethnic and religious segregation was a basic characteristic of oasis towns. Furthermore, typology and architecture were normally consistent within

14

Fig. 7: The vernacular morphology based on courtyard houses.

one settlement, wherein the courtyard house was usually the smallest building block. Its open-air court and rooftop were important living areas for the family, whose privacy was protected from outside views due to standardised heights. The architecture itself usually made use of local materials such as adobe and incorporated structural features intended to accommodate the hard climatic conditions, for example, wind towers. Because of a collective mindset and Islamic traditions, buildings had a unified exterior design within one settlement, which emphasised a feeling of affiliation and unity. Consequently, the vernacular oasis town was formed by the rules of the desert environment and Islam and thus it became a fundamental part of the landscape and culture. Humans had to adjust their living conditions according to the nature of the given surroundings and could not afford to cause any harm to the fragile desert ecosystem which they profoundly depended on.

1.2. The Oil City as the Product of the Oil Boom

1.2.1 The Socio-economic Transformation during the Oil Production

Due to a high deposit of sedimentary rock, huge amounts of oil accumulated about 100 million years ago along the Persian Gulf and its coasts. Oil production first began on the Arabian Peninsula in 1938 when oil fields were discovered in the Eastern Province of Saudi Arabia and on the island of Bahrain during explorations conducted by the British. No other region of the world is so rich in oil as the Gulf as about 50% of all currently known oil resources are located there[44]. In addition, processing this oil is simple and fast because the deposits are close to the earth's surface. Although the first oil fields to be exploited were already allocated to British and American oil companies in the 20s and 30s, the oil boom and, along with it, inevitable change in the economy and society only began after the second World War[45]. The largest single oil field called Ghawar, with a

15

length of 240 and a width of 35 kilometres, was found in the Eastern Province of Saudi Arabia, which has remained in possession of the largest amount of oil resources of all the Gulf states until the present day[46]. Since the 60s, more and more offshore oil production started to be carried out along the coast, from which the smaller Gulf states in particular benefited profoundly. While the oil fields in Oman and Bahrain were comparatively small, large oil resources were detected on the territories of Kuwait and the emirate Abu Dhabi within the UAE[47]. In addition to oil, huge amounts of natural gas was found, particularly on the Qatari peninsula[48].

At the same time as the oil boom began, the Gulf states began to develop into independent nations. The precise area and borders of the countries' territories were established over the course of the middle of the 20th century. At the beginning of the century the western part of the Arabian Peninsula and thus the majority of the population had been still under the reign of the Ottoman empire. Only the harbour towns along the trading route to India had been under the protectorate of Great Britain. When the Ottoman empire lost the first World War on the side of the German Reich in 1918, the tribe Al Saud succeeded in uniting several Arab tribes under its umbrella to free the peninsula from Turkish hegemony. The Allies, in particular Great Britain, supported this struggle for independence, which was also made possible by the spread of the Islamic reform movement known as 'Wahabism'[49]. In 1932, the kingdom of Saudi Arabia was founded under the leadership of the tribe of Al Saud. Nevertheless, the attempt to unify the whole Arabian Peninsula under the flag of Saudi Arabia failed due to the contracts that were maintained between the leading tribes that lived along the Gulf coast and Great Britain as their protectorate. Although the territorial demands of Saudi Arabia have not stopped over decades, the small Gulf states were founded. In 1961, Kuwait became the first emirate to become independent, followed by Bahrain, Qatar and the UAE in 1971 after their attempt at a possible unification failed. The UAE itself came into being through the consolidation of the seven smaller emirates known as the 'Trucial States', which are located along the southern coast of the Persian Gulf, of which Abu Dhabi (almost 80% of the territory), Dubai and Sharjah constitute the leading political, economic and cultural centres[50].

The Sultanate of Oman in the south east is the only sovereign country in the Gulf that has a rather long history of independence. While Yemen was until 1990 divided into the national republic of North Yemen and the socialist republic of South Yemen, all the other Gulf states have been monarchies without initially any constitution but nevertheless with strong economic and political ties to the West[51]. The monarchies were built on the tribal hierarchy of each country and therefore future rulers have usually been decided by succession. While Saudi Arabia, Qatar and Oman have developed as more or less absolute monarchies, Kuwait, Bahrain and the UAE can currently be considered as semi-constitutional monarchies, although the participation of the inhabitants has been generally very limited. The sole right that these populations received was the right to vote a representative parliament, whose political influence was normally limited to an advisory function. Neither the foundation of opposition parties nor the creation of labour unions have been made possible in order for these reigning monarchies to retain control. Despite various limitations, Yemen, as an Islamic republic, can be regarded as the only democratic political system on the Arabian Peninsula. Economic and political cooperation among the Gulf countries did not use to be a strong political priority until Iraq invaded Kuwait, following which all the Gulf countries apart from Yemen joined the newly founded Gulf Cooperation Council (GCC) in 1992 in order to unify foreign policies and defend common interests[52]. However, economic cooperation and the coordination of domestic policies never developed in the manner of the example of the European Union that the GCC modeled itself on because of old tribal rivalries and a general wish among the GCC members to retain their autonomy, particularly in the case of the smaller countries.

In the connection with the oil boom, a process of industrialisation began in the Gulf states, which was initially limited to representative and generally oversized projects. In addition to aluminium and copper converting industries, numerous dry docks and petrochemical plants have been constructed. Furthermore, the building industry required vast investments from the states, local entrepreneurs and shareholders. In particular, major infrastructure projects such as streets, energy plants, desalination plants, airports and harbours have been the focus of public investment[53]. After this first stage of industrial development, investment became more directed toward establishing industries, which helped to decrease the need for the import of basic commodities such as food, furniture and building materials. The industrial production of other, non-essential consumer goods followed as the last stage in the development of what was a relatively short industrial revolution in the Gulf, taking place over no more than three or four decades. Since its beginning, industry was mainly based directly or indirectly on the oil and gas production in the region. Almost all the other elements needed by industry had to be imported, for example, labour, various raw materials and licences[54]. Consequently, the industry in the Gulf often relied on public subsidies, which again derived mainly from oil exports. In conjunction with increasing oil production, the liquid assets of the Gulf states began to grow and along with this the development of the local financial infrastructure in the form of banks. Nevertheless, for a long time the Gulf states preferred to put their investments in foreign financial systems rather than their own. Regional financial centres have developed mainly in small countries or emirates such as Bahrain, Kuwait and Dubai. At the same time as the financial markets began to grow, trade itself started to flourish, particularly along the Gulf coast because of its fortunate geopolitical location, the establishment of import monopolies and the abandonment of taxes[55].

The transformation from a predominantly agricultural region to oil exporting national states changed the structure of society within a few decades. The urban population increased rapidly because of four main factors, namely, the massive immigration of guest workers, the move of rural populations to the towns and higher birth rates and lower mortality rates because of better living conditions. At the beginning of the oil boom guest workers came primarily from neighbouring Arab countries such as Egypt, Jordan and Syria. For example, guest workers formed 20% of the entire population of the town of

17

Riyadh in 1969[56]. Over the following decades labour was needed in the manufacturing as well as increasingly in the service sector. Consequently, millions of South Asian and South-east Asian workers, predominantly from India, Pakistan and the Philippines, have immigrated on the basis of short-term working contracts to the Gulf. In addition to lower paid labour, many Europeans, Americans and Australians immigrated to the Gulf to work as businessmen or consultants. The general preference of governments to focus on the development of their capital cities has led to an acceleration of urbanisation due to migration from rural areas. In the cases of some towns such as, for example, Riyadh, the capital of Saudi Arabia, the annual population growth has in some years at the beginning of the oil boom been more than 10%. In Kuwait it has even reached 18%[57].

The political attempt to make bedouin tribes settle down has erased the millennium-old nomadic culture on the peninsula. The general challenge of domestic policies was to increase national consciousness in order to become free of tribal conflicts and unify the nation. One method that was widely used was to appease sheikhs by giving them money or land. The end of the dominance of tribal structures led to the end of the traditional division of the population into nomads, oasis-farmers and townspeople, as well as fishermen and seamen in the case of coastal regions. The traditional horizontal differentiation of society into various tribes changed into a vertical differentiation based on income and a division between the local and expatriate population. While the equalisation of all native groups to nationals has led to the dissolution of the social tribal network, it has also opened the doors for the development of the individual without predetermination of profession and social rank. Although it was not possible to suppress favouritism completely between related tribes or clans, recruitment practices in public sectors, the allocation of land and access to public education have to a large extent been freed from tribal thinking[58].

In all six GCC countries the majority of the local working population has been employed in the public sector while the private sector has been predominantly occupied by guest workers, whose general intention has been to earn as much as possible within the short amount of time of their limited contracts. The manufacturing sector and the lower positions within the service sector have mostly been dominated by South Asian and South-east Asian labour, and due to their low incomes and the strict conditions of their working contracts they are not permitted to immigrate together with their families, who remain in their homelands relying on the salaries that are sent back. In some cases such as Dubai in the UAE, most jobs are occupied by foreign workers while the income of the local population on the other hand largely comes from the rent they receive by leasing accommodation to the continuously changing expatriate population[59].

1.2.2 The Introduction of Central Administration

The new order of national states led to the rapid development of governmental institutions, which had to regulate and administer the rapid urban growth. Because of the authoritarian political order, the government and administration has been developed from above with a clear, centralised structure. Therefore, the King or Amir and his ministers held the greater responsibility for all decisions made regarding the country's development. It has often been the ruler's initiative to develop new strategies such as five-year plans, budget plans or even single projects and new laws in cooperation with the Council of Ministers, which has always been directly appointed by the ruler himself. The ministries themselves have developed plans and programmes within their own planning departments while the final decision about any plan or programme has generally remained in the hands of the Council of Ministers. In many countries, including, for example, Saudi Arabia, a state-run Central Planning Organisation (CPO) had the responsibility of consulting ministries as to the technical aspects of the realisation of their plans. Furthermore, it was usually the CPO that developed socio-economic studies as the basis for certain planning projects and it normally performed the function of coordinating and acted as consultants for the development of any five-year plan. The Ministry of Finance usually had the important role of determining whether projects being planned were financially feasible. In some cases it was in charge of generating all economic statistics[60].

In Saudi Arabia, the Ministry of the Interior for Municipal Affairs, the later MOMRA[61], was put in charge of physical planning itself. It was responsible for spatial planning at the national, regional and local levels in addition to the provision and management of infrastructure[62]. Below MOMRA, there were general directors, each in charge of the municipalities of their respective region, as well as regional directors in charge of town planning who were responsible for so-called 'guide plans', which were done in the 50s and 60s and were valid for about 20 years[63]. In addition, the regional director of town planning fulfilled the task of cooperating with consultants and coordinating studies and projects. The municipalities took care of building permits and were also in charge of initiating the construction of public buildings and infrastructure projects. In 1977, the municipalities gained more financial and administrative independence and were put under the direct control of MOMRA[64]. The responsibilities of the ministries were divided according to different sectors such as, for example, industry and business, agriculture and water, finance and macroeconomics, labour or communication[65]. If a ministry developed an idea for a project, it would usually have to be analysed by means of a preliminary feasibility study. After an open discussion and declaration of intent, a second feasibility study would have to follow before the CPO would start to inspect the project. After this examination the Council of Ministers would listen to the proposal and the contract negotiations would start. Finally, engineers would design the project and the construction would begin[66]. This standard procedure only concerned large projects of economic importance for the public or private sector. Small-scale projects such as single buildings had to follow the general building code and laws in order to obtain building permission from the local municipality. During this time in the 70s, new building laws were established in Saudi that were mainly based on similar rules and regulations in effect in neighbouring Arab countries where they had been developed over a longer period of time under European influences. Traditional building laws were not included in this new legal framework due to their

19

divergent concept of town building using a different form of organisation and traditional construction methods. Laws were generally introduced to solve predominantly technical problems such as the development of a sufficient infrastructure as well as to support the private ownership of land in the form of lots. For example, in the case of land to be sold, the general rules in Saudi Arabia in the 70s insisted on the importance of specifying the precise area of such lots of land in accordance with official plans. With regard to the heights of private homes, the latter were restricted to 8 metres. Furthermore, there were regulations as to what percentage of one lot could be built upon (60%) and what distance a house had to be from the street (1/5 of the width of the street and no further than 6 metres)[67]. In addition, it was forbidden to build on land which was reserved for future public use. Part of the legislation provided for the establishment of municipalities that were to have jurisdiction over the approval of land issues related to private ownership, which, from an administrative point of view, was dealt with separately from the granting building permits[68]. These initial building laws were generally limited to strict but simple rules in order to cope with the development of infrastructure and the creation of lots for private or public use in the context of the rapid urban growth taking place and the import of new building types and construction techniques.

1.2.3 The Historic Development of Oil Urbanisation

In conjunction with the economic and social transformation taking place, a new way of living was introduced due to the import of a new form of mobility. It was the car which essentially changed life because it made it possible to move quickly in an air conditioned environment. British and American oil companies introduced this new way of mobility and, in doing so, also introduced a new type of housing when they built the first settlements for their employees. One of the first of these oil settlements was Aramco[69], a settlement in the Eastern Province of Saudi Arabia that was built in 1938[70]. The general typology used was the detached one- or two-storey villa on a square plot within an orthogonal grid of streets, organised so as to provide every single house with access to the grid by car[71]. Supermarkets or public buildings such as schools were built along central roads. These oil settlements initially developed separately from traditional settlements and thus there was almost no economic and personal interdependence between traditional and modern settlements. In the case of the Aramco oil settlement, it later grew to become the town of Dhahran, containing the headquarters of Aramco, a university and an airport[72]. Many other oil settlements never developed into towns and remained suburban housing areas[73]. After the Second World War and the rising oil boom, the new forms of mobility and housing became the new standards of living for the majority of the local population within only a few decades.

Particularly in the 50s and 60s, many major infrastructure projects were carried out by most Gulf countries, for example, road networks and harbours. During this time, the planning instruments used were limited to the guide plans previously mentioned, which

20

were in effect reduced land-use plans (concerning, for example, use, number of storeys or density of housing) that were difficult to implement because of a lack of an effective legal framework for their enforcement. When the process of oil urbanisation first began, the traditional oasis towns were starting to grow beyond their former clear borders that were outlined by walls. These were demolished because they were no longer needed as protection. In some cases, the old settlement was completely replaced by an orthogonal street grid and square building lots such as in, for example, the city of Abu Dhabi. In most cases however the substitution of old structures was less radical. In some towns such as, for example, Manama, Riyadh and Dubai, parts of the old street structures continue to exist because old buildings were simply replaced by new ones on the existing lots without necessarily changing the street layout. The modern streets were built on top of the old road system but were broadened and straightened. These new streets connected the towns and changed the traditional built environment from a compact and densely built unity into an urban landscape divided by streets into blocks. The new system was based on a grid of large main roads and streets that enabled every single block to be accessed by car. The consequence of this has been the disappearance of the traditional system of main roads and labyrinthine sidestreets that constituted the densely built housing areas with their traditional courtyard houses, which have been the victims of both modern building techniques and the absence of maintenance.

In the context of the intense immigration of guest workers and the exodus of the local population to the outskirts of towns, the old centres became dense areas of mixed use. The old markets along the former central main roads were often replaced by modern buildings. At the fringes of the centre, large apartment blocks were constructed mainly by private companies to house their workforce. Apart from the urban centres and their peripheral areas, guest workers were accommodated in large working camps that were often located close to the industrial areas in the outskirts. Because of the vast immigration from rural areas, many families had to live temporarily in shanty towns on the fringes of large towns such as, for example, Riyadh, where such settlements were built out of cement blocks, cardboard and corrugated metal, often surrounded by enclosures of livestock such as sheep, goats or camels[74]. While most of the poor population consisting particularly of immigrants moved into the old centres and its fringes, the middle and upper classes were moving into newly developed suburbs. The new development areas were often defined by the general direction of urban growth caused by the topographical situation and certain infrastructure projects. The new 'central business districts' (CBDs) were often positioned linearly along large roads due to the convenience of easy access by car and the capacity for future, steady expansion. In particular, administrative and office buildings, shopping malls and banks were built along those roads, which often led from the city centre to airports or large highways that further connected economic centres. Examples of such roads are the old airport roads in Dubai and Riyadh that connect the city centres to the airports. However, the old traditional markets in the centres have nevertheless continued to function as business centres for particularly lower-income groups.

Fig. 8: Satellite image of Riyadh.

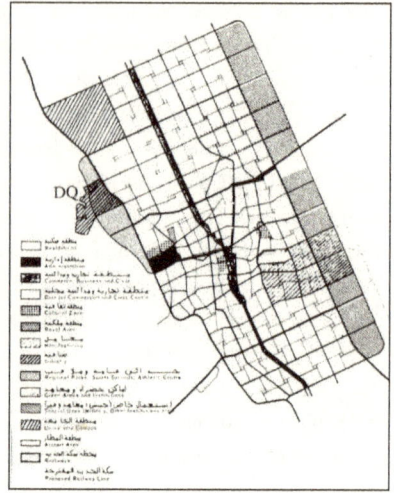

Fig. 9: The masterplan by Doxiadis.

New housing areas for the middle and upper classes were often built several kilometres away from the old centres. One of the very first modern housing areas to be built in the suburbs was the Al Malaz project in Riyadh, which was built during the 50s. When government ministries were moved to Riyadh in 1954, a 500-acre satellite suburb was developed in order to house the state employees. In this way, 754 single houses and 340 apartment units as well as supporting facilities were created[75]. This project marked a turning point away from the previous laissez faire attitude and led to the institutionalised and standardised use of the grid and, in particular, the detached villa as the dominant housing typology, which was preferred by the local population[76]. From then on suburbanisation continued to develop due to the widespread subdivision of land into standardised plots within a built network of orthogonal streets. Governmental support in the form of land provided for free or favourable long-term loans helped to reduce the exponentially growing need for housing. A factor hindering the rapid supply of lots was speculation with unbuilt land because of its quickly growing value[77]. The consequence was a rapid urban sprawl with a built urban area of very low density. The growing problem of managing this development trend led to the government initiative of replacing outdated planning methods with the introduction of master plans in the 70s. These master plans were developed by foreign consultants but because of a lack of appropriate local regulations they were difficult to implement, particularly with regard to the aspects of the planning that were intended to cater to the needs of whole communities, for example, parks[78]. Nevertheless, the first generation of master plans had an impact on the built environment in certain respects, particularly in relation to the road grid[79].

An example of one of the first master plans in the region is the 1973 Doxiadis master plan for Riyadh. The Greek consulting company Doxiadis Associates International developed a plan with a linear growth concept as the basic development idea. A so-called 'central

22

spine' was introduced as the new CBD extending in a north-south direction avoiding the encroachment of the city onto the Wadi Hanifah in the west. This master plan had to be extended over time in order to accommodate an ever growing urban sprawl driven by land speculation. A city-wide gridiron network of highways circumscribed 'super blocks' of 2 by 2 kilometres in size and land use was divided according to function, leading to a lower-income, service oriented, industrial area in the south-west set against an upper-income, residential, commercial and administrative area in the north-east[80]. In addition, the plan regulated zoning, which included definitions of the type and density of residential development, for example, minimum lot sizes. Some elements of the plan such as the design of individual 'action areas' in order to revitalise the city's centre initially had less impact on development than the new modular order of the grid and its equally sized plots or the functional division of land use. The plan was projected up to the year 2000 but due to an underestimation of the annual population growth, the plan had to be reviewed as early as 1976[81], when 250 square kilometres of land were added for further subdivision. The 80s and 90s were marked by a growing, leapfrog rate of development, which included the expansion of commercial strips and scattered, low-density housing developments[82]. Neither the master plan nor the 'urban growth boundary' (UGB) policies of the 80s could prevent the continuing urban sprawl in almost every direction, including the area of the Wadi Hanifah in the west[83].

The oil boom has led to the transformation of the former traditional built environment into a completely new type of town on the Arabian Peninsula – the so-called 'oil city'. The process of urbanisation related to oil has been particularly focused on the new capitals of the young national states and new economic centres located close to oil fields or harbours. The development of modern infrastructure has led to a rapid transformation process, wherein the old, dense town with its former clear borders has been replaced by an overwhelming, rapid urban sprawl. The inland or coastal topography has had a major impact on general land-use decisions. Furthermore, the direction of the wind has usually led to the establishment of industrial areas in the south and along with them poor residential areas and work camps have been developed. On the other side of the towns, large areas have been transformed into suburbs, predominantly for the local population and upper-income groups. The airports have developed into important regional and global hubs and therefore business as well as administrative buildings have lined up along main roads leading to new traffic junctures. Due to rapid urban growth, these early CBDs were later outshone by large, linear business centres along new growth corridors such as, for example, Sheikh Zayed Road in Dubai and King Fahd Road in Riyadh. Because of limited accessibility by car, the old centres lost their function as the biggest market place. Shopping malls in the outskirts soon become favoured, particularly by higher income groups. Due to their air conditioned environment and good location along main roads, shopping malls have developed to become not only new market places but the most important public arena and meeting point of society as well. Parks and public squares have been generally very rare due to the climatic circumstances and the initially low demand of

23

the local population for them as a result of the former urban culture that fostered a very restricted public life defined by Islamic rules.

1.2.4 The resulting urban structure and its characteristics

Generally, the oil city can be divided into three major areas – the old city core, the new CBD and the suburbs in the outskirts. The biggest part of the urban area of an oil city is occupied by the suburbs, which are typically structured within a system of streets and highways arranged in a strong geometrical grid. Apart from some public facilities such as schools, supermarkets and shopping malls, land use is predominantly occupied by housing. The most common typology used in oil cities is the two-storey villa on a square plot with walls. The old city core has normally remained as a business centre with the expansion and reconstruction of the old market for lower-income groups. In addition, most foreign workers have been accommodated in the city core or in fringe areas nearby, where multi-storey apartment buildings have often been built. As a result, the most dense areas of oil cities tend to be found in these areas. The mix of high-rise and low-rise typologies is generally very restricted, particularly in the suburbs due to privacy concerns. In addition to their function as a housing area for labour and their continued use as business districts, the city centres have often become the last remaining cultural identification points for society due to the more recent restoration of old mosques, palaces and souqs that has been carried out since the 90s. The main CBD is located outside of the city core and beyond its close fringes along main axes in the direction of new airports or economic centres, which has had a profound impact on the nature of urban growth.

Apart from the rather small city core, the density of the overall built area has generally been very low due to land speculation and the space occupied by streets and parking lots. While public transportation has only been introduced in the form of buses for lower-income groups, the car became one of the essential parts of everyday life, not only due to the climatic circumstances but also because of the nature of the functional division of land use, which involved people having to travel long distances between home, work, leisure and shopping. In addition, the disappearance of traditional neighbourhoods, where all the members of extended families settled together, has further increased the need for cars in order to be able to visit the often scattered family members in the suburbs. Developing urban centres by creating pedestrian zones has not been part of oil urbanisation mainly because of the dominance of shopping malls, which widely took over the role of public centres. The urban area was generally designed to accommodate the car and its needs while pedestrians have been rather neglected. Cheap petrol and the absence of taxes made cars affordable for the large majority of the population, leaving out only low-income groups, who have had to rely on buses, mopeds or bicycles. In the case of Saudi Arabia, where women are generally restricted from going out on their own without being accompanied by a male relative, women have not gained the permission to drive cars and thus rely on male drivers.

24

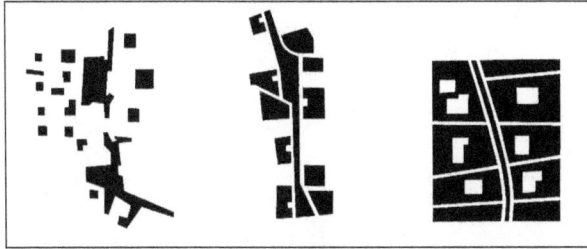

Fig. 10: The transformation from courtyard houses to detached villas.

Three major typologies can be distinguished in the oil city – the two-storey villa, the multi-storey block and the high rise. While villas are generally situated in the suburbs, blocks have been constructed in the core of the city and its borders while high-rise buildings have normally been restricted to certain areas such as the CBD. Villas are built with cement blocks in the middle of a square lot with surrounding walls that protect the occupants from outside views due to the need for privacy that has persisted in the context of the Islamic culture. The villa, which has replaced the traditional courtyard house, is typologically almost exactly its antithesis. Instead of the traditional courtyard, the house itself is the core of the lot surrounded by walls. Climatically and culturally, the new, imported typology was not ideally suited to the requirements of living in the Middle East but due to air conditioning and the use of high walls, the villas have been able to accommodate these requirements. Because of the missing courtyard, all activities take place within one block. Usually, a large salon forms the core of the ground floor as an alternative inner courtyard which connects to the kitchen and living rooms. One or two majlis remain as traditional living rooms where guests are entertained and separated between women and men[84]. At the back of the villa, often close to the kitchen, the maid is given a small room while the family's living area is located on the first floor. While the front façades of the villas are often designed with various ornamental elements that have a European or international rather than local origin, the other façades are often neglected from a design point of view. The local population mainly lives in such villas in the suburbs while guest workers with a higher income are generally accommodated in special compounds with walls and guards.

The first multi-storey block buildings were built in the 50s in order to house lower-income groups as well as to function as office buildings for administration or business. Apartment blocks were designed with a purely functional approach and consequently have the appearance of massive concrete buildings. While the ground floors are occupied by either parking or shops, the upper floors consist of apartments, often rented by companies for their employees. An open-air ducting chamber is usually placed in the centre of the building in order to supply bathrooms and kitchens with little windows and to be used as a convenient space for installations such as water pipes. Apartment blocks have been built either as stand-alone buildings on square plots or in groups side by side as a perimeter block development surrounding a central space that is used as a parking

25

Fig. 11: The suburban morphology.

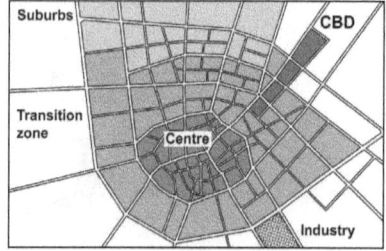

Fig. 12: The typical structure of an oil city.

area, as is the case in, for example, Abu Dhabi. In contrast to the predominantly technical design of many apartment and office blocks, public buildings have often been built with a design statement in mind using more representative architecture, commonly involving an attempt to integrate local ornaments. For a long time, such buildings have been the major landmarks and identification points of towns as have been large monuments placed on roundabouts until high-rise buildings took over this important representative role within cities. Due to the vast urban sprawl with predominantly low-rise villas, the new high-rise buildings have become important orientation and identification points. Not every town has granted permission for the construction of this new typology because of its interference with the privacy of estates and their subsequent loss of value. Nevertheless, places such as Dubai, but even Riyadh, did not want to refrain from constructing high-rise buildings, although areas of possible sites for these were initially very restricted to the CBDs only.

Oil urbanisation began in the 50s and was particularly intense during the 70s and 80s when growth reached new peaks and in most countries around one fifth of the whole population lived in towns[85]. Due to the political independence that they had already achieved and their vast oil wealth, Saudi Arabia and Kuwait could start their development earlier than Qatar and the UAE, which became independent in 1971. Saudi Arabia, the largest country in the Gulf with about 4 million inhabitants in 1950 and more than 22 million in 2000, developed three major urban areas during this time of growth in the 70s and 80s[86]. These three were, firstly, the Eastern Province, where because of the regional oil production three major cities developed – Dammam, Khobar and Dhahran. This large new cluster of oil cities had a population of about 552,000 in the 70s[87]. Secondly, the development of Riyadh led to it becoming the capital city and thus the administrative centre of the whole kingdom. Because of this new function, the city grew from about 80,000 inhabitants in 1950 to more than 4 million at the beginning of the 21st century, occupying an area of 1,554 square kilometres, making it the largest oil city in the world as well as one of the least dense cities[88]. It is a classic example of an oil city with vast suburban areas, a 'central spine' constituting a linear CBD and a city core with mixed use. Thirdly, in the western part of Saudi, Jeddah was developed from only about 30,000 inhabitants in 1940 to over 1 million in 1986. It is Saudi's second largest city and its most important harbour on the Red Sea[89]. Inland, Mecca and Medina expanded as cultural and religious centres of the region and of the Islamic world.

26

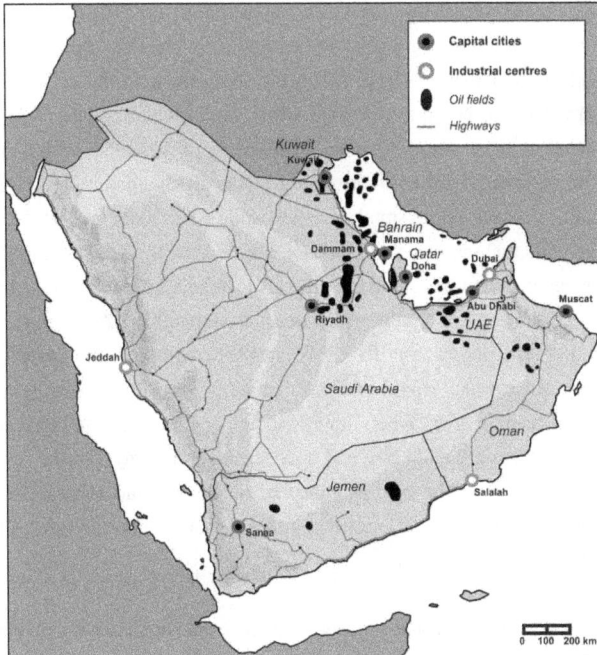

Fig. 13: The major urban centres during the oil boom.

Along the Persian Gulf six major oil cities developed in the second half of the 20th century. In the north, Kuwait grew from about 100,000 inhabitants in 1950 to almost 2 million in 2000. In contrast to many other cities, officials in Kuwait tried to prevent development from focusing on the capital, Kuwait City, and enforced the extension of the satellite towns around it. Due to vast demolition and a period of rebuilding, the capital's population dropped from 96,860 to 60,365 between 1961 and 1980 while the overall population of the emirate grew to 1,358,000[90]. The Gulf War in 1990 led to the emigration of about 500,000 inhabitants and the loss of Kuwait City's position as the most important harbour in the Persian Gulf[91]. In Bahrain, because of less immigration, population growth was slower in contrast to other countries, with the population numbering 61,726 in 1965 and growing to an estimated 148,000 in 1995[92]. At the end of the 20th century, local Bahrainis were the majority, constituting almost 75% of the population on an island of just 665 square kilometres[93]. Manama on the main island and Muharraq on its own island in the north are the main urban centres by virtue of being connected by bridges. In Qatar's capital city Doha, the population grew from just a few thousand inhabitants in the middle of the 20th century to over 340,000 inhabitants in the late 90s[94].

In the southern part of the Persian Gulf, Dubai, the second largest emirate of the UAE, became an important traffic juncture due to its early investments in infrastructure projects in the form of two harbours and an international airport in the 70s. The population grew from about 183,187 in 1975 to 826,387[95] in 2000, with the proportion of the local

27

population consistently decreasing, constituting just 30% of the whole population in the 80s[96]. Development in Dubai was carried out around the old centres Deira and Bur Dubai and the new harbour Jebel Ali in the south-west. In the north, the emirate of Sharjah spread to the borders of Dubai, which has resulted in one large urban agglomeration including the emirate of Ajman, the territory of which lies within the emirate of Sharjah. The capitals of the northern emirates Ras al-Khaimah, Umm al-Quwain and Fujairah have remained rather small oil cities while Abu Dhabi City in the south has, with more than 650,000 inhabitants in the late 90s, become the second largest city in the UAE and its development has spread beyond its own small island to the mainland[97]. In the inland of the emirate Abu Dhabi, which occupies more than 80% of the territory of the UAE, Al Ain, with a population of 120,000, has become one of the largest cities[98]. In the south-east, Muscat, the capital of the Sultanate of Oman, has been developed within the so-called 'capital area', where many settlements have been founded separately in the hilly terrain around the old city and connected by highways. In the south of Oman, Salalah was developed to become an industrial centre of Oman. Because of a lack of oil, internal political problems and long-term political isolation on the peninsula, there has been no oil urbanisation in Yemen.

The socio-economic change brought about by the production of oil that began in the middle of the 20th century meant the end of the traditional oasis town. The wealth they derived from the oil allowed the newly independent Gulf states to invest in the industrialisation and modernisation of their countries. In order to achieve the goal of catching up with the developed world, many infrastructure projects in the form of harbours, airports and roads were undertaken, marking the beginning of a process of modern urbanisation. This development focused mainly on the capital cities, which took on the new role of administration centres, and the industrial areas, which were predominantly either oil wells, oil related industries or harbours. In the context of this economic change, the former rural society became urban, but because they were lacking in education, guest workers were needed to achieve a rapid rate of modernisation. Low-cost labour has generally been recruited from South Asia and South-east Asia. Many immigrants from neighbouring Arab countries such as, for example, Egypt started to work in middle-class jobs as teachers, doctors or engineers, etc. Westerners have been the minority and have generally been engaged as major consultants. Consequently, the new planning organisation was based on an adaptation of foreign models with some minor adjustments. All major decisions were centralised, resting on the King or Amir and his close Council of Ministers. The ministries themselves had the responsibility of developing plans, which were definitively decided on from above, for all sectors, including urban development plans. After all, the municipalities had to administer the plans within their separate local departments.

The first consequence of the modern urbanisation was the loss of the compact form of the previous oasis settlements. Because of the exponential growth of the urban population and the introduction of cars, the majority of the locals started to settle in the outskirts while most of the foreign labour moved into the old centres. This has marked the beginning of a new kind of social segregation in the cities on the peninsula. In the 60s and 70s, governments started to replace the old approach of sectoral project-based planning with the development of new master plans designed by foreign consultants. These new master plans have been predominantly technical documents, creating a grid of hierarchical streets and establishing a strict functional division of land use. It was important that any plan was comprehensible and expandable in order to cope with the rapid annual growth. The consequence has been linear CBDs along main multi-lane roads and endless subdivisions of land into equally sized plots. The dominance of the two-storey villa as the housing typology has been mainly due to the demand of the local population, who refused to live in apartments. The sprawl of suburbia was in addition accelerated through land speculation and widespread public subsidies in order to cope with the increasing housing shortage. The need for privacy based on Islamic culture restricted the possibility of mixing different housing typologies, which has only been possible in the old centres and their fringes. The result has been an urban landscape with one of the lowest densities in the world.

Like the town planning, the architecture has followed the principles of Western modern design of the middle of the 20th century. The old traditional adobe architecture has been almost completely replaced by cement constructions. The only landmarks of the oil cities have mainly been public buildings such as ministries and roundabout monuments. In the 90s, many governments started to restore or rebuild certain buildings and areas such as old palaces and markets in order to preserve their cities' last remaining architectural heritage. The oil city as a whole can bee seen as almost the exact antithesis to the oasis town, and similarly, the villa to the traditional courtyard house. While oasis towns are among the most ecological cities in the world due to their sensitive adjustment to the desert environment, oil cities have become among the most unecological cities in the world due to the fact that they involve one of the highest rates of energy consumption per person. This high level of energy waste is closely connected to a vast waste of area because of urban sprawl and the functional division of land use into housing, working and supply areas. The consequence has been a high dependency on the car, which has become the most convenient mode of transport because of its air conditioning and free mobility. Furthermore, it has remained affordable because of cheap petrol[99]. That the car so rapidly became an essential part of practical life, as well as a status symbol, has made it difficult to introduce and develop public transport, which has generally been reduced to buses for lower-income groups. Another reason for the increasing energy waste has been the highly energy-consuming technique of desalinating water, which has remained largely subsidised by governments[100]. The third reason that oil cities have the biggest ecological footprint in the world is the buildings themselves, which have generally not been adjusted to accommodate the desert climate, which has caused extreme dependency on air conditioning.

In addition, the constant decrease in liveability of towns has been predominantly caused by the car, leading to traffic jams, pollution, high rates of accidents and the overwhelming occupation of urban land in the form of streets and parking lots. Another negative side effect of rapid urbanisation has been the society's extreme dependency on guest workers and the local population's growing dependency on public subsidies, which has led to an underdevelopment of human resources and the risk of increasing unemployment rates. The oil city was built within a few decades and it is one of the world's singular cases of a city that is basically an import. Oil wealth made it possible to buy labour, know-how, technologies and materials. The Western planning ideas of the 50s and 60s shaped the oil city into an exemplary car town with hardly any relationship to the previous urban form. The result has been the loss of the traditional identity within the built environment and a growing conflict between the two opposing cultural forces of Islamic traditions on the one hand and Western consumption industry on the other. The local population found itself caged within only a few decades in a modern environment without being an active participant in the development itself. The question of identity has been a rising conflict in addition to general concerns about the nature of the ecological, social and economical development that has been taking place. The foreseeable end of oil production, which is the fundamental basis of the oil city, would have to lead to a profound economic diversification, which again means a new transformation of the built environment and thus the creation of a new type of town – the post-oil city.

1.3. The Post-oil City as a Future Service Hub

1.3.1 The Initiation of Economic Diversification
It is an undeniable fact that the remaining oil wealth of the Gulf states will decline in the 21st century. All the GCC countries apart from Bahrain and Oman are members of OPEC[101], which they joined in the 60s. In Saudi Arabia, the oil peak is predicted for the year 2014 and in the emirate of Kuwait for 2013[102]. The Sultanate of Oman already reached its peak in 2000[103]. In the case of Bahrain, one of the oldest oil producing countries in the Middle East, the peak lies even further behind with 75,000 bbl/d in the 70s. Today, Bahrain is only producing less than half of that – about 35,000 bbl/d[104]. In the UAE, the emirate of Abu Dhabi has increased its possible production up to 1.8 Mbbl/d and the peak is expected in 2026[105]. The emirate of Dubai, the second largest oil producer of the UAE, has decreased its production since the 90s from 230,000 to 170,000 bbl/d and reserves are expected to be exhausted within the next 20 years[106]. Furthermore, oil production is decreasing in Qatar, which was 780,000 bbl/d in 2004 and is expected to be 270,000 bbl/d in 2020, while its liquified natural gas production is increasing to an expected 1.4 Mbbl/d in 2011, making it to the world's largest producer[107].
The impending end of the oil and natural gas reserves is leading to a profound impact on the economies and societies of all the Gulf states, which are still highly dependent on the export of fossil energy resources. At the same time, domestic demand for fossil fuels is exponentially increasing due to the high energy consumption demanded by, for example,

Fig. 14: The Sheikh Zayed Road in Dubai.   *Fig. 15: The Bahrain Financial Harbour.*

industries, car use, desalination plants and air conditioning in private and public buildings. During the late 80s, the Gulf states gradually fell into a situation of increasing foreign indebtedness due to inflation, the stagnating profits of oil exports and the high costs of major industrial and infrastructure projects[108]. Therefore, an attempt to diversify the economy has become more and more a part of national development strategies. Although efforts at building up an economy independent of oil were already being made in the 70s, major steps in post-oil development were undertaken during the 90s and at the beginning of the new millennium. Due to their limited oil resources, smaller Gulf countries and emirates such as Bahrain and Dubai began to invest in their post-oil future several decades earlier than other countries, whose wealth of natural resources engendered less pressure to rapidly develop economic diversification. Nevertheless, since the beginning of the new millennium, all the GCC countries have been developing oil-independent economies, particularly in trading, banking, tourism and high-end technology sectors.

Because of their declining wealth of resources, the Gulf states have started to invest in the development of economic sectors other than oil such as, for example, tourism and more traditional sectors such as trade. In the context of their history and geopolitical location, the Gulf states have been trying to establish their function as trading centres between Asia, Europe and Africa. Especially along the Persian Gulf many harbours have been built in order to create a new global trading hub. In addition to harbours, many international airports have been constructed in order to create an air cargo industry. The development of trade as an essential part of a future post-oil economy has been accelerated through the introduction of the concept of 'free trade zones' (FTZ) in the Gulf by the emirate of Dubai. In 1985, the first FTZ was established in Jebel Ali, which attracted many companies because of few taxes and modern infrastructure. Furthermore, very minor bureaucratic requirements and limited labour legislation increased the interest of international entrepreneurs and investors in building up businesses in Dubai. Over the following decades, several FTZs were founded in the emirate Kuwait, the state of Bahrain and particularly in the UAE. The size of FTZs, which have normally been positioned at airports or harbours, varies from single ports such as, for example, in Bahrain or airports such as, for example, in Sharjah to large industrial areas such as in Dubai[109].

In the context of the profits of oil production and the Gulf states' increasing function as trading hubs and market places, the financial sector has become a growing economic factor. In 1975, Bahrain overtook the role Beirut once had as a leading centre of offshore banking in the Middle East parallel to the rise of Kuwait and Dubai as important financial centres[110]. In the late 90s, 163 banks and 58 insurance companies made Bahrain the largest financial hub in South-west Asia. In Kuwait, about 150 finance companies have established their position in the international banking world[111]. The emirate of Dubai has become the most important financial centre of the UAE and in recent years it has gained major regional and global attention by outshining all its neighbours as the fastest growing investment hub in the region. One particularity of financial business in the Gulf has been the establishment of so-called 'Islamic banks'. In 1975, the Islamic Development Bank (IDB) was founded and the Dubai Islamic Bank became one of the first Islamic banks in the Gulf[112]. Although the most important centre of Islamic banking has remained in Kuala Lumpur (Malaysia), Bahrain, Dubai and Kuwait have recently started to expand this future-oriented banking sector[113]. The most important characteristics of Islamic banking are the prohibition of interest, speculation and gambling according to shariah rules[114]. Apart from creating their own banking system, all the GCC countries have been major investors, joint partners or shareholders in Western companies, real estate holdings, banks or states[115].

In some Gulf states the development of electronics industries, including, for example, communication and software production, has been part of the diversification process. The increasing attraction of the Gulf for high-tech producing companies has been particularly based on the ample supply of highly educated labour from the Indian subcontinent and the advantages of almost no taxes and few restrictions due to an absence of strict bureaucracy and labour laws. Another advantage in this respect has been the major effort expended to attract science and research industries to the Gulf by developing new universities and technology parks such as Education City in Doha. Furthermore, the telecommunications industry has been growing in recent years, especially in small countries such as Bahrain and Dubai, due to the exponential growth of this sector in the region itself, as is the case in Saudi Arabia, for example, which is planning to build up an information technology centre in the north of Riyadh[116]. Apart from the electronics industry, the development of the media has had a growing role in establishing the Gulf as a business area of global importance. The Gulf media began to gain worldwide recognition through Qatar when the broadcasting channel Al Jazeera was founded in 1996[117]. Over the following years, representatives of international media have increasingly settled in the Gulf. For example, CNN and the BBC have presences in Dubai Media City, the first media FTZ in the world, in order to use it as an important base in the Middle East[118]. Along with media, town marketing has begun to play a major role in increasing the prospects of another future post-oil industry – tourism.

Tourism has played a major role in the economies of the smaller Gulf states and Oman in recent years. In Saudi Arabia, international tourism has never been part of an economic development strategy due to the influence of conservative Islam on the society, the extent of the impact of which is greater than in all other GCC countries. The emirate of Dubai in particular has been developing since the 90s into an international tourist destination, offering high-class hotels, beaches and shopping possibilities. In addition to attracting international tourists from Europe, North America and Russia, Dubai has established itself as a major tourist destination for visitors within the Gulf region[119], with a growth from about 3 million[120] visitors in 2000 to over 7 million visitors in 2007 coming from the neighbouring GCC states. This high number of visitors from neighbouring countries can be attributed to the restrictions in these countries and general lack of entertainment and leisure in the region. Vast numbers of hotels, large shopping malls and entertainment parks have sprung up as a consequence of this growing regional demand. Apart from Dubai, which is the most major tourist centre in the Gulf, Bahrain has become another attractive destination, particularly for visitors from the Eastern Province of Saudi Arabia. Sports events such as Formula 1 in Bahrain and the Asian Games in Doha have marked a new step in development in terms of the provision of entertainment centres, attracting both regional and global attention. More recently, investments have been made in developing the Gulf as an international cultural destination. Examples of such investments are the Guggenheim Museum in Abu Dhabi and the opera house in Dubai[121].

The transformation from the oil economy, which has been based on the export of fossil resources and oil-related industries, to the diversified post-oil economy has gradually taken place over the last decades and can be considered in most countries as being still at the beginning. In this process, the evolution of society has involved local populations being challenged by changing conditions. Traditionally, the majority of the local inhabitants of the Gulf oil cities have worked in the public sector while guest workers have overtaken the private sector, particularly the service sector. Furthermore, the new types of economies have created a growing need for more educated and specialised workers from abroad. The lack of local workers in various economic fields has been due first of all to the speed and extent of economic development. In addition, continuing public subsidies have led to minor efforts on the part of the local population to play a dominant role in all economic fields, particularly the private sector. In many towns, the majority of the locals has become used to working in the public sector. Another factor has been a lack of education since the first universities in the Gulf have generally not been specialised in the teaching of typically post-oil professions. The result has been a growing unemployment rate in recent years, which has led to major public investments being made in the development of educational centres in order to integrate the local population into the future economies. Due to increasingly globalised conditions, local labour has started to face the challenge of increasing competition with foreign labour, particularly foreign labour from South Asia, who have been preferred by companies because of their education and willingness to accept hard working conditions and lower salaries.

1.3.2 The Decentralisation of Urban Governance

In order to devise new economic strategies, many countries and towns have founded new independent authorities such as, for example, the EDB (Economic Development Board) in Bahrain, the DED (Department of Economic Development) in Dubai and the ADA (Ar Riyadh Development Authority) in Riyadh. Consequently, the changing development goals have led to a new form of urbanism and a new role for cities in the Gulf. While rapid modernisation and the development of major infrastructure through public investments was given priority in the time of the oil boom, the new era of post-oil economies has been dominated by the growing impact of private investments in urban development. In previous decades, the participation of the private sector was limited to single projects located on plots created by monotone subdivisions. The new attempt to attract more investors has led to a new era of an investor-driven urban development in the Gulf and thus the creation of major real-estate developers. Furthermore, the legal introduction of freehold properties has become a growing factor in ever increasing speculation. The private sector's interest in investing in the building of cities in the Gulf has been particularly influenced by changing investment conditions in the West and high expectations of future revenues in the GCC countries as an up-coming global market. Regional investors in particular have started to invest in local projects instead of overseas ones. The effect has been a major push in urban development, particularly in Dubai followed by the neighbouring emirates and Bahrain and Qatar.

1.3.3 The Transformation of Urban Structures

As result of the growing role of the private sector in urban development, major developers have started to operate as managers of large-scale developments in the form of landmarks, housing areas, business parks and small cities within already existing cities or as satellites while the public sector has taken over the function of organising and developing the infrastructural supply of these projects. All decisions about the major planning of developments and the distribution of land have remained in the hands of the rulers, who have become direct or indirect associates of the developers. Although the planning authority has remained central, each real estate company has received more freedom to plan the master plans of their developments individually with only few restrictions. This new decentralised form of organisation between the private and public sectors has led to a more diversified urban landscape with mixed land use and typologies, which has begun to put an end to the former structure of the typical oil city with its clear functional land use division and one of the lowest urban densities worldwide. Like the social transformation, the change in the built environment began recently at the start of the new millennium and it is only in the cases of the UAE, Bahrain and Qatar that already built or almost completed projects have started to transform the built environment visibly. In other countries and emirates such as Saudi Arabia, Kuwait and Oman, this new urbanisation trend has just begun in the form of ambitious plans and projects.

The new development goals of post-oil urbanism have had an impact on both the urban structure and architectural development. Modern architecture has been imported to the Gulf since the 50s and has had a major impact on the outer appearance of the oil cities with its functional design and cement construction. Due to the interdependence of international architects and the traditionally minor participation of local architects, who have had a visible influence on architectural development in Saudi Arabia only, recent architecture in the Gulf has developed along post-modernist lines. One consequence of the more commercially oriented design has been the widespread integration of oriental forms and decorations in order to create a new connection to the local culture, diminish the problems of identity caused by the vast mixture of designs and last but not least create some kind of individuality. Architecture itself has started to become a major factor in town marketing and therefore an economic factor as well, such as, for example, the Burj Al Arab in Dubai, which has become a globally recognised local landmark. The new goal of urban design and architecture has changed in the direction of fostering international recognition and attracting investors and tourists, replacing the previous reduced role of representing the newly founded nations through public monuments.

The development of a post-oil economy has led to a new way of understanding towns as a future asset that can sustain and widen the economic prosperity of the region. The consequence of this has been an attempt to open markets and interweave with global business. The transformation of the built environment that is beginning is the direct expression of this economic change and the new trend of globalisation in the Gulf. While Dubai has been the major centre of this new tendency because of its introduction of methods for creating vast urban growth and international attention, other countries such as Bahrain and Qatar have either followed in Dubai's footsteps in recent years or, in the case of the remaining Gulf states, have only just begun to develop projects and strategies. Currently, the biggest and fastest buildings sites of the world are to be found in the Gulf, and the resulting transformation of the built environment has become more and more visible. The growing economic competition between the Gulf states has increased the speed and extent of the developments. In order to achieve this growth, planning organisation has been decentralised by granting more liberties to and imposing less bureaucratic restrictions on the private sector. The economic success of these private developers, driven by growing speculation and based on an enormous volume of investment, has increased land prices and the pressure of capitalistic competition in the whole region. Nevertheless, the current model of post-oil urbanisation has remained at the heart of widespread discussions about new controversial development strategies based on the fact that in addition to the issues of future identity and liveability, a conflict is beginning to rise between the new built environment and a growing need for ecological, economic and social balance in the Gulf region.

1 Heck and Wöbcke, 2005, p. 15.
2 Heck and Wöbcke, 2005, p. 19.
3 Wadis are dry riverbeds which carry water only in the case of heavy rains.
4 Heck and Wöbcke, 2005, p. 15.
5 Heck and Wöbcke, 2005, p. 31.
6 Hijab denotes the standard of dress prescribed by Islam, which especially affects women and their obligation to cover their hair. The style of hijab can differ from region to region.
7 de Montequin, 1983, p. 48.
8 de Montequin, 1983, p. 55.
9 Sayrafi, 1981, p. 157.
10 Hakim, 2007, p. 153.
11 Hakim, 2007, p. 154.
12 Fiqh literature consists of answers to legal questions based on the Islamic sources of the Quran and hadith (historical narratives) by a scholar.
13 Hakim, 2007, p. 160.
14 Hakim, 2007, p. 154.
15 de Montequin, 1983, p. 51.
16 de Montequin, 1983, p. 48.
17 de Montequin, 1983, p. 51.
18 The fina was a gazebo which could extend upper floors by one metre.
19 Sayrafi 1981, p. 156.
20 de Montequin, 1983, p. 50.
21 Hakim, 2007, p. 154.
22 Majlis are simple meeting rooms either within private homes or in separate buildings.
23 Fadan, 1983, p. 295.
24 Diener, Gangler and Fein, 2003, p. 20.
25 Hakim, 2007, p. 156.
26 Dayaratne, 05.10.2008.
27 Korn, 2003, p. 27.
28 Diener, Gangler and Fein, 2003, p. 19.
29 Diener, Gangler and Fein, 2003, p. 20.
30 Heck, 2004, p. 90.
31 Heck and Wöbcke, 2005, p. 86.
32 Heck and Wöbcke, 2005, p. 31.
33 Heck and Wöbcke, 2005, p. 102.
34 Heck and Wöbcke, 2005, p. 130.
35 http://en.wikipedia.org/wiki/Sana%27a, 05.10.2008.
36 Sayrafi, 1981, p. 143.
37 Pape, 1977, p. 16.
38 Pape, 1977, p. 46.
39 Garba, 2004, p. 597.
40 The palace in the latter had mainly representative functions, while the ruler's home was normally outside of the city centre (de Montequin, 1983, p. 55).
41 Al Hathloul, 2002, p. 1.
42 'Macro-level' means the large-scale distribution of land according to ethnic or familial affiliations (Hakim, 2007, p. 153).
43 Hakim, 2007, p. 153–154.

44   Heck and Wöbcke, 2005, p. 17.
45   Scholz, 1999, p. 32.
46   Blume, 1976, p. 267.
47   Scholz, 1999, p. 62.
48   Scholz, 1999, p. 188.
49   Wahabism began as a movement of reform founded by the scholar Abdul Wahab (who died in 1787) and which aimed to abolish the different Islamic schools and create unity by introducing a literal interpretation of the Quran (Pape, 1977, p. 17).
50   Heck 2004, p. 14.
51   Heck and Wöbcke, 2005, p. 16.
52   Heck and Wöbcke, 2005, p. 33.
53   Scholz, 1999, p. 70.
54   Scholz, 1999, p. 71.
55   Scholz, 1999, p. 70.
56   Pape, 1977, p. 50.
57   Reichert, 1978, p. 46.
58   Scholz, 1999, p. 74.
59   Scholz, 1999, p. 77.
60   Reichert, 1978, p. 106.
61   Municipality of Municipal and Rural Affairs (Mubarek, 14.10.2008, p. 9).
62   Garba, 2004, p. 600.
63   Reichert, 1978, p. 106.
64   Garba, 2004, p. 600.
65   Reichert, 1978, p. 109.
66   Reichert, 1978, p. 110.
67   Reichert, 1978, p. 114.
68   Reichert, 1978, p. 115.
69   Arabian-American Oil Company
70   Reichert, 1978, p. 69.
71   Reichert, 1978, p. 76.
72   Reichert, 1978, p. 87.
73   Reichert, 1978, p. 96.
74   Mubarek, 14.10.2008, p. 9.
75   Mubarek, 14.10.2008, p. 10.
76   Al Naim, 2008, p. 138.
77   Mubarek, 2004, p. 588.
78   Reichert, 1978, p. 113.
79   Mubarek, 14.10.2008, p. 11.
80   Mubarek, 14.10.2008, p. 11.
81   The 1976 revised master plan was developed by the French consulting company SCET International (Mubarek, 14.10.2008, p. 12).
82   Mubarek, 14.10.2008, p. 12.
83   Mubarek, 14.10.2008, p. 13.
84   Rahman, 1991, p. 50.
85   Melamid, 1980, p. 473.
86   http://www.populstat.info/Asia/saudiarc.htm, 14.10.2008.
87   Reichert, 1978, p. 60.
88   http://www.arriyadh.com/En/Ab-Arriyad/index.asp, 14.10.2008.
89   Al Hathloul, Mughal, 2004, p. 611.

90   Scholz, 1999, p. 80.

91   Scholz, 1999, p. 81.

92   Scholz, 1999, p. 80.

93   http://www.cia.gov/library/publications/the-world-factbook/geos/ba.html, 13.10.2008.

94   Scholz, 1999, p. 80.

95   http://www.ameinfo.com/16342.html, 13.10.2008.

96   Scholz, 1999, p. 229.

97   Scholz, 1999, p. 236.

98   Scholz, 1999, p. 236.

99   Al Mosaind, 1998, p. 263.

100   Melamid, 1980, p. 476.

101   Organisation of Petroleum Exporting Countries.

102   http://www.abc.net.au/4corners/special_eds/20060710/, 19.10.2008.

103   http://www.indexmundi.com/energy.aspx?country=om&product=oil&graph=production+c
onsumption, 19.10.2008.

104   http://www.eia.doe.gov/emeu/cabs/Bahrain/Oil.html, 19.10.2008.

105   http://www.theoildrum.com/story/2006/10/5/215316/408, 19.10.2008.

106   Butt, http://www.uaeinteract.com/uaeint_misc/pdf/perspectives/11.pdf, 19.10.2008.

107   http://www.theoildrum.com/story/2006/10/5/215316/408, 19.10.2008.

108   Scholz, 1999, p. 43.

109   http://www.kishtpc.com/Freetrade%20ZONES.htm, 19.10.2008.

110   Sassen, 1997, p. 44.

111   Scholz, 1999, p. 41.

112   Gafoor, http://users.bart.nl/~abdul/chap4.html, 19.10.2008.

113   http://de.wikipedia.org/wiki/Islamic_Banking, 19.10.2008.

114   http://www.islamic-banking.com/shariah/index.php, 19.10.2008.

115   Scholz, 1999, p. 41.

116   ADA Report, 2004, p. 65.

117   http://www.independent.co.uk/news/media/aljazeera-the-new-power-on-the-small-
screen-512562.html, 21.10.2008.

118   http://www.dubaimediacity.com/, 21.08.2008.

119   http://www.arabianbusiness.com/518541-dubai-tourist-numbers-set-to-hit-10mn?ln=en,
21.10.2008.

120   http://www.uae.gov.ae/Government/tourism.htm, 21.10.2008.

121   http://www.guggenheim.org/press_releases/release_159.html, 21.10.2008.

2 Dubai as the Primary Model of Post-oil Urbanism

2.1 Historical Background of Urban Development

2.1.1 The Traditional Settlements of Dubai

The very first settlement founded at the location of the current Dubai was most probably a small fishing village founded in the 18th century. Its first remarkable development step was when about 800 members of the tribe Bani Yas moved from the neighbouring oasis settlement of Abu Dhabi to this village, which led to a doubling in the number of its inhabitants[1]. This event took place in 1833 after an argument took place between clans within the tribe of Bani Yas, which was a powerful tribal conglomeration whose influence stretched along the coast as far west as the Qatari peninsula and far inland including the chain of Liwa oases[2]. The most prominent people of the new settlers in Dubai[3] was the Al Bu Falaseh clan including the family of Al Maktoum. Its sheikh Maktoum bin Buti took over the leadership of Dubai after the death of his uncle in 1836, which marked the beginning of the Al Maktoum dynasty in Dubai[4]. During the 19th century the British influence in the region was increasing due to the trade routes to India and led to the establishment of the Trucial States in 1892 in order to curb increasing piracy in the Gulf. Consequently, Dubai became one of these states along with its six neighbouring emirates[5]. The settlement sustained itself through various basic economic activities such as fishing, animal husbandry, basic re-export trading and some limited agriculture. More exceptional kinds of income included the trade of pearls, which could be found along the coast, in addition to boat building and rope manufacture[6].

The early settlement was located at the mouth of a prominent natural water inlet in the lower Persian Gulf known as Dubai Creek[7]. At the beginning of the 20th century, the settlement was divided into three major built areas around the creek – Deira in the north, Bur Dubai in the south and Al Shindagha, which was where the residence of the royal family was located on a small strip of land between the coast and the creek[8]. From the 1870s, the harbours on the southern coast of the Gulf benefited from increasing political instability in Persia, which led to a large immigration of Persian merchants to Dubai, consequently establishing the settlement as one of the most important harbours in the region. Immigration grew at a remarkable rate after 1902, when the Persian harbour town Bandar Lingeh increased its customs duties and many merchants began settling in Dubai due to its liberal tax policies[9]. Over the following years the Persian merchant families settled in Deira or in their own district called Bastakyia in Bur Dubai. Apart from these Persian immigrants many merchant families arrived from various other regions such as Belutchistan or Iraq, particularly during the peak of the pearl trade in the 20s. This led to an early mixture of various ethnicities and nationalities, which formed an ideal basis for the development of Dubai as a trading hub due to the exchange of knowledge and contacts.

At the beginning of the 20th century, the settlement consisted of about 10,000 inhabitants

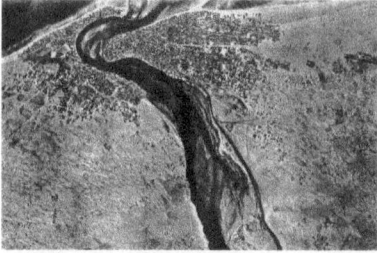

Fig. 1: Dubai in the 1950s.   Fig. 2: The district Deira.

living in courtyard houses made of coral stone, gypsum and date trunks or simple barasti huts, which remained the traditional housing of the lowest income groups[10]. The houses of the wealthier inhabitants often had an air conditioning system of wind towers, a system that was imported by the Iranian immigrants. The Friday mosque, the fort and the palace were built on the southern bank of the creek in Shindagha and Bur Dubai. The markets were located on both sides of the creek while the main souq developed in the district of Deira in the north due to its harbour and the fact that it was located at a closer distance to the emirate of Sharjah. At the beginning of the 20th century, the district of Deira consisted of about 350 souq shops and 1,600 houses. In Al Shinadagha the royal family and its related clans lived in around 250 houses segregated away from the remaining town. The smallest residential area was in Bur Dubai with only about 200 houses, which was mainly the home of Persian and Indian merchants, as well as containing a small souq with 50 houses. Until the 50s the size of the urban area remained modest at about just 3.2 square kilometres[11]. Until the middle of the 20th century, Dubai was a typical harbour town with the traditional town structure, typology and architecture of the region, namely, a strictly tribal and ethnically segregated community lived in similarly built two-storey houses in a dense vernacular set-up with narrow labyrinthine streets, roadside markets and central cores with mosques.

2.1.2 Oil Urbanisation and National Independence

In the 30s, economic development stagnated due to a global economic crisis and the invention of cultured pearls in Japan, which led to a breakdown in the pearl trade in the Gulf. After the Second World War, Dubai's economy focused on trading gold to India in response to India's growing demand for the metal and Dubai subsequently established itself as the biggest transshipment centre for gold in the region because of its liberal customs policies[12]. In 1966, oil was discovered for the first time off the coast of Dubai, which led to a remarkable and rapid transformation of the economy and society within only one decade[13]. The population quickly increased from about 20,000 inhabitants in the middle of the 20th century to about 59,000 in the late 60s[14]. In 1971, Dubai joined the United Arab Emirates to become its second largest emirate with the second largest oil resources after Abu Dhabi, which became the UAE's leading emirate by providing the major part of the national budget and developing its role as an administrative centre. Nevertheless, Dubai was able to expand its political influence within the federation by becoming a trading and banking centre with a remarkable economic growth over the following years. While external policies were unified through the founding of the UAE, domestic policies and administration within the seven emirates remained decentralised and in the hands of each Amir, each of whom became the indisputable authority within his own territory and neither alternative political parties nor labour unions were permitted to be set up to allow any form of public participation[15]. In addition to these political changes, the society of Dubai became increasingly multicultural and multi-ethnic during the second half of the 20th century due to a vast influx of immigrants, particularly from South Asia.

The urban modernisation of Dubai first began in the 50s, when the first basic electricity, communication and water infrastructure was built. Furthermore, the waterway in the creek was developed for modern shipping and the first provisional airfield was put into operation. These early investments, including the construction of the first hospital and schools, were made possible by the financial support of Great Britain and loans from Bahrain, Qatar and Kuwait, where oil production had already started. Apart from this external support, it was Dubai's own economic success based on its high volume of re-export (for example, in 1950 85% of all imports were re-exported) that made the emirate one of the most important trading hubs of the region and therefore one of the most attractive locations for business[16]. Two basic factors for this successful economic development can be distinguished, namely, the local merchant community with its network of international contacts and the attractive trading policies provided by the government[17]. At the end of the 60s, the high revenues generated from of its own oil production enabled the government of Dubai to invest in large-scale industrial and infrastructural projects such as Port Rashid, the dry docks, the aluminium smelter and Jebel Ali Port with its industrial area about 30 km to the south-west of Dubai. Although the population came to just over 60,000 inhabitants at the beginning of the 70s, the government decided to construct Dubai's first international airport with a terminal for more than 1,200 people and several extensions over the following decades[18]. These risky investments developed into the key stones for further economic development and growth.

As in many Gulf countries, the first phase of oil urbanisation in Dubai was dominated by individual major developments without major comprehensive planning. Nevertheless, after the establishment of a modern public administration in the late 50s, a first development plan was published in 1960 by the British architect John R. Harris, which was called 'Survey and Plan – Capital City of Dubai'. This early guide plan included five major development goals. These were the provision of a road system which would be appropriate for the anticipated volume of traffic, the zoning of town areas suitable for industry, commerce, schools and public buildings, the choosing of areas for new residential quarters outside the boundaries of the existing town, the choosing of sites for school buildings, open spaces and local centres within the new residential areas and last but not least the creation of a town centre in Dubai. The main focus of this plan was the development of the infrastructure on a still relatively small budget[19]. Over the following years, the grid of the main roads was constructed and the settlement spread mainly at the borders of Deira toward the north and east[20]. In accordance with Arab-Islamic traditions, land ownership in Dubai was based on two principles – within a settlement any plot of land that had been occupied by a homestead for a long period of time belonged to the inhabitant. Elsewhere, unbuilt land was at the disposal of the ruler. These principles were legally applied at the beginning of the 60s and led to considerably centralised control of later urban development due to the free use of unsettled land by the ruler[21].

In 1971, it was again the architect John R. Harris and his office who were engaged to develop a new master plan, the so-called 'Dubai Development Plan'. The new plan reacted to the unexpected rapid urban development of the 60s and the new financial possibilities that arose due to the beginning of oil production. The plan included the development of the Shindagha tunnel beneath the creek and two bridges in order to connect Deira and Bur Dubai. The detailed land use plan arranged new housing areas along the coasts in the north and south. Furthermore, the harbour Port Rashid and the road grid were to be expanded under the new plan[22]. Over the following years, most parts of the plan were executed, particularly in relation to the network of hierarchical streets consisting of 'main traffic routes' with up to six lanes and 'primary and secondary access roads[23]'. The subdivision of land led to suburbanisation that had a standard typology of the two-storey villa. This process was increased due to land speculation and public subsidies provided to the local population, wherein every man over 20 years of age received a plot of 15,000 square feet[24]. Almost all major infrastructural projects, including the Shindagha tunnel, the two bridges, the harbour extensions and the airport, were completed in the 70s according to the plan and its goals. However, the rapid growth of the population from about 60,000 inhabitants in 1967 to over 120,000 in 1973 and then to more than 240,000 in 1980 was again unexpected and thus the master plan became outdated[25].

The consequence was an increasing urban sprawl in the outskirts on the basis of additional subdivisions and the expansion of the existing road network. In spite of the exodus of the main housing areas into the suburbs, the business centre of Dubai remained at the creek in the second half of the 20th century because of its unique function and form as mentioned in the Dubai Development Plan: 'Additionally to the support given to the economy by activities using the Creek as a facility, there is an undoubted attraction in the presence of water and of boats near the centre of the town. Both from a functional and an aesthetic viewpoint, therefore, developing Dubai should still be centred on the Creek, where marine activity continues to flourish. The clear implication of this is that there shall be future development inland on both banks of the Creek[26]'. In the north, the Bani Yas Road in the direction of the international airport became the location for the modern CBD along the shoreline of the creek. Apart from many banks, all main administration and governmental institutions such as the Dubai Municipality settled along this CBD in the south-east of Deira. While the old traditional houses in the old districts were mainly

Fig. 3: The Sheikh Zayed Road in the 1980s.   *Fig. 4: The CBD in 2005.*

replaced in the first phase of the modernisation in the middle of the 20th century, many new multi-storey apartment and office blocks were built in the 70s, creating a dense urban centre with mixed land use. Within only a few years, the downtown of Dubai was almost completely occupied by guest workers with lower incomes due to the development of apartment blocks and the increasing move of the local population to the outskirts along the coasts and close to the border of the emirate of Sharjah. At the end of the 70s, Dubai was an exemplary oil city with the special feature of an urban centre spreading along the creek.

2.1.3 The Introduction of Post-oil Strategies

In 1979, the Dubai World Trade Centre became the first high-rise and landmark of Dubai. With 39 storeys, it was the highest building in the entire Middle East and thus it became a symbol of Dubai's attempt to become the largest economic centre in the region[27]. Furthermore, it marked the beginning of a new CBD in the south of Dubai along the main development corridor toward Jebel Ali along a multi-lane road called Sheikh Zayed Road. This CBD became more and more developed due to further expansion of the urban area toward the south. Based on the exponential growth of the population, the Greek consulting firm Doxiadis was engaged at the beginning of the 80s to develop the third major development plan, which, while possessing relatively minor structural changes, included several adjustments that were to be made to accommodate the population growth. A factor of the increasing development toward the south was the installation of the first free trade zone (FTZ) in Jebel Ali. In 1985, the 300-hectare FTZ was developed close to the harbour for customs-free trade and industrial use, for example, aluminium production. The development of a free economic zone enabled Dubai's rulers to circumvent a federal law of the UAE called 'Commercial and Companies Law' (1984), which stipulated that all emirates adhere to a local sponsorship or kafil system that required all registered companies to be at least 51% owned by a UAE national[28]. Consequently, many international companies settled in Jebel Ali and started to increase the diversification of Dubai's economy, which therefore became more and more independent from its own oil production. Further factors of Dubai's rapid economic diversification were the two Gulf wars during the 80s and at the beginning of the 90s, which made Dubai the most

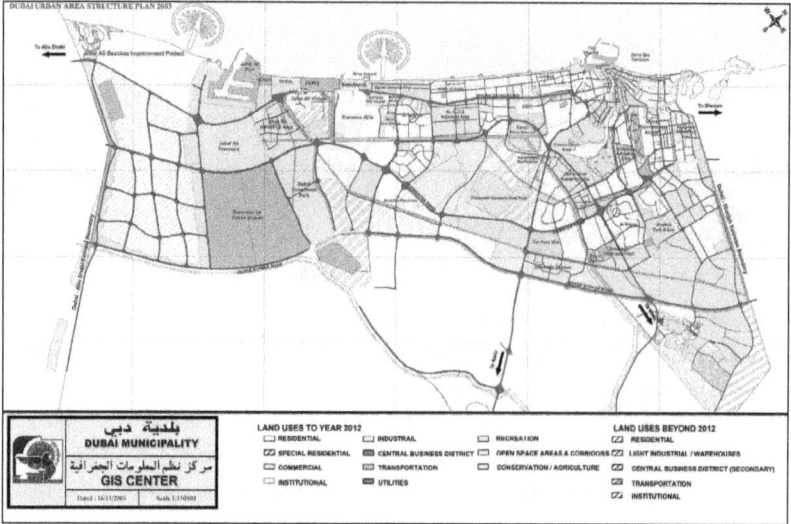

Fig. 5: The Dubai Urban Area Structure Plan.

important harbour in the region due to the disruption of trade in the north. Moreover, the civil war in Beirut led to the setting up of many Lebanese banks in Dubai, making it to a growing financial centre in the Middle East[29]. Apart from these external factors, huge investments made in the development of its first airline 'Emirates' brought Dubai closer to its goal of becoming a global economic centre.

In the early 90s, the government commissioned the 'Dubai Urban Area Structure Plan' for the physical and economic development of Dubai between 1993 and 2012. The plan was developed by the consultants Parsons Harland Bartholomew & Associates and it integrated major physical development goals. First of all, it accommodated urban expansion by allocating additional land in a phased planned process to meet current and future needs for residential, industrial and commercial use. It also extended the existing transport network and infrastructure facilities. Apart from general physical planning, the strategic plan included major goals concerning the rapid economic development. It intended to promote continued economic growth and support and attract private investment by ensuring sufficient land supply, adequate infrastructure, simplified administrative and planning procedures and the carrying out of publicly funded feasibility studies for major development proposals to minimise the risk to private capital. In addition, it encouraged expatriates to reinvest capital and profits in local enterprises and the development of an inter-departmental planning framework capable of reviewing, monitoring and implementing the Structure Plan. Last but not least, the plan proposed the introduction of a regulatory environment capable of operating within a strong market economy and incorporating the needs and interests of a large number of agencies and organisations[30].

The end of the 20th century was marked by a period of rapid urban expansion with the total urban area of Dubai increasing at a rate of more than 3% a year[31]. The population grew from 370,788 in 1985 to 862,387 in 2000. This was due to a huge immigration influx of expatriates, who in 2000 constituted about 53% of the total population[32]. Consequently, one of the biggest challenges for the urban development was the growing need for housing. This need was increasing due to the continuing practice of giving free land to the local population and subsidising their living expenses, including the provision of the costly infrastructure. The result was an automobile-dependent city with a very low density of less than 1,400 people living on one square kilometre. On the one hand, the low-density suburbs occupied a larger and larger area with their widespread typology of two-storey villas and massive road infrastructure. On the other, the centre of Dubai and its fringes became more dense during the 90s due to the vast construction of multi-storey apartment buildings by companies in order to accommodate their foreign labour. Generally, there was almost no state provision for housing expatriates in order to avoid unnecessary competition with the growing private sector[33].

At the end of 1999, the construction of the hotel Burj Al Arab was completed and along with it the beginning of a new era of landmarks in Dubai. The hotel not only claimed to be the first '7-star hotel', it also still holds the record for being the tallest hotel in the world at 321 metres and 60 storeys[34]. The unique design gave a new face to Dubai at the beginning of the millennium with a major emphasis on luxury and tourism, which became an important factor in the town's increasing marketing of itself and further economic growth. The Burj Al Arab and other landmark projects such as the Emirates Towers and the Jumeirah Beach Hotel were direct investments of the Amir himself which were intended to develop the attraction of Dubai as a high-class tourist destination as well as a centre of various future economic possibilities[35]. An important basis for this direction in development was the entrepreneurial foresight of Amir Mohammed bin Rashid Al Maktoum[36] and his general liberal policies that imposed almost no Islamic restrictions, thus allowing those who wished to to lead a Western lifestyle. This liberalism in Dubai is mainly based on the diversity of its own society comprising different ethnic and religious groups, which is over a century old. The economic strategy of developing as a tourist centre was intended on the one hand to fill the vacuum of leisure and entertainment between the two Islamic conservative countries of Saudi Arabia inland and Iran on the other side of the Gulf, and on the other, tourism itself was used to market Dubai as a business centre and growing future market on an international level. Consequently, the number of annual visitors increased to over 7 million in 2007 and the population grew from just about 961,000[37] inhabitants in 2002 to around 1.5 million in 2007, with a growing majority of expatriates at more than 83%[38].

The events of 11 September 2001 and the following 'war against terror' led to a new investment situation in the Gulf, wherein regional investors started to search for new opportunities beyond the old strategy of investing in the West. Consequently, Dubai began to develop new free trade zones and in 2002 it became the first place in the region to officially open the local real estate market to foreigners[39]. The consequence of this new strategy was an unexpected surge in private investment in the development of so-called

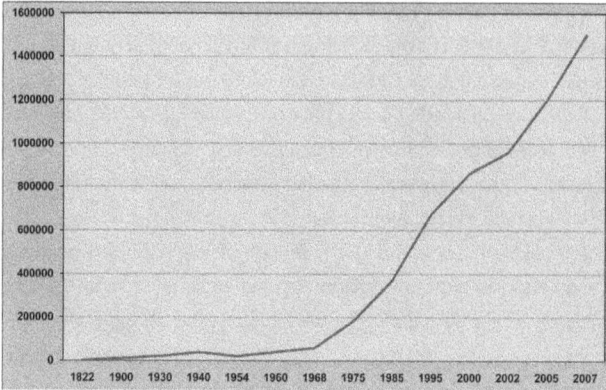

Fig. 6: The population growth in Dubai.

'freehold properties'. In order to cope with this large pressure of developments being carried out by the private sector, the old centralised urban management structure of one municipality developing the infrastructure and allocating building permits according to the Structure Plan of 1993 had to be changed under a more decentralised and flexible approach. This led to a significant change in urban governance wherein real-estate companies, which were typically public joint stock companies, played a new role as the main developers shaping urban growth in Dubai. The three main real-estate companies in Dubai – Nakheel, Emaar Properties and Dubai Holding – were founded in the 90s. These private enterprises have the special feature of being under the control of the royal family, who initiated their founding and became either their largest shareholder, as in the case of Emaar, or a private owner of the whole company, as in the cases of Nakheel and Dubai Holding, by means of direct investments or the provision of land for developments[40]. This privatisation of urban management has caused Dubai's physical and economic development to become more decentralised and entrepreneurial in nature while the basic control and power of veto have continued to remain in the hands of the ruling Amir.

The project 'Emirates Hills' marked the actual beginning in 1999 of the real-estate boom of recent years by offering freehold properties for sale for the first time to foreign investors on a limited basis of plots on lease for 99 years[41]. Although this project contradicted existing laws in the UAE, under which only locals could possess property, it became the precursor to many subsequent developments because Sheikh Mohammed was determined to foster this type of change in his emirate in order to accommodate the growing interests of investors[42]. The first project of this kind that attracted more widespread international recognition was the 'Palm' by Nakheel, which was proposed in 2001 and developed over the following years with overall completion estimated to be in 2010[43]. The project consisting of a palm-shaped island in front of the coast in Jumeirah constituted a new benchmark in real-estate developments sold as freehold properties. The Palm was a new landmark showing the future possibilities of property developments in the Gulf by using large-scale forms as pictorial branding for successful marketing. In the following

46

years many similar developments intended predominantly for housing and tourism use followed both inland and offshore. Furthermore, many new FTZs were either planned or built in order to increase economic diversification and attract future investors. In 2000 and 2001, the FTZs Dubai Internet City and Dubai Media City were established as new economic sub-centres of Dubai[44]. Both 'cities' were constructed within the area of many new developments in Jumeirah, which became the first large-scale city extension in the new millennium including the housing developments Dubai Marina, Emirates Hills and Jumeirah Islands. A characteristic of the new FTZs has been their set-up as 'cities' within the city with their own particular theme and specialisation in an economic sector such as, for example, ICT[45], media or health care.

The result of the construction boom was a remarkable rise in land prices due to growing speculation and a considerable shrinkage in the amount of unbuilt area in the small emirate. The consequence has been the development of several offshore projects, made possible by reclaiming land, and the construction of many multi-storey blocks and high-rises in order to create higher densities. Because of increasing apartment prices, many foreign inhabitants have opted to move to the neighbouring emirates of Sharjah or Ajman, which have become dormitory towns. In spite of the rapid transformation of the urban structure, the centre of Dubai has remained the home of foreign labour in addition to its new function as a tourist attraction due to the development of the Creek and several restoration projects such as, for example, the traditional souqs or the area of Bastakiya. In order to create a local atmosphere, several projects have integrated oriental designs and decorative elements such as traditional wind towers. The best example of this is the development Madinat Jumeirah, which includes hotels, a spa, theatres and commerce in a traditional looking set-up[46]. The increasing competition and growing investments have led to a growing variety of different architecture, from catalogue designs to landmark projects by renowned architects. Consequently, Dubai has become a laboratory of post-modern architecture and construction superlatives such as the highest tower in the world, the Burj Dubai. This over 800-metre skyscraper is set to be completed in 2009 and will mark the centre of the development 'Downtown Dubai' located along Sheikh Zayed Road, which has established a role for itself as the main CBD of Dubai in recent years[47].

2.2. The New Concept of Urban Governance

2.2.1 The Vision of Developing a Global Service Centre

The high economic growth of recent years has affirmed the ambitious goal of Dubai's Amir to develop the first global city in the Gulf[48], which can be compared more easily with other economic success stories in Asia such as Singapore rather than the traditional global cities of New York, London or Tokyo. The basis of this extraordinary economic development was the entrepreneurial skill of the rulers of Dubai, who initiated important decisions and investments at the right moments in time by understanding how to use

Dubai's fortunate geopolitical location in combination with outer circumstances such as, for example, regional conflicts. This strategy began at the beginning of the 20th century and continued into the time of the oil boom as well as recently at the beginning of the post-oil era. As a traditional harbour town, it was trade which became the key stone of Dubai's post-oil economy in addition to various new economic sectors ranging from tourism to media and ICT, which have been recently promoted by huge investments, low taxes and few restrictions. In addition, at the beginning of the new millennium, the development of freehold properties led to a global awareness of Dubai as an important up-coming market and haven for investment. Since 2000, GDP has been growing at a compounded rate of 13% with an annual population increase of 6.9%[49]. In 2005, the non-oil sector contributed 95% of the GDP in comparison to only 46% in 1975. The key driver of the economic diversification has been the service sector with an annual growth rate of more than 21% since 2000[50].

The general aims of the economic development today are summarised by the Dubai Strategic Plan 2015 as follows – sustaining economic growth through the establishment of a GDP annual growth rate of 11% for the next 10 years and an increase in the real GDP per capita to USD 44,000 by 2015 as well as the enhancement of labour productivity and sector development by expanding existing sectors and creating new ones with a focus on competitiveness and innovation. In order to reach these goals, the future economic strategy is based on six main sectors – tourism, financial services, professional services, logistic services, trade and construction. In order to develop these sectors, there are seven basic factors which have to be established in parallel, namely, human capital, productivity, science, costs for living and business, quality of life, economic policy and the legal framework[51]. Consequently, an estimated 882,000 additional workers will be required by the year 2015, bringing total employment to 1.73 million with a significant shift toward higher skilled employment[52]. The most important foundation is the concept of a liberal market with only few restrictions in order to attract both regional and global investors and to strengthen public-private partnerships, which have become essential for the economic diversification.

The social development challenge of the future is the integration of the national population within all sectors of the economy in order to preserve national identity and culture and to become more independent from foreign labour[53]. Today, less than 17% of the population is local, facing a growing immigration of foreign guest workers. Most of the around 300,000 jobs that are created in Dubai each year are in the private sector and it is thought that only 1% of this workforce actually comprises nationals[54]. The majority of the population of Dubai is formed by South-Asian labour with more than 63% of the overall population working mainly in the lower service sector and construction[55]. In 2008, more than half a million construction workers were engaged on building sites, forming almost one third of the overall population. A large part of these low-cost labourers earn less than USD 160 a month, facing hardly any labour laws and living in labour camps at the outskirts[56]. The small salaries are usually sent back home to sustain the life of the families left overseas and thus millions of dollars are not reinvested within Dubai's economy. This does not only concern the construction sector but also the largest part of the overall service sector, which is run by expatriates who benefit from very low taxes. Furthermore, the majority of foreign labour works only temporarily in Dubai in order to save money for their return to their homelands. The result has been an ever-changing, multi-cultural society that develops very few roots within Dubai itself, another contributing factor of which is the fact that this foreign segment of the population is almost completely excluded from political integration. Now, this issue is being addressed by the encouragement and support of the national population to become educated and take over the key roles within Dubai's future post-oil economy. In addition, the integration of educated foreign labour in particular within society is being encouraged by providing them with more attractive living conditions than is available in their home countries.

Dubai's current overall goal is to develop into a high-ranking global city with the best international standards by becoming independent from the export of its own fossil resources. In order to achieve this in a rather short period of time, the structure of urban governance has started to become more decentralised and increasingly dominated by the private sector. However, the central decision making remains in the hands of the ruler, who has started to govern Dubai more and more as a corporate manager in addition to his genuine function as political ruler. This has particularly affected the development of real estate with the creation of private corporations initiated by the Amir himself in order to develop freehold properties on a large scale. The general aim is to develop a total urban area that can house a projected population of about 3.7 million by 2015[57]. Furthermore, it is expected that an estimated 15 million tourists will visit Dubai in 2010 with their numbers growing in the future[58]. Therefore, the real estate sector has been rapidly growing in recent years, which has led to an increasing need for infrastructure in order to achieve consolidation and sustain liveability.

2.2.2 The Parties Involved in Urban Governance

At the beginning of 2006, the Amir Sheikh Mohammed bin Rashid Al Maktoum became the ruler of Dubai after the death of his older brother Sheikh Maktoum bin Rashid Al Maktoum, who had been in charge since 1990. While their father Sheikh Rashid bin Saeed Al Maktoum led Dubai through the decisive time of the oil boom and the foundation of the UAE from the 50s to the late 80s, it was his third son, Sheikh Mohammed, who took over the initiative to develop Dubai into a global city. This was made possible during the rulership of Sheikh Maktoum, who left Dubai's economic development to his younger brother and mainly focused on the political day-to-day business of the UAE. Consequently, since the early 90s, Sheikh Mohammed was already able to start creating his vision of a modern Dubai by introducing new economic strategies[59]. Sheikh Mohammed, who was educated at Cambridge in Great Britain, was introduced into the political leadership at a very young age of just 23 years, when he became Minister of Defence, and thus the youngest Minister of the UAE, at the beginning of the 70s[60]. His entrepreneurial skills were first proven in 1985, when he decided to develop the first airline of the UAE, Emirates, involving an investment of USD 10 million. In the mid-90s, he initiated the expansion of the tourism sector by founding the Shopping Festival, which soon became one of the biggest regional attractions. It was this ambition which led to his appointment as Crown Prince by Sheikh Maktoum instead of his older brother Sheikh Hamdan[61] in 1995.

In the 90s, Sheikh Mohammed created the key stone of the rapid urban growth of the following years by instigating the establishment of large real estate companies in order to develop freehold properties for sale to non-UAE nationals. In cooperation with other investors, the joint stock company Emaar was founded in 1997 with an equity capital of 2.65 billion dirhams[62]. Because of the allocation of land for the development of real estate, the Maktoum family became the biggest shareholder with 33% of the stock. The Amir appointed Mohammed Al Abbar as chairman of Emaar, who has proven his skills as a financial expert in previous projects in Dubai. Emaar's first project in 1999, 'Emirates Hills', was also the first freehold property project in Dubai. In 2007, the company had an annual net profit of more than 6.5 billion dirhams that continued to grow over the following years due to the company's expansion within the region and globally[63]. In parallel with Emaar, Sheikh Mohammed initiated the establishment of a second large real estate company, Nakheel, named after its signature project of reclaimed islands in front of Dubai's coast in the form of a palm. The intention of Sheikh Mohammed was to create a real estate company wholly owned by the Al Maktoum family unlike Emaar, and due to this advantage that Nakheel was the private property of the ruler himself and the fact that Emaar was already running at full capacity, Nakheel soon became the biggest real estate developer of Dubai. The appointed chief executive of Nakheel has been Sultan Ahmad Bin Sulayem, who already had previous senior management experience in relation to his leadership role at Jebel Ali Harbour and its FTZ[64]. The net profit of Nakheel in 2007 was 4,688 billion dirhams[65].

Although since the 70s all ministries have been established in Abu Dhabi in order to govern the major domestic and external affairs of the UAE, several governmental institutions were founded in Dubai that have been playing a major role in the emirate's economic development. The Dubai Municipality, which had already been established in the 50s, was followed by the Department of Economic Development (DED) in 1992, the Dubai Ports Authority (DPA) in 1991, the Dubai Commerce and Tourism Promotion Board in 1989, which was later replaced by the Department of Tourism and Commerce Marketing (DTCM) in 1997, and furthermore the Dubai Development and Investment Authority (DDIA) in 2002[66]. The DED was mainly developed in order to analyse the economy unlike the DDIA, which was put in charge of executing the outcome of these analyses within the sectors of infrastructure development, project management and investment raising. Thus, the DDIA was one of the most important institutions and consequently put under the direct supervision of Sheikh Mohammed himself, who appointed Mohammed Abdallah Al Gergawi as its chairman[67]. In the following years these public institutions were often

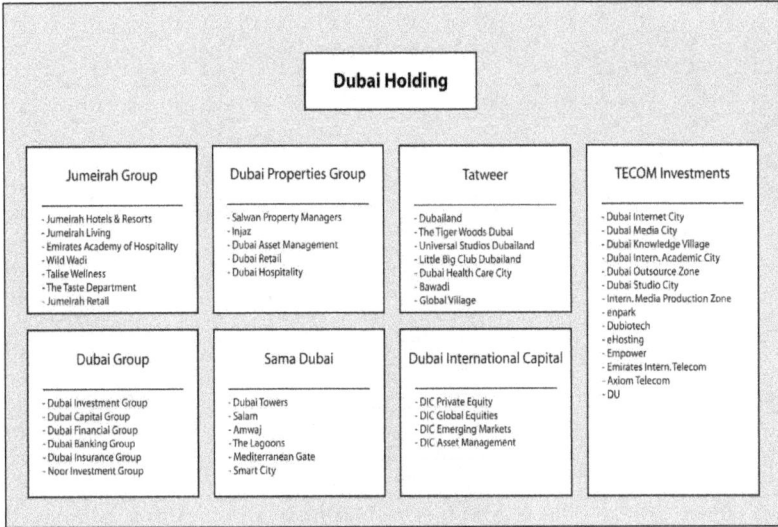

Fig. 7: The companies of Dubai Holding.

reformed and restructured, particularly the DDIA. After four years of its existence it was completely privatised and subordinated to the private Dubai Holding (DH), which was developed in parallel at the beginning of the millennium as a private version of the DDIA, and due to DH's success and future perspectives as a private cooperation, DH was able to establish a strong position for itself[68].

DH was established as a holding of seven company conglomerates of the Al Maktoum family, covering 13 different economic sectors including finance, real estate, tourism, communication, technology, industrial manufacturing, education and research. The Jumeirah Group, one of the oldest conglomerates of DH, includes companies working in hospitality, education, leisure and entertainment. In addition, the conglomerate Tatweer was established as the parent company of various companies in the entertainment industry, including, for example, Dubailand[69], as well as additional companies involved in real estate, health care, industrial manufacturing and the energy sector. A further parent company, TECOM Investments, has integrated all the technology, communication and research branches by being the head of several FTZs, including Dubai Internet City, Dubai Media City and Dubai Knowledge Village. In addition, there are the conglomerates Dubai Group and Dubai International Capital, which are umbrella organisations that cover mainly investment and insurance companies. Last but not least, two real estate companies, Dubai Properties and Sama Dubai, were founded in order to develop real estate projects in Dubai itself and abroad. Although the DDIA was taken over by DH, most leading positions within the private holding were given to former leaders of the DDIA. Consequently, Mohammed Abdallah Al Gergawi became the chairman of Dubai Holding, which has established itself as the most important developer of Dubai's economy beside a second recently founded holding called Dubai World.

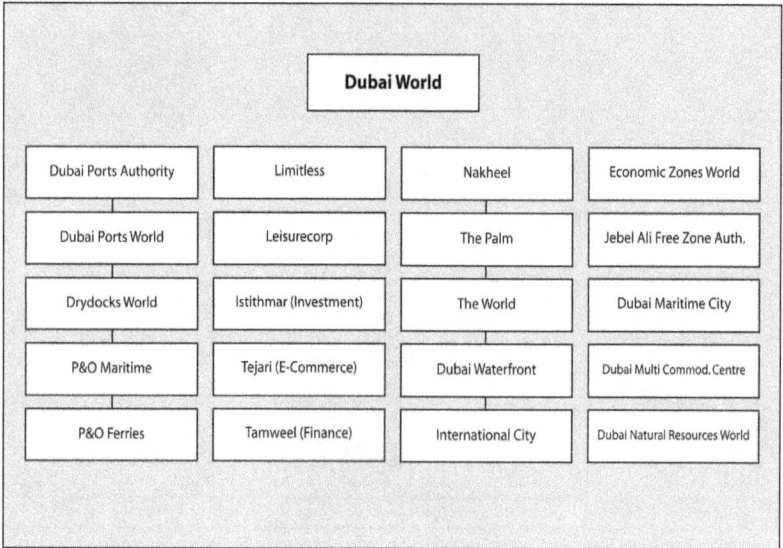

Fig. 8: The companies of Dubai World.

The new holding Dubai World was founded in 2006 after the restructuring of the Ports, Customs & Free Zone Corporation (PCFC), which was founded in 2001 as an umbrella organisation covering the Jebel Ali FTZ and the Dubai Ports Authority (DPA). Because of its remaining public functions, the DPA, which had already been founded in 1991 as a public institution, did not become part of the holding Dubai World despite the fact that it was financially independent and profit-orientated, unlike other companies that joined the new holding such as, for example, Dubai Maritime City (DMC)[70] and, most importantly, the real estate giant Nakheel. Today, the Dubai World conglomerate covers over 13 different economic sectors with a focus on the maritime sector. The former chairman of the PCFC, Sultan Ahmad Bin Sulayem, has also been appointed as new head of the board of directors of the new Dubai World group. Furthermore, he has remained the chairman of the companies Nakheel, Dubai Ports World (DPW) and Jebel Ali Free Zone Authority (JAFZA) within the holding. In addition, he has also been put in charge of the semi-public DPA and has thus gained an important governmental role. However, he is not the only case of appointment to a leading position within the private and public sectors as both Al Gergawi (DH) and Al Abbar (Emaar) have been appointed to important public functions in addition to their chairmanship of private corporations. While Al Gergawi has become Minister of State for Cabinet Affairs, the position of the General Manager of the Department of Economic Development has been given to Al Abbar. Consequently, all three have become important members of the Dubai Executive Council and thus some of the politically most influential persons in Dubai[71].

The Dubai Executive Council (DEC) was founded in 2003 in order to create development plans for Dubai and decide on the phrasing and implementation of new laws. The deputy

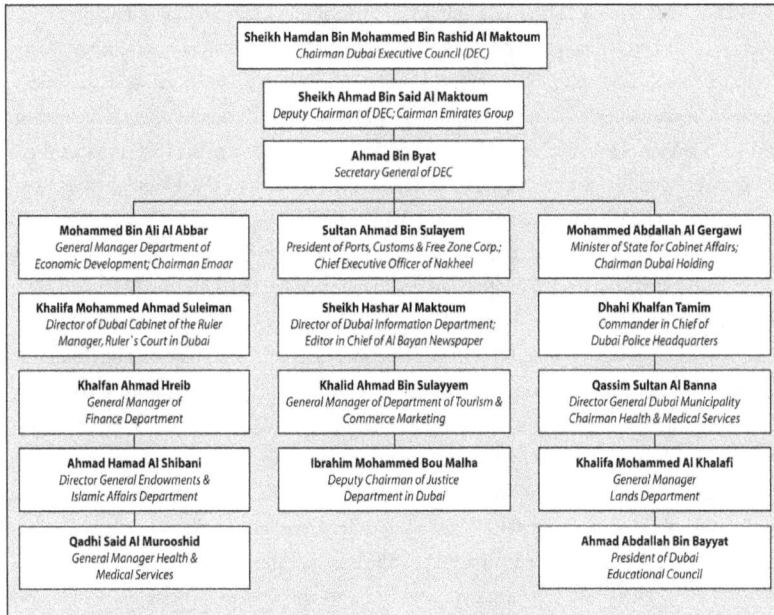

```
┌─────────────────────────────────────────────────────────────┐
│          Sheikh Hamdan Bin Mohammed Bin Rashid Al Maktoum     │
│                 Chairman Dubai Executive Council (DEC)         │
└─────────────────────────────────────────────────────────────┘
                Sheikh Ahmad Bin Said Al Maktoum
          Deputy Chairman of DEC; Cairman Emirates Group

                     Ahmad Bin Byat
                  Secretary General of DEC
```

Mohammed Bin Ali Al Abbar General Manager Department of Economic Development; Chairman Emaar	Sultan Ahmad Bin Sulayem President of Ports, Customs & Free Zone Corp.; Chief Executive Officer of Nakheel	Mohammed Abdallah Al Gergawi Minister of State for Cabinet Affairs; Chairman Dubai Holding
Khalifa Mohammed Ahmad Suleiman Director of Dubai Cabinet of the Ruler Manager, Ruler's Court in Dubai	Sheikh Hashar Al Maktoum Director of Dubai Information Department; Editor in Chief of Al Bayan Newspaper	Dhahi Khalfan Tamim Commander in Chief of Dubai Police Headquarters
Khalfan Ahmad Hreib General Manager of Finance Department	Khalid Ahmad Bin Sulayyem General Manager of Department of Tourism & Commerce Marketing	Qassim Sultan Al Banna Director General Dubai Municipality Chairman Health & Medical Services
Ahmad Hamad Al Shibani Director General Endowments & Islamic Affairs Department	Ibrahim Mohammed Bou Malha Deputy Chairman of Justice Department in Dubai	Khalifa Mohammed Al Khalafi General Manager Lands Department
Qadhi Said Al Murooshid General Manager Health & Medical Services		Ahmad Abdallah Bin Bayyat President of Dubai Educational Council

Fig. 9: The Dubai Executive Council.

chairman of the DEC is the uncle of Sheikh Mohammed, Sheikh Ahmad Bin Saeed Al Maktoum, who is the president of the Civil Aviation Department[72]. Furthermore, he is in charge of the airline Emirates as chairman of the Emirates Group. Apart from Al Abbar, Al Gergawi and Bin Sulayem who have already been mentioned, there are eleven other appointed members of the DEC including Khalifa Mohammed Al Khalafi, the General Manager of the Lands Department, and Qassim Sultan Al Banna, the Director General of the Dubai Municipality (DM).

The Dubai Municipality itself has been organised into five sectors, namely, the Corporate Service Sector, the General Service Sector, the Environmental and Public Health Services Sector, the Health Service and Environmental Control Sector and last but not least the Planning and Engineering Sector. However, all central development decisions and initiatives have originated from Sheikh Mohammed and his four main economic leaders. Thus, the governance of Dubai has been largely formed by corporate interests and the general growth and profit orientation of the private sector. This new model of corporate-driven governance has led to a more decentralised administration and thus to an accelerated development.

One consequence of this privatised form of governance has been a decrease in the administrative authority of the Dubai Municipality caused by its loss of responsibility for all of the urban area due to the newly founded FTZs, which have been run as private companies and not as public institutions. Therefore, the current urban development has been mainly determined by the decisions of the main corporations with the permission of Sheikh Mohammed as the last authority. Nevertheless, the Dubai Municipality has retained its urban planning responsibility in certain areas. Its Planning and Engineering Sector is subdivided into six different departments, namely, the Planning Department, Survey Department, GIS Department, Building Department, General Projects Department and Architectural Heritage Department. The Planning Department itself is structured into three main sections, which are the Planning Studies Section, the Planning Execution Section and the Advertisement Section. While the Planning Studies Section (PSS) is responsible for comprehensive and detailed planning, the Planning Execution Section (PES) has to deal with land allocation and zoning ordinance. The PSS's main function is to formulate and periodically review the general planning regulations. It is also in charge of formulating, reviewing and implementing the urban area structure plan and has to design zoning plans per administrative area. The PES is mainly occupied with the formulation and administration of planning ordinance and the issue of planning permits. Furthermore, it exercises planning control by following up the implementation of the planning zoning ordinance. While the Planning Department is in charge of land use planning and building permission, the independent RTA[73] (Road and Transport Authority) is responsible for developing the general road network and public transport projects.

The Planning Department operates on the basis of the five-year strategic plan of the Dubai Municipality, which currently covers the development period from 2007 to 2011. There are three main general strategic directions regarding the urban planning section. The first involves complying with the existing rules and legislation that govern urban development concerning both projects and institutions. The second involves general planning development, including conservation projects related to architectural heritage wherein the planning is particularly focused on the development of existing communities and their supply with infrastructure and the provision of residential plots for citizens. The last main strategic direction is the creation of a suitable legislative environment, including the maintenance of steady updates to the planning zoning ordinance and ensuring comprehensiveness in the update of planning criteria for public services. The zoning ordinance is one of the biggest responsibilities of the Dubai Municipality and is based on three kinds of plans with different time frames. These are the Dubai Emirate Strategy Physical Plan (1993–2050), the Dubai Urban Area Structure Plan (1993–2015) and the Five-year Plan for the Dubai Urban Area (2001–2005). Although the Planning Department has retained its function as the main urban planning institution of Dubai, it has lost its planning influence on most of the urban area, which has become occupied by large-scale real-estate projects of the private sector. Consequently, the role of the Planning Department has been restricted to mainly the development of the former urban area and the macro-infrastructure of the overall urban area.

The transformation of the form of Dubai's urban governance from a centralised administrative form to a more and more corporate-driven model began in the 90s when the Amir Sheikh Mohammed became responsible for the economic development of the small emirate. One of the first consequences of this transformation was the foundation of the DED in 1992 as a governmental institution intended to create economic strategies for rapid economic growth and diversification. Furthermore, the Sheikh initiated major investments in order to establish new economic sectors in Dubai. This particularly concerned the tourist industry, when hotels, beaches, cultural heritage sites, theme parks and events such as the Shopping Festival were subsidised by the government in order to market Dubai as an up-and-coming tourist centre in the Middle East. In this context, landmarks such as the Burj Al Arab have become driving forces in the creation of the image of a modern Dubai. Along with tourism and growing international attention, many other economic sectors started to develop such as trade and financial businesses. Other important factors in the move of many companies to Dubai include liberal policies with regard to freedom from taxes and legal and bureaucratic burdens as well as Dubai's interesting geopolitical location between growing global markets and the availability of relatively cheap energy, labour and modern infrastructure. This successful attraction of many new economies led to a growing interest in investment in real estate, which was expected to gain rapidly in value because of economic growth and thus exponential demand. Consequently, Sheikh Mohammed initiated the founding of large real estate companies in the form of Emaar and Nakheel in order to develop projects for the new investment market, which became more and more driven by speculation.

Although the legal process of opening the real estate market to foreign investors took several years until 2002, many projects were already announced and in the process of realisation. While the first freehold properties were predominantly housing developments, there was an early attempt to use the growing investment possibilities to create specialised business parks such as Dubai Internet City and Dubai Media City in order to encourage rapid economic development and attract future investments. Consequently, more and more companies were founded and subordinated under the umbrella of two new holdings, Dubai Holding and Dubai World. While the real estate developers Dubai Properties and Sama Dubai became part of DH and Nakheel joined DW, Emaar has remained independent due to its specific characteristic of not being 100% owned by the royal family. Nevertheless, Sheikh Mohammed announced who was to be the chairman of Emaar as he did in the case of his holdings. In addition to their leading positions in the private sector, these businessmen gained public functions and became the most important figures in the Dubai Executive Council, which is currently the highest authority after Sheikh Mohammed himself. Consequently, four leading businessmen, including Emaar's chairman Sheikh Ahmad Bin Saeed Al Maktoum, run a large part of Dubai under the supervision of Sheikh Mohammed from a purely business oriented perspective, which led to an enormous growth and speed in economic development in recent years. This joint venture between the ruler and the private sector is profoundly dependent on the fact that most unbuilt land legally belongs to the royal family. The result has been an increasingly diminishing influence on the part of the public sector on urban development due to the large-scale distribution of land to private developers.

55

While the Planning Department of the Dubai Municipality has remained in charge of the implementation of the urban area structure plan within already existing built areas, most new developments are planned individually by each real estate company itself. In many projects, one master developer creates a master plan of the development and many further sub-developers execute the project under his supervision. In order to guarantee rapid development, the infrastructure is usually developed for the whole project by the master developer and a less bureaucratic system of building permissions has been established through a system in which initial approval is granted by the master developer before being passed on for approval by the Dubai Municipality as well as other authorities such as, for example, JAFZA, which allocate building permits within their areas. This decentralised way of planning and developing Dubai has been the biggest factor in the exponential growth and profound transformation of the former urban structure. Consequently, Dubai can claim itself to be one of the fastest building sites in human history and an embodiment of a new concept of a city governed as a profit oriented company speculating on future growth. This particular form of corporate governance with few restrictions and burdens for the single investor in combination with a powerful Amir with the power to make ultimate decisions together with Dubai's geopolitical location give Dubai great economic potential and thus highly attractive for real-estate investment.

2.3. The Urbanisation Driven by Private Developers

2.3.1 The Four Main Areas of Development

New Dubai – the First Urban Expansion in Jumeirah:
After the Emirates Golf Club was founded in 1988 and several hotels including the Burj Al Arab were built during the 90s along the coast, western Jumeirah became the first address for freehold property developments. The convenient infrastructural supply due to the expansion of the already existing road network and Jumeirah's location along the coast prompted Emaar to develop its very first freehold property project, called Emirates Hills, in Jumeirah in 1999. Over the course of the following years the project was expanded through several developments including the Springs, the Meadows, the Lakes, the Views and the Greens into one large development area called Emirates Living, which will be home to approximately 40,000 people. The whole area is about 12.3 sq. km and is predominantly occupied by two-storey villas[74]. Multi-storey apartment buildings are being built within the projects the Views and the Greens as well as within several smaller projects in the north of the development. In the case of Emirates Hills, the master developer Emaar has built the basic infrastructure and provided freehold properties in the form of undeveloped large-scale plots of about 40 x 100 sq m, which have been designed and constructed individually by each investor[75]. Most areas of the neighbouring projects however were developed by the master developer itself by constructing a reduced set of different types

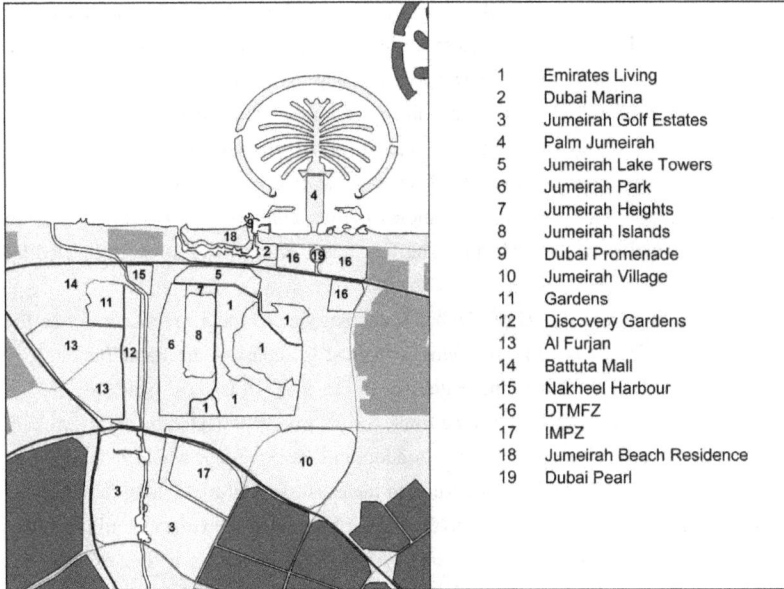

1	Emirates Living
2	Dubai Marina
3	Jumeirah Golf Estates
4	Palm Jumeirah
5	Jumeirah Lake Towers
6	Jumeirah Park
7	Jumeirah Heights
8	Jumeirah Islands
9	Dubai Promenade
10	Jumeirah Village
11	Gardens
12	Discovery Gardens
13	Al Furjan
14	Battuta Mall
15	Nakheel Harbour
16	DTMFZ
17	IMPZ
18	Jumeirah Beach Residence
19	Dubai Pearl

Fig. 10: The developments in New Dubai.

of villa. The special feature of Emirates Living is the vast area of unbuilt land in the form of more than 20 artificial lakes, many green areas and a golf course occupying more than 35% of the whole development. Furthermore, instead of an orthogonal road network, an ornamental layout was chosen for the developments, creating an individual structure with winding streets.

Apart from these housing areas, which have mainly been set up as gated communities, several new 'free economic zones' have been developed in order to create business areas close to the new suburbs. In 2000, Dubai Internet City (DIC) was established as the first free zone in Jumeirah providing optimised business opportunities for international technology, software and internet companies such as HP, Microsoft, IBM and Siemens[76]. In 2008, around 850 companies were already settled in DIC, creating more than 10,000 jobs[77]. In addition to DIC, the company TECOM, which was established as a subsidiary of Dubai in order to invest in the knowledge-based economy, launched Dubai Media City (DMC) in 2001. The main goal of DMC has been to attract large regional and international media organisations ranging from news agencies, publishing companies, online media, advertising, production and broadcasting facilities[78]. More than 800 media groups including the likes of CNN, BBC and Reuters have decided to settle in DMC. Apart from three main buildings, DMC will be expanded by the development of 84 high-rise buildings that will mainly contain apartments[79]. In 2003, TECOM initiated the development of Dubai Knowledge Village (DKV) as part of the free zone conglomerate. In recent years, many private schools and universities moved to DKV, for example, the British University in Dubai and the American College of the Emirates. In addition to the

57

American University of Dubai, DKV has made Jumeirah one of the biggest educational centres of Dubai. Instead of having clear borders, DIC, DMC and DKV have been merged together into one free zone called Dubai Technology and Media Free Zone, which is situated between Sheikh Zayed Road and the coast with an area in the south-east covering about 3.5 sq. km to allow for future expansion. In the centre of the conglomerate, the construction of the project Dubai Pearl has been started on a circular area to form a commercial centre and a residential complex of four connected high-rise buildings to house approximately 9,000 residents. The first phase is expected to be completed by 2011[80].

A bypass leading from TECOM's Dubai Technology and Media Free Zone forms the starting point of what is currently Dubai's biggest landmark – the Palm Jumeirah. A 300-metre long bridge leads to the beginning of the artificial islands, which are shaped in the form of a palm with a 2-km long trunk and 16 fronds protected by an 11-km long crescent functioning as a breakwater[81]. In addition to a monorail, which runs from the crescent over a bridge and down the trunk to the coast, an 800-metre long tunnel at the top of the palm connects the crescent to the palm. Since 2001, more than 92 million cubic metres of sand have been needed to create the whole land mass on an area of about 5.5 sq. km, which has added about 78 km of new coast line[82]. In 2008, about 1,400 villas were already been built on 11 fronds, and 20 multi-storey apartment buildings providing about 2,500 housing units have been developed on the east side of the trunk. More than 25 multi-storey buildings, predominantly apartment blocks and hotels, are currently under construction, including the highest landmark, the Trump Tower, with a height of 270 metres, located in the middle of the Palm trunk[83]. The whole project covers an area of 5 x 5 km and mainly consists of luxury freehold properties in the form of two-storey villas along the fronds and apartment blocks along the trunk in addition to 32 hotels and resorts, of which as of 2008 eight were under construction[84]. As one of the biggest landmarks, the 5-star hotel Atlantis has already been completed in the middle of the crescent. After a development period of about nine years, Nakheel's signature projects are expected to be completed by 2010. From the start, the Palm Jumeirah was planned as a freehold property project with additional tourist attractions in the form of hotels, resorts, marinas, beaches and shopping centres. Instead of the previously planned 4,500 housing units, there will be approximately over 8,000 in the whole development[85].

In addition to the growing number of residence units on the Palm project itself, it has attracted new housing projects along the coast offering views on the artificial islands. One of these developments is Emaar's Dubai Marina, a conglomerate of residential high-rise buildings along one of the largest man-made marinas in the world. Since 2003, the project has been developed in different stages on an area of more than 4 sq km for about 100,000 future residents[86]. The overall completion of the whole project, including about 200 high-rise buildings and one shopping mall, is expected in 2011[87]. After a 3.6-km long artificial channel was dug, the first six residential towers were built in the east of the project as well as the promenade along the marina. Most of the towers have an average height of between

130 and 200 metres and are generally designed as freehold properties offering various sizes of apartments for the upper real estate market[88]. About nine high-rise buildings are currently being developed with a height of over 300 metres, including the 516-metre tall Pentominium, which has been launched as the future tallest residential building in the world[89]. The Dubai Marina Mall, located centrally along the promenade, will form the commercial centre of the project.

Between the coast and Emaar's Dubai Marina, a second residential high-rise development for about 30,000 residents was completed by Dubai Properties in 2007 – the Jumeirah Beach Residence includes 36 residential towers, four hotel towers and four beach clubs spreading along the 1.7-km long shoreline[90]. The third and second largest development of a high-rise conglomerate in Jumeirah is Nakheel's Jumeirah Lake Towers stretching over an area of 1.8 sq. km on the opposite side of Sheikh Zayed Road along the Dubai Marina. After completion in 2014, the whole development will be the first mixed-use free economic zone of Dubai, including 87 towers, which will predominantly be residential towers for more than 60,000 people and office towers for more than 120,000 working visitors[91]. About 78 towers with 35 to 45 floors will be clustered in groups of three, surrounded by four artificial lakes covering an area of about 180,000 sq m[92]. In addition to the artificial waterways and lakes, the overall unbuilt area of the whole project will be more than 730,000 sq m, including landscaped parks[93]. The centrepiece of the project will be the Almas Tower with a height over 360 metres[94]. In the south of the project, Nakheel has started a smaller high-rise development called Jumeirah Heights offering about 2,300 residences within four high-rise buildings and six multi-storey apartment blocks[95]. Apart from these two developments inland, Nakheel has launched another high-rise project on reclaimed land at the entrance of Emaar's Dubai Marina. The Dubai Promenade will include eight residential towers, one office tower and two hotels including the Icon Hotel as a landmark in the shape of a vertical doughnut[96]. After completion by 2012, the built-up area will be over 1 million sq m, providing apartments for approximately 9,700 future residents[97]. All five high-rise developments cover an area of more than 6 sq. km, providing housing units for a total future population of approximately 200,000 residents.

The project Jumeirah Heights marks not only the end of what is currently Dubai's biggest high-rise agglomeration, it is also designed to be part of another signature project of Nakheel in Jumeirah – the Jumeirah Islands. The 300-hectare development consists of 46 clusters of man-made islands surrounded by artificial lakes. The 736 villas have been developed in different sizes and designs in order to create individual looking clusters[98]. Along the borders of this development Nakheel has launched the project Jumeirah Park, which will include 2,000 villas and about 10 mid-rise residential buildings in the centre on an area of more than 350 hectares and is due to be completed in 2009[99]. Three different architectural designs and nine different sizes of villa have been developed for the whole project, which will in addition offer small public spaces in the form of gardens and commercial areas[100]. In the east of the project along Emirates Living, there will be a sixth Nakheel project called Jumeirah Village, which will cover an area of about 811 hectares providing mixed-use areas with mainly housing units in the form of more than 7,000 villas and apartments with additional commercial areas[101]. The development is divided into two districts called the Triangle and the Circle, which both follow an ornamental layout of streets, plots, small public spaces, artificial lakes and commercial areas[102]. At the centre of the Circle, eight boulevards with four rows of multi-storey buildings will end at a park, surrounded by 12 towers.

Next to Jumeirah Village Leisurcorp, one of Dubai World's subsidiaries, has started to develop Jumeirah Golf Estates on an area of almost 11.2 sq. km. When completed by 2009, the first phase will occupy an area of 3.75 sq. km and in addition to four golf courses it will integrate 13 residential developments offering about 1,000 individual properties, mainly in the form of villas[103]. Between Jumeirah Golf Estates and Jumeirah Village, TECOM has started to develop a new free economic zone called International Media and Production Zone (IMPZ), which aims to attract companies from the media production sector, for example, media production, graphic design, publishing and packaging[104]. After completion by 2010, it is expected that more than 375 companies will offer jobs for about 75,000 people[105]. In addition to its function as a free economic zone, the 4-sq. km IMPZ will integrate residential developments such as the already launched project Dana Gardens, which will comprise 11 residential towers with about 3,500 apartments on an area of around 280,000 sq m[106]. In addition to its six developments in Jumeirah, the real estate giant Nakheel has begun to develop several projects on areas further west of Jumeirah. A large part of the new development area will be occupied by the project Al Furjan, which is one of Nakheel's biggest inland developments apart from Jumeirah Village. In its first phase, which will cover around 50% of its 5-sq. km project area, over 4,000 homes will be developed comprising low-rise dwellings in addition to several multi-storey apartment blocks along the borders of the project. The master plan involves two development phases and four residential areas with a 'village centre' as commercial area and identification point[107].

While Al Furjan is still at the beginning of its development and is expected to be completed in 2010, the projects the Gardens and Discovery Gardens were already completed in 2008. The Gardens project is split into a bigger area in the north with 129 three-storey apartment blocks and eight multi-storey apartment blocks offering 5,354 housing units and a smaller area of about 208 villas in the south[108]. Along Sheikh Zayed Road on the east side of the Gardens development, the Ibn Battuta Mall was established by Nakheel as one of the biggest shopping malls of Dubai in 2005. The project Discovery Gardens stretches along an area of a length of over 4.5 km and a width of about 500 metres. The whole development comprises 290 multi-storey apartment buildings, which have been mainly grouped into clusters of three within six themed residential areas, providing more than 26,000 housing units[109]. In the north of the Garden developments and close to Emaar's Dubai Marina, Nakheel has launched a new project called Nakheel Harbour and Tower. The development will be divided into four districts covering an area of 2.7 sq. km with a built area of around 1.5 sq. km, which will include 250,000 sq m of hotels, 950,000 sq m of commercial and retail space and large public spaces in the form of landscaped parks and canal walks. The centrepieces of the project will be the inner-city harbour and the over 1,000-metre tall Nakheel Tower, which will be made up of four individual cores with over 200 floors offering space for about 14,000 apartments and 14 hotels. The whole development will include more than 40 towers and several mid-rise buildings with a total of about 19,000 housing units for more than 55,000 future residents. Moreover, the project will be the workplace for around 45,000 people after the expected completion of all its development stages by 2018[110].

After the completion of more than 18 development areas in New Dubai, including Nakheel's projects in the south-west, there will be homes for more than half a million residents on an area of about 70 sq. km. The business districts will be located in TECOM's free economic zones in the north and south in addition to office towers and commercial areas within several projects, providing approximately more than 250,000 future jobs. Apart from the main goal of developing freehold properties, the majority of projects are intended to expand Dubai's tourist industry through the construction of new hotels, resorts and shopping malls. In 2010, when most developments are expected to be completed, the Dubai Metro will connect New Dubai with the old centre of Dubai in combination with further public transport projects.

The Developments along Coast and Creek:

Since 2005, the Mall of the Emirates with its famous ski hall has formed a centrepiece along Sheikh Zayed Road at the crossing point to the famous hotel cluster along the coast that comprises the Burj Al Arab and Medinat Al Jumeirah. Further north, the Dubai International Financial Centre (DIFC) has been launched on a site close to the high-rise agglomeration that lines Sheikh Zayed Road, which currently consists of more than 55 towers containing mainly offices, including Dubai's landmark, the Emirates Towers. In addition to the 15-storey landmark 'the Gate', more than 50 buildings are currently being developed in the DIFC district, which will cover an area of about 450,000 sq m[111]. Apart from financial business, which will offer about 10,000 jobs, the development will integrate hotels, apartments and cultural institutions. The biggest landmark of the project will be the 400-metre 'Lighthouse' tower, which is designed to be a low-carbon commercial high-rise using photovoltaic panels in combination with wind turbines at the upper floors[112]. In addition to the DIFC, a new project called Downtown Burj Dubai has been initiated by Emaar Properties in order to extend the CBD. The 190-hectare project will contain the world's tallest building, the Burj Dubai, and the world's largest shopping mall, the Dubai Mall. In addition, there will be 30,000 housing units in the form of apartments within 19 residential towers and about 38 low- and mid-rise apartment buildings, which will be located around the 12-hectare Burj Dubai Lake[113]. The 162-storey Burj Dubai reaches a height of 818 metres, including its 200-metre tall spire. In addition to 779 private apartments and 43 offices, the first 37 floors of the tower will be occupied by the Armani Hotel, which will be one of in total nine luxury hotels within the whole development[114].

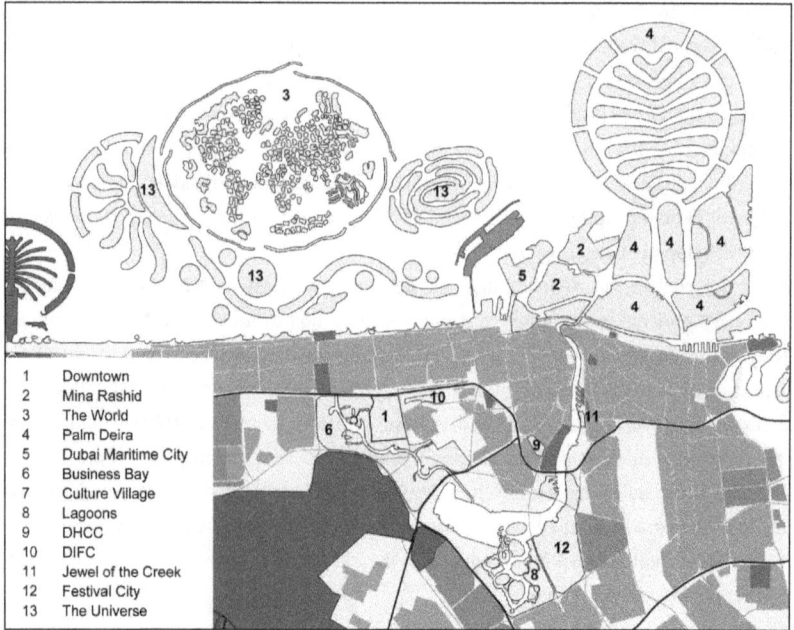

1	Downtown
2	Mina Rashid
3	The World
4	Palm Deira
5	Dubai Maritime City
6	Business Bay
7	Culture Village
8	Lagoons
9	DHCC
10	DIFC
11	Jewel of the Creek
12	Festival City
13	The Universe

Fig. 11: The developments along coast and Creek.

While the Burj Dubai is expected to be completed in 2009 along with many other projects within the development, the Dubai Mall was already opened in November 2008. The three-storey shopping mall provides 350,000 sq m of sales area for about 1,200 shops within a total floor area of 550,000 sq m including an underground floor[115]. The second tallest building of the development is the already completed 63-storey Burj Dubai Lake Hotel with a height of 306 metres located close to the Dubai Mall facing the Burj Dubai on the opposite side of the artificial lake[116].

In addition to the downtown project, Dubai Properties has started the large-scale mixed-use development Business Bay leading from Sheikh Zayed Road along the Downtown Burj Dubai project area toward the end of Dubai Creek. When completed in 2015, the development will consist of 240 towers on an area of 6 sq. km with a total gross floor area (GFA) of about 7.2 sq km, offering apartments for about 191,000 residents as well as offices and shops for a further 110,000 employers and shoppers. While about 22% of the total GFA will comprise residential developments, about 60% will be built as mixed-use developments and about 18% will be developed for commercial use[117]. The high-rise projects will be grouped around the Creek extension and its waterways, which will occupy more than 1 million sq. km of the whole development. Furthermore, there are plans to expand the artificial canal by a length of 2.2 km, thus connecting the coast in Jumeirah by crossing the Sheikh Zayed Road[118]. In 2008, the first project within the Business Bay – the Executive Towers – was completed by Dubai Properties. The project consists of 10 residential towers, one commercial tower, the Aspect Tower and one hotel called Business

Bay Hotel[119]. In addition to this high-rise conglomerate, the 51-storey Vision Tower with a height of 260 metres has already been developed. While Bahrain Bay will be developed along a 10-km long system of canals linked to Dubai Creek, several projects have been launched along the still unbuilt shorelines of the lower Creek.

One of these projects is the Lagoons, which is being developed by Sama Dubai, another real-estate subsidiary of Dubai Holding. The 6.5-sq km development will include seven different districts combining residential units, corporate and commercial facilities and cultural attractions such as the Dubai Opera House. Similar to the Business Bay, the Lagoons project will extend the Creek in the direction of the inland in order to offer more waterfront developments. The artificial canals will create islands for each district, which will be connected by over 40 bridges[120]. The biggest landmark apart from the opera house will be a high-rise cluster called Dubai Towers within the district Al Dana in the centre of the whole development. The four high-rise buildings, which will have heights from 57 to 94 floors and a unique design inspired by a moving candle light, will include apartments, offices, a retail area and a hotel[121]. After the completion of the Lagoons in 2010, about 53 residential buildings will offer about 4,166 housing units within the mixed-use development, which is rather fortunately centrally located close to Dubai's CBD, its old centre and the wildlife sanctuary at the end of Dubai Creek[122]. Along the northern border of the Lagoons, a further 5.2 sq. km mixed-use development has been established by the Al-Futtaim Group. The Dubai Festival City will stretch more than 3.8 km along Dubai Creek and will be divided into three districts which will be connected by a 30-km internal road network. The centrepiece of the development will be the Festival Centre with over 600 shops, 100 restaurants and cinemas. Upon completion in 2010, 50,000 people will live in 20,000 villas, town houses and apartments. In addition, 50,000 office personnel will be accommodated in 930,000 sq m of commercial area. Another attraction of the development will be the Marsa Automotive Park, set to be Dubai's biggest centre of car showrooms and the Four Seasons Golf Club[123].

On the opposite side of the Creek, a further development of Dubai Properties, the mixed-use project Culture Village, will be developed in three phases. Each phase will be approximately 1.2 sq km although the GFA will differ within each phase. The first phase is being established along 1 km of Dubai Creek, offering housing units for about 15,000 future residents within low-, medium- and high-rise apartment buildings. Apart from residential use, phase one will be mainly for commercial use in the form of retail areas and entertainment sub-districts. Furthermore, it will integrate cultural centres such as museums and art galleries. Similar to the already built Medinat Jumeirah, the development will use traditional design elements such as the wind towers in order to create a reference to the locality. Similar to the Lagoons and Festival City projects, the Creek will be extended by a canal and some reclamation will be undertaken to create five extensions into the Creek. Halcrow has planned the 60-storey residential Tower D1 as the highest landmark of the development and an equivalent to what is currently the highest residential tower Q1 at the Gold Coast in Australia[124]. Inland behind Culture Village,

Tatweer has launched a new free economic zone called Dubai Healthcare City (DHCC) at two different sites on an overall area of about 2.2 sq. km. The first phase of the total of two development phases comprises a medical cluster including hospitals, hotels and apartment buildings on an area of about 380,000 sq m between the Wafi City complex and Dubai Creek Park. The second phase consists of a wellness cluster on an area of about 1.8 sq km, which will spread along the lower Dubai Creek facing the Lagoons project on the opposite shoreline[125]. In the future, over 350 hospitals will be established within the Medical Community and the Wellness Community. Furthermore, there will be many research institutions such as universities, wellness centres and residential buildings for employees. The completion of the first phase is expected by 2010[126].

Further up the Creek, the real estate company DIRE[127] has started the small-scale mixed-use development Jewel of the Creek on an area of about 800,000 sq m close to the famous Creek Golf Club[128]. After completion in 2011, the conglomerate of 20 high-rise buildings with 20 to 23 floors will offer a GFA of about 1 million sq m for offices, apartments and two hotels. Apart from the six developments along Dubai Creek, there are currently two main reclamation projects underway in order to extend the urban area in Bur Dubai and Deira along the coasts. While the Dubai Maritime City (DMC) will be built on an area of about 2 sq km between Mina Rashid and the Dubai Dry Docks in Bur Dubai, the Palm Deira of Nakheel will expand the city district of Deira in the north by a surface area of about 46 sq km. Dubai Maritime City belongs to the holding Dubai World and it will be a mixed-use development providing six districts, namely, the Marina District, Marina Centre, Harbour Residences, Harbour Offices, Dubai Maritime City Campus and Industrial Precinct. The development will consist of 14 residential towers, 19 office towers and 12 mixed-use towers within the Marina Centre including four hotels and the landmark of the project in the form of the 229-metre Landmark Tower. The Industrial Precinct will provide all facilities necessary for ship repair and maintenance, including the manufacturing of yachts. Furthermore, a maritime educational campus for about 1,300 students will be built next to the Industrial Precinct and close to the Marina District in the north[129]. While the DMC project is expected to be completed in 2012, Nakheel has launched a plan to develop and expand the port Mina Rashid as a residential area[130]. Thus, the existing harbour area will be expanded by land reclamation to more than 7.2 sq km in order to establish a mixed-use development for about 200,000 future residents[131].

While Nakheel has almost completed its first Palm project in Jumeirah, it has already begun to develop two further Palm projects along the coast in Jebel Ali and in Deira, where the Palm Deira will be a city in itself located close to the old city centre. After the master plan was reviewed in 2007, the length of the project has been reduced from about 15.5 km to 12.5 km while the width of 7.5 km has remained unaltered. If completed, the third Palm project will be the largest man-made island ever with a size seven and a half times bigger than the Palm Jumeirah and thus it has been promoted as being as big as the centre of Paris[132]. The project will be a leisure, residential and tourist development featuring villas, a number of hotels and large marinas. Furthermore, the development will be divided into nine major districts in the form of man-made islands, which will add 226 km of shoreline including 121 km of beach front to Dubai. Over 40 bridges will connect the total of 25 islands with each other and the coast[133]. At the end of 2008, 36% of the reclamation was completed including the first districts in the form of the five main islands Deira Island Front, Deira Island Mamzar, Deira Island Central, Deira Island South and Deira Island North. In the next phases the reclamation of the four remaining districts, namely, Deira Island Trunk, Palm Crescent, Palm Fronds and Palm Crown, are expected to follow. About 55% of the GFA will be occupied by residential developments in addition to around 28% of mixed-use projects, which together will offer housing units for about 1.3 million future residents. The remaining 17% will mainly comprise commercial buildings, which will be particularly developed within the central business districts on the Deira Island Trunk and Deira Island South[134].

Apart from the waterfront developments close to the centre, Nakheel has built another large-scale reclamation project about 4 km off the coast of Jumeirah. The World development was already started in 2003 and its reclamation was completed in 2008. It consists of about 300 islands in the shape of the world, adding 232 kilometres of beach front to Dubai's coastline. The project offers investors the possibility of developing residential, leisure or tourist projects[135]. The individual islands can only be reached by ship or helicopter and thus a system of canals, marinas and transportation hubs have been developed. While a centralised infrastructure system will be established to provide the residents with the basic services, most islands will have to be connected individually to the distribution and collection mains. Four central water plants will be located at the hubs and will deliver water via a network of underwater pipelines. Power will be brought from the main land and distributed to the islands using underwater cables. Waste-water and reuse systems will be located on each island and together with the whole utility system will be placed under water. All utilities within the project will be distributed by way of primary corridors, which will later branch out into a second utility corridor based on the development patterns[136]. Each island has a size from about 14,000 sq m to 42,000 sq m and can be developed, according to the master plan, as either low-, mid- or high-density residential projects. In addition to about 240 residential islands, there will be 12 commercial islands, four transportation hubs, three service hubs and nine resort island-clusters covering 40 islands[137]. The entire project will cover an area of about 931 hectares with a length of 9 km and a width of 7 km, surrounded by an oval breakwater[138].

Nakheel has recently published its plans for a further island development around the World project and along the coast in the shape of the sun, planets and the Milky Way galaxy. The actual master plan of the 3,000-hectare Universe development has not yet been published as the project is still in the phase of engineering, feasibility and environmental studies[139]. Including the new project, the five waterfront developments will add about 90 sq km to Dubai and housing units for about 1.5 million residents. The five projects along Dubai Creek will accordingly add about 18.3 sq km to the urban area and dwellings for more than 90,000 residents. Moreover, the extension of the CBD will add residential units for

65

about 290,000 future inhabitants spreading across a newly developed area of about 8.5 sq km. After the completion of most developments by 2015, Dubai's former urban area will have expanded by about 117 sq km, which all in all will comprise housing units for almost 2 million new residents in addition to offices and commercial units for more than 200,000 employers. Apart from Nakheel's Universe and Mina Rashid projects, the construction of most developments has already been started in recent years and apart from the Palm Deira it is expected that most of them will be completed by 2015 in spite of the current financial crisis.

The Inland Developments and Dubailand:
While the first projects have been developed close to the coast and the already existing infrastructure, the inland of the emirate generally remained untouched during the first years of the recent building boom. In 2003, Nakheel's International City was launched as one of the first inland developments located in Dubai's Al Warsan area about 12 km from Dubai International Airport and 6 km from Dubai Creek. The project covers 800 hectares and is divided into five districts, namely, the Residential District, Central District, Dragon Mart Complex, Lake District and Forbidden City. The development will consist of 413 buildings in total, of which 387 buildings were already delivered in 2008. After completion in 2010, it is expected that the current population of about 60,000 residents will double to a total population of around 120,000 living in 23,847 housing units. The 5-storey apartment blocks have been designed in accordance with 10 themed clusters depending on their location within the Residential District. In the coming years, the development, which mainly houses mid-income guest workers and their families, will be expanded by the addition of the Forbidden City district as a further residential area. In addition, there will be 5,254 retail units mainly located within the Central District, over 65% of the 34 plots for which have already been built by private investors in 2008. Apart from the commercial area in the centre of the project, there has been the development of a large-scale shopping mall at the border of the development called Dragon Mart Complex, which has been fully operational since 2004[140]. Emaar Properties has recently started another development located close to the International City project and Mushrif Park, which was already established in the 80s. The project Mushrif Heights will be a mainly low-rise residential development with an integrated commercial centre on an area of about 4 sq km[141].

In 2002, a huge technology park named Dubai Silicon Oasis (DSO) was launched by Sheikh Mohammed in order to create one of the world's leading centres of advanced electronic innovation, design and development[142]. 'The main objectives behind DSO are to establish an industry based on micro- and nano-electronics and to attract companies that are globally competitive in manufacturing these types of products' says Jurgen Knorr, CEO of DSOA (Dubai Silicon Oasis Authority)[143]. The FTZ was established on an area of about 7.2 sq km just about 2 km to the south of Nakheel's International City and includes a master-planned residential, commercial and educational district along Emirates Road.

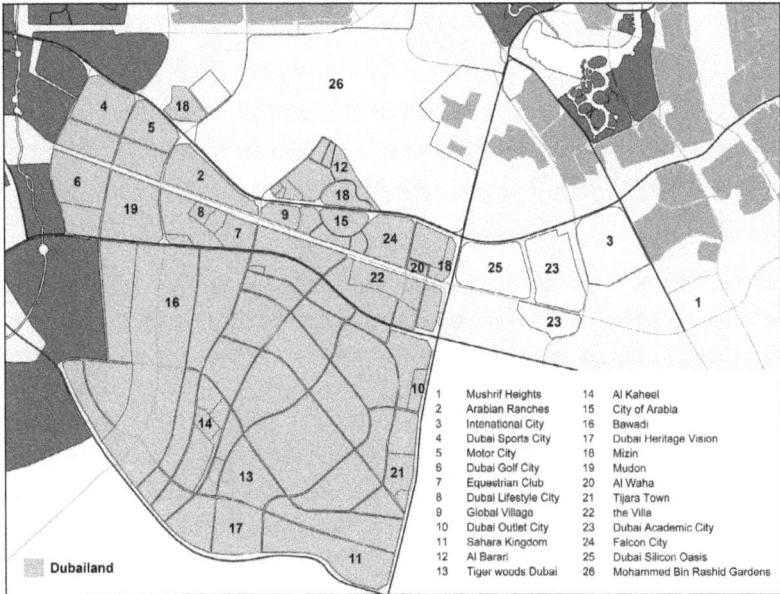

1	Mushrif Heights	14	Al Kaheel
2	Arabian Ranches	15	City of Arabia
3	International City	16	Bawadi
4	Dubai Sports City	17	Dubai Heritage Vision
5	Motor City	18	Mizin
6	Dubai Golf City	19	Mudon
7	Equestrian Club	20	Al Waha
8	Dubai Lifestyle City	21	Tijara Town
9	Global Village	22	the Villa
10	Dubai Outlet City	23	Dubai Academic City
11	Sahara Kingdom	24	Falcon City
12	Al Barari	25	Dubai Silicon Oasis
13	Tiger woods Dubai	26	Mohammed Bin Rashid Gardens

Fig. 12: The inland developments and Dubailand.

The project will consist of about 2.37 sq km of residential area split into about 185,000 sq m of high-density area, about 884,000 sq m of medium-density area and 1.3 sq km of low-density area. Furthermore, there will be a commercial area of about 366,000 sq m located in the centre and in the north in the form of a shopping mall along Emirates Road. The technology park itself will be split into two areas, together covering more than 2.5 sq km. The conglomerate of universities and schools will be established in the centre on an area of over 300,000 sq m. In addition, there will be an area of offices covering about 270,000 sq m, which already contains the headquarters of the DSOA as the landmark of the overall development on a central area of more than 60,000 sq m[144]. After its completion between 2010 and 2015, DSO will offer residential units for about 100,000 inhabitants and more than 20,000 jobs[145].

Close to DSO there will be another conglomerate of private schools and universities called Dubai Academic City (DAC), which will include TECOM's free zone for tertiary institutes called Dubai International Academic City (DIAC). DAC will cover an area of about 12 sq km and will be completed by 2012. While 19 academic institutions have already decided to move to DAC, there are over nine institutions, including colleges and universities currently based in DKV in Jumeirah, which are expected to move in the future. DIAC itself will be developed on an area of about 2.3 sq km and it is planned that more than 40 academic institutions with almost 40,000 students will be established by 2015[146]. Today, TECOM's 'knowledge free zones', including DKV, are already catering to over 10,000 international students from various countries[147].

Dubai's largest inland development is the project Dubailand of Dubai Holding's subsidiary

Tatweer, which will cover an area of more than 279 sq km stretching between Dubai Silicon Oasis and Nakheel's Jumeirah Village. In order to develop such a large area, the development has been divided into 45 main projects and 200 sub-projects. Furthermore, the development will be built in four phases, each of which will last about five years. During the first phase, all initial infrastructure such as the main roads will be developed between 2005 and 2010. In the following years, the main projects will follow, which will comprise leisure and entertainment developments in order to establish Dubailand as the biggest entertainment park in the world, about twice the size of the Walt Disney World Resort in Florida, by the year 2020[148]. At the present time, the development of more than 21 projects of different categories on an area of about 30 sq km is underway[149]. After these 21 and the remaining projects have been completed they will form seven themed zones known as 'Worlds' which will each focus on a different aspect of the development, namely, Attraction & Experience World (13 sq km), Retail & Entertainment World (4 sq km), Themed Leisure & Vacation World (29 sq km), Eco-tourism World (75 sq km), Sports & Outdoor World (19 sq km), Downtown (1.8 sq km) and Science & Planetariums[150]. Most projects however will be mixed-use and integrate various aspects in addition to the central aspect of their theme. Thus, almost every project will include a large number of residential units in the form of multi-storey apartment buildings or villas. While a large number of projects are currently being built by Tatweer's master developer Dubailand, other developers are building other projects, for example, Emaar, which is currently building a project called Arabian Ranches[151].

The Arabian Ranches development is part of Sports & Outdoor World as it includes the Arabian Ranches Golf Club, Equestrian Centre and Dubai Polo Club. The project will cover an area of about 6.7 sq km providing around 4,000 housing units in the form of villas located in 22 districts surrounding the Golf Club and two artificial lakes[152]. Most of the developments within the project will be completed in 2009. It will thus be one of the first large-scale projects opening within Dubailand. Next to the Arabian Ranches along its western border, the Dubailand master developer has started two further signature projects for Sports & Outdoor World – Dubai Sports City and MotorCity. On an area of more than 4.6 sq km, Dubai Sports City will be the world's biggest and first purpose-built sports city with four stadia as the centrepiece of the development, which will include vast residential and commercial developments in addition to sporting facilities such as a golf club and a 3,000-sq m gymnasium. In addition to housing units for about 65,000 people, there will be international schools, hotels and entertainment venues[153]. While the first phase, which involves the golf course and sports academies, was completed in 2007, the remaining developments will follow by 2010[154]. Similarly to Dubai Sports City, MotorCity will be a mixed-use development with integrated sports facilities, in this case the Dubai Autodrome. In addition to one low-rise and one high-rise residential project, there will be a Business Park MotorCity that will function as MotorCity's main commercial area and be divided into three different parts, namely, a retail district, the Auto Commercial district and Applied Technologies district. Furthermore, there will be an 'F1 Theme Park' on an area of about 300,000 sq m. The whole development will be completed on an area of about 3.5 sq km by 2009[155]. Additional projects of Sports & Outdoor World will be Dubai Golf World (including five themed golf courses, a golf academy and a hotel on 5 sq km), the Plantation Equestrian & Polo Club (including three polo fields, a polo academy and a hotel on 1.85 sq km) and Dubai Lifestyle City (including sports academies and hotels on more than 150,000 sq m)[156].

Attraction & Experience World will be a conglomerate of entertainment parks within about 16 projects covering more than 13 sq km. In addition, it will include one of the oldest projects in the area, called Global Village, which was already established as part of the Shopping Festival in 1996. In recent years, Global Village has been expanded as an entertainment centre of Dubai, where several countries participate in the organisation of various cultural events at their pavilions and promote themselves as tourist destinations. In 2008, more than 4 million visitors were attracted by the development, which has become one of the subsidiaries of Tatweer[157]. One of the biggest developments within Attraction & Experience World will be the Falconcity of Wonders on an area of about 4 sq km in the north-eastern part of Dubailand. The project will be developed by Salem Al Moosa Enterprise and completion is expected within five overlapping phases by 2020. The mixed-use development is promoted as a self-contained and multi-faceted tourist and recreational project. The Falconcity project has its name due to its master plan shaped in form of a falcon, which is the emblem of the UAE. While a shopping mall will form the head of the falcon, residential areas with villas divided into different themes will shape the wings and body of the falcon. The centrepieces of the development will be a number of projects trying to imitate famous buildings from around the world that will be used as hotels, shopping malls, entertainment parks or residential developments. Apart from the Dubai Eiffel Tower, which will be the highest landmark in the centre, there will be imitations of the Taj Mahal, the town of Venice, the tower of Pisa, the Great Wall of China, the Hanging Gardens, the pyramids and the lighthouse of Alexandria[158]. In addition to the Falconcity development, there will be more than 12 theme parks within Attraction & Experience World including Legoland, Aqua Dunya, Hit Entertainment, Six Flags, Universal Studios Dubai, Warner Brothers Movie World and Dreamworks Studios Theme Park. Furthermore, there will be science museums within Science & Planetariums and retail projects within Retail & Entertainment World, including Dubai Outlet City on an area of almost 1 sq km, which will include the first shopping mall for factory outlets in the region.

Themed Leisure & Entertainment World will comprise about six projects on an area of almost 30 sq km. Apart from a large number of hotels, spas and resorts, this 'World' will include several residential areas. Very similarly structured will be Eco-tourism World, which will cover with about 75 sq km, the largest area in Dubailand. Apart from a Safari Park and the Al Sahra Desert Resort, a number of residential areas will be included in the projects. For example, Al Kaheel will offer about 400 residences and equestrian facilities on an area of almost 1 sq km similarly to Emaar's Arabian Ranches[159]. A further large-scale residential low-rise development will be Dubai Property's The Villa, stretching over 3.2 sq km along the southern border of Falconcity of Wonders[160]. Furthermore, Dubai Properties has launched another residential development called Mudon, which will provide five themed districts on an area of about 6.8 sq km with housing units for around 50,000 future residents after its completion in 2012[161]. On an area of about 3.8 sq km in the eastern part of Dubailand within the Nad Al Sheba area, Dubai Heritage Vision will be established as an authentic desert village oasis in order to attract tourism and exhibit the regional culture[162]. Close to Dubai Outlet City, Dubai Properties has started to develop the mixed-use project Tijara Town on an area of about 2.3 sq km, offering about 7,700 residential units in the form of apartments[163]. In addition, the development Al Barari, with its 330 villas surrounded by botanical gardens and leisure facilities, has been launched on an area of about 1.3 sq km along the north-western border of central Dubailand[164]. The centrepiece of Dubailand apart from all its attractions and residential developments will be the Downtown project, which is mainly occupied by the City of Arabia. This development will be the main gateway into Dubailand on a circular area of about 1.85 sq km and in addition to its function as one of the biggest retail centres of Dubai it will integrate residential and tourism projects.

All in all, the City of Arabia will consist of four projects including the theme park Restless Planet, the world's biggest mall The Mall of Arabia, the commercial and residential district Wadi Walk and the high-rise project called Elite Towers. All in all, there will be a total of about 8,200 residential components upon completion for an expected population of about 32,800 future residents. Apart from mid-rise apartment buildings within the project Wadi Walk, there will 34 commercial and residential high-rise buildings within Elite Towers[165]. With a gross leasable area of almost 1 sq km, the Mall of Arabia will replace Dubai Mall as the world's biggest shopping mall[166]. Moreover, there are plans to develop a monorail that will be connected to the Dubai Metro in order to transport residents and visitors[167]. The City of Arabia will have a phased opening that will start in 2010. Tatweer, as master developer of Dubailand, is currently involved in three projects in addition to its already established Global Village. In addition to Tiger Woods Dubai, which is a mixed-use development with a golf course on an area of around 5.1 sq km in the north-east of Dubailand, there will be a large-scale mixed-use development called Mizin, which will be divided into four themed cities at four different locations. The mixed-use high-rise development Majan will form, together with the City of Arabia, the downtown of Dubailand[168]. Together with the three remaining districts Liwan, Arjan and Remran, the project will offer residential units for more than 187,000 people[169]. One of Tatweer's largest projects within Dubailand will be Bawadi – a development on an area of around 12.5 sq km stretching 10 kilometres from north to south. The centrepiece of the mixed-use development, which was inspired by the Las Vegas strip, will be a boulevard comprising a commercial district and about 51 hotels, which will establish the development as the world's longest shopping and entertainment project[170]. While Tatweer's Bawadi will be completed by 2014, several future developments of Dubailand will follow by 2020, making it to the world's biggest tourism and entertainment project in combination with many residential developments, creating an urban area for an estimated population of 2.5 million people including tourists, workers and residents[171].

In 2008 a new major inland development was launched by Dubai Properties – Mohammed Rashid Garden City – which will be located between Dubailand and the expanded CBD district along Sheikh Zayed Road. The development will occupy an area of almost 82 sq km and it is expected to cater to about 200,000 future residents in addition to approximately 150,000 daily visitors[172]. Over 70% of the project area will be developed as landscaped parks including waterways in the form of a 25 km-long system of canals that will connect both Dubai Creek and the coast in Jumeirah. Within the ornamental layout of the project there will be three main centres of the development in the form of the House of Humanity as the home of humanitarian establishments, the House of Nature as a family resort with an integrated wildlife park and the House of Commerce as a commercial and retail district. All three centres will surround the House of Wisdom as the future centrepiece of the whole development, which will include a university campus with an integrated library and mosque[173]. While the completion of the main infrastructure is expected in 2015, the construction of the entire project will follow within different phases beyond 2020.

The Developments in and around Jebel Ali:

Since the 80s Jebel Ali has been an important economic centre as the world's largest man-made harbour and the first free trade zone of Dubai. While most recent developments have been completed or started close to former urban areas further north, Nakheel has begun to develop its second Palm project in Jebel Ali. Land reclamation for the Palm Jebel Ali, which is about 50% bigger than the Palm Jumeirah, began in 2002 and was completed in 2008. The project area will cover more than 7.7 sq km and the development will add 70 kilometres of beaches to Dubai's coastline. After completion by 2020 the project will include residential units for about 250,000 people in the form of more than 1,950 villas, around 680 town houses and a large number of apartments[174]. Apart from freehold residential developments, there will be various commercial and tourism projects such as the theme-park cluster The Worlds of Discovery, comprising SeaWorld, Aquatica, Busch Gardens and Discovery Cove. This agglomeration of theme parks will be developed on an island in the shape of an orca whale forming the central island of the palm's breakwater crescent. Other special features of the project will be a Sea Village consisting of six marinas and 1,060 Water Homes, which will be built on stilts along the crescent in form of an Arabic poem[175] by Sheikh Mohammed himself[176]. In addition to the Palm Jebel Ali, Nakheel has launched its biggest project in Jebel Ali – the Waterfront – which will be a self-contained city on an area twice the size of Hong Kong island. Thus, the total area will cover around 130 sq km, of which 20% will be on reclaimed land in the form of five islands in the shape of a crescent surrounding the Palm project as well as one detached island between the two projects[177].

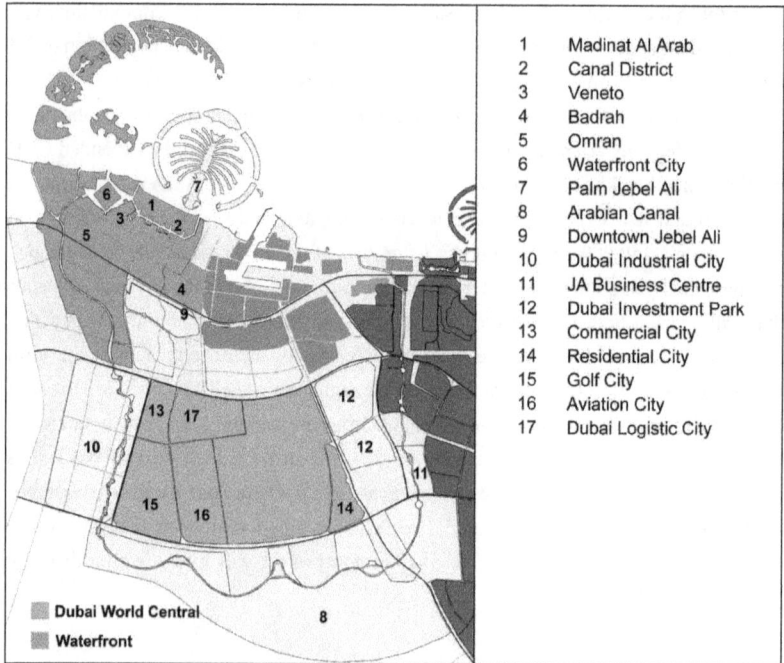

1	Madinat Al Arab
2	Canal District
3	Veneto
4	Badrah
5	Omran
6	Waterfront City
7	Palm Jebel Ali
8	Arabian Canal
9	Downtown Jebel Ali
10	Dubai Industrial City
11	JA Business Centre
12	Dubai Investment Park
13	Commercial City
14	Residential City
15	Golf City
16	Aviation City
17	Dubai Logistic City

Dubai World Central
Waterfront

Fig. 13: The developments in Jebel Ali.

The Waterfront project will be mixed-use and developed in different stages starting at the coastline and stretching to the south. The first phase of the project will be the development of Madinat Al Arab, which will be developed along the coastline from the beginning of the Palm Jebel Ali toward the south-west on an area of about 7 sq km. The mixed-use development will comprise a built-up area of more than 12 sq km and a future population of approximately 251,000 by its completion in 2015. Further developments will be the Palm Canal Towers, a conglomerate of six mixed-use towers, and a 138-hectare park called Boulevard Park stretching along Madinat Al Arab and the second phase of the Waterfront, which will be developed in the form of a medium-density residential area called Canal District on an area of almost 1.8 sq km. Its built-up area of about 4.3 sq km will be mainly occupied by residential units housing about 75,000 future residents by 2016. In addition to its location at Boulevard Park, the project will stretch along the Palm Cove Canal, which will be about 140 metres wide and 8 kilometres long. The third phase of the Waterfront project will comprise the development Veneto, which will offer low-to-medium density residential areas along the opposite side of the canal on an overall area of about 2.1 sq km. Thus, the built area of more than 850,000 sq m will be mainly occupied by dwellings in the form of villas and town houses for an estimated future population of about 14,000 people. Next to Veneto on its south-eastern border, a medium-density project called Badrah has been launched. The project targets middle-income groups and will offer residential units in the form of apartments for about 120,000 residents by 2017.

Furthermore, there will be integrated commercial areas on an area of about 5.75 sq km and within a built area of around 5.7 sq km. Close to the centre of the Waterfront on an area of almost 2.4 sq km, there will be a further residential development called Omran offering accommodation for about 60,000 workers by 2010[178].

Apart from the Islands, which will mainly comprise residential projects on an area of 2,400 hectares, the Waterfront City, which will be the CBD of the whole development, will be completed by 2018. Its 3.3-sq km project area will consist of an almost square island in the waters of the Palm Cove Canal in addition to development areas along the opposite shorelines. The project will be developed as a high-rise agglomeration with a built area of 12 sq km providing housing units for around 80,000 future residents. Furthermore, it is expected that more than 340,000 daily visitors, including a large number of commuters, will add to an overall population of about 425,000 people[179]. After completion of all development stages in 2025, more than 1.5 million people are expected to live at the Dubai Waterfront, which will form the starting point of a further large-scale project by Nakheel called Arabian Canal[180]. Around 23% of the overall 75-km long man-made canal will be part of the Waterfront development, where it will be connected to Waterfront City and the Palm Cove Canal[181]. The Arabian Canal, which will be the largest and most complex engineering project ever undertaken in the Middle East, will lead from the Waterfront development to the Dubai Marina in Jumeirah by passing through the desert. Nakheel launched its first phase at the beginning of 2008 and is expecting its completion in 2013. The second phase overlaps the first and was begun in the second half of 2008 with an expected completion date of 2014. The development of the projects along the canal will take at least 15 more years. The canal's waterway navigation width will be established between a minimum of 75 metres and a maximum of 150 metres. Due to the waterway depth of 6 metres, there will be places where contractors will have to excavate to a depth of 70 metres to keep the canal at sea level. The digging of the canal will lead to an estimated 1 billion cubic metres of excavated earth, which will be used for landscaping along the shorelines of the man-made creek[182]. Many residential developments will follow along the Arabian Canal creating homes for approximately 2 million people.

In addition to the above, several commercial developments have been initiated along the future canal, for example, Tatweer's Dubai Industrial City and Dubai Investment Park. On an area of about 52 sq km in the south-east of the Waterfront, Dubai Industrial City will be developed as a self-sustaining city, specifically designed for the needs of manufacturing companies with a downtown and residential areas, including low-cost housing solutions for about 500,000 future residents[183]. Although the project will not be established as a free zone, there will be certain benefits to companies such as duty exemption on imports, tariff-free access to the GCC and MENA markets and no taxes on personnel and corporate income. After completion in 2015, the industrial sectors will include the manufacture of machinery and mechanical equipment, transport equipment, base metal, chemicals and mineral products[184]. Similarly to Dubai Industrial City, the company Dubai Investments has developed its project Dubai Investment Park (DIP), which was launched in the 90s, as a self-contained city where future inhabitants can live and work after its completion by 2010[185]. On an area of approximately 32 sq km close to New Dubai and along the Arabian Canal, the DIP project will integrate industrial, commercial and residential areas. The development will offer more than 19,000 housing units and several hotels within the residential and recreational zone. Furthermore, there will be a commercial district in the form of a business park, where each company will be able to design and construct their own facilities. The central feature of DIP will be its industrial complex, including the High-tech Industrial Zone, Light Industrial Units and Staff Accommodations[186]. Today, more than 800 companies are located in DIP including cosmetics, food, light engineering, logistics and distribution industries. On the opposite shoreline of the Arabian Canal, several projects will be built next to the DIP development, including the Jebel Ali Business Centre.

In close proximity to the Jebel Ali Free Zone, Limitless, a company of the holding Dubai World, has launched the development of a new CBD project called Downtown Jebel Ali within four separate zones stretching 11 kilometres along Sheikh Zayed Road, which leads to Abu Dhabi. The 2-sq km project area is expected to be developed by the year 2012[187]. All in all, 237 of a total of 326 buildings will be occupied by residential use, offering apartments for about 30,000 people. In the daytime, more than 200,000 people will be living and working in the Downtown project, which will be centred in between three main development areas including New Dubai, the Waterfront and the new airport more inland[188]. Each zone will have its individual Urban Centre with a plaza surrounded by a cluster of high-rise buildings. Furthermore, there will be so-called Trellis Districts for low-rise residential buildings and Medina Districts at the periphery of each zone for mid- to low-rise residential buildings[189].

In addition to the current expansion of Dubai International Airport with a third terminal, which will become the world's single largest terminal building with an annual capacity of about 43 million passengers after its completion by 2011, a new airport has been launched in Jebel Ali. Al Maktoum International Airport will be ten times the size of Dubai International Airport and will be the centrepiece of a new development cluster called Dubai World Central (DWC), located close to Jebel Ali Port and surrounded by the Arabian Canal project. In addition to its future position as the world's largest passenger airport, accommodating an estimated 120 million annual passengers, the new airport is expected to become the world's biggest cargo hub with an annual capacity of 12 tonnes of cargo[190]. Although the overall completion of the airport is expected by 2015, the construction of the first runway with its terminal has already been started and is expected to operate by the end of 2009[191]. The airport will cover about 50% of the DWC, which will cover a total area of about 140 sq km including DWC Aviation City, Dubai Logistics City (DLC) and three real estate developments within the DWC Real Estate cluster. The 6.7-sq km large DWC Aviation City will host end-to-end manufacturing, maintenance, research, training and other facilities related to aviation services. Its construction has already commenced and it is expected that the first operations will start in 2009.

Furthermore, the DLC development has been launched on an area of about 25 sq km between the new airport and the Jebel Ali Port, to which it will be connected by a bridge in order to reduce sea-air lead times for cargo to a few hours in comparison to the current time frame of one to three days[192]. DLC has been designed as an integrated multi-modal logistics platform handling over 12 million tonnes of annual air cargo in up to 22 air cargo terminals. In addition to terminals, office buildings, land plots for industrial business, trading companies, distributors, logistics service providers, warehouses and air-side cargo handling facilities, there will be an integrated staff village with accommodation for over 40,000 workers[193]. Phase one of DLC will start to operate in 2009 and 40% of its area has already been reserved. The contract logistics and forwarding areas will be completed by 2010 and 2012[194].

DWC Real Estate consists of three main projects, namely, Residential City, Commercial City and Golf City, on an overall area of about 39 sq km. The 8-sq km large Residential City will provide homes for approximately 250,000 people and working places for about 20,000 employees. The focus of the mainly residential project will be the development of affordable dwelling units to accommodate the future employees of DWC[195]. Close by, Golf City will offer over 5,000 residential units for the upper real-estate market in addition to two integrated golf courses[196]. The CBD of DWC will be formed by Commercial City, consisting of 850 towers with six to 75 floors. The commercial cluster will be established along the Arabian Canal and it will offer offices for more than 130,000 future employees[197]. Like Residential City and Golf City, it will integrate a retail district and its development will progress in overlapping phases with the first expected to be completed between 2009 and 2010[198]. After completion of the whole DWC project, including its additional infrastructure and service projects Global Technologies and DuServe, more than 1 million people will live and work within the mixed-use developments on an area almost twice the size of Hong Kong Island by 2020[199].

2.3.2 The Facts and Figures of the Future Urban Structure
The main developers Emaar, Nakheel and Dubai Holding's real estate companies have launched or completed more than 55 large-scale projects covering a development area of around 660 sq km. While Emaar, as one of the first real-estate companies producing freehold properties, has been in charge of just five masterplanned projects, including Emirates Hills, Dubai Marina and Downtown Burj Dubai, Nakheel, which is the biggest real estate developer, has been in charge of over 19 developments, mainly on reclaimed land, for example, the palm projects, the World and the Waterfront. Furthermore, Nakheel has initiated the biggest engineering project, the Arabian Canal, and several inland housing projects, for example, International City. Like Emaar, Nakheel has started to develop projects in Jumeirah, where it became the largest developer with projects such as the Palm Jumeirah, Jumeirah Lake Towers and Jumeirah Islands, covering an overall area of more than 35 sq km. In recent years, Nakheel has focused on an area toward the south-

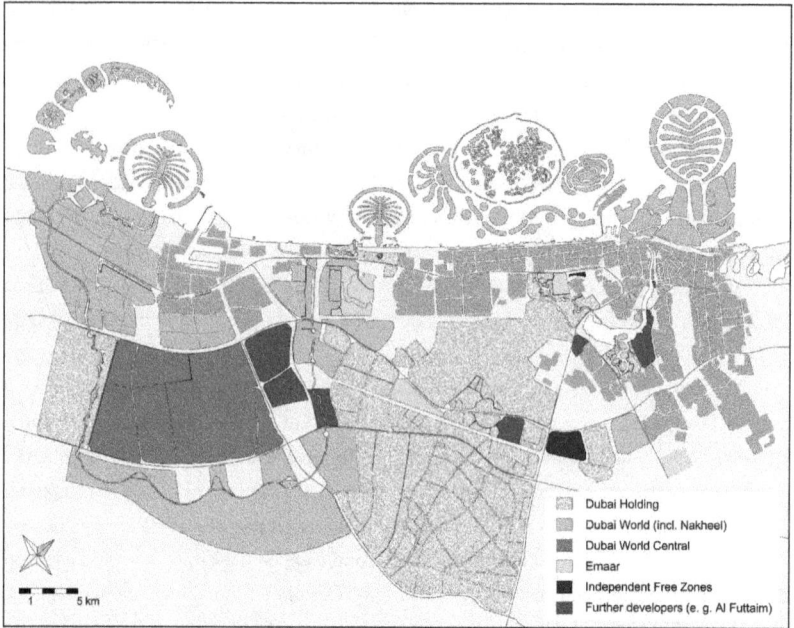

Fig. 14: The project areas of major developers.

west of Jumeirah, where it has launched three housing projects in addition to a mixed-use, high-rise development called Nakheel Harbour & Tower, which will include the over 1-km tall Nakheel Tower, which will outshine Emaar's Burj Dubai as the world's tallest tower. While Nakheel gained its leading position mainly through large-scale reclamation projects covering an area of almost 74 sq km, the subsidiaries of Dubai Holding have been in charge of most free economic zones and some of the biggest inland developments. Apart from its two real-estate companies Dubai Properties and Sama Dubai, which are in charge of several projects such as Business Bay and the Lagoons, the subsidiaries TECOM and Tatweer have become real-estate developers in addition to their role as leading forces of major economic sectors in Dubai. These are the technology, communication, research, education, healthcare, industrial production and entertainment industry sectors. Thus, TECOM has developed more than eight companies including the free economic zones Dubai Internet City, Dubai Media City, Dubai Knowledge Village and Dubai International Academic City. Tatweer on the other hand has been in charge of the biggest entertainment projects, for example, Dubailand, which will comprise around 45 projects after its completion by 2020. All in all, the companies of Dubai Holding are currently involved in the development of projects covering more than 156 sq km, almost 32% of the total development area of around 493 sq km of Dubai at the time of writing. The holding Dubai World, including its main real-estate developer Nakheel, will be in charge of a larger development area of about 167 sq km. Emaar has been responsible for the development of just about 6% of Dubai's overall project area. Several other investment

76

groups or companies such as the Al Futtaim Group have launched projects on an area covering about 14%. In addition, the DWC development will add another 70 sq km of real estate and free economic zones as well as the new airport. The remaining project areas within Nakheel's Waterfront and Tatweer's Dubailand in addition to the project area of the recently announced Mohammed Rashid Garden City will add a future development area of about 390 sq km with an expected completion beyond 2025.

While the first projects have been mainly developed in the south-west of Jumeirah, which is today known as New Dubai, on an area of more than 70 sq km, many additional developments have been launched along the CBD and Dubai Creek on unbuilt areas along and within the former urban areas of Dubai City, which will be expanded by almost 100 sq km. Additionally, some developers have started projects in the inland of Dubai on an area of about 310 sq km stretching from Warisan to New Dubai, which is mainly occupied by Tatweer's Dubailand covering about 279 sq km. Close to Jebel Ali in the south-west of the emirate, there are two main development areas: Nakheel's Waterfront project including the Palm Jebel Ali, covering an area of about 130 sq km, and the 140-sq km Dubai World Central situated inland, which will comprise five large project areas. In addition, there are plans to connect the main development areas with the already launched Arabian Canal, which will lead from its starting point at the Waterfront development around the DWC project toward New Dubai. Several other projects, for example, Dubai Industrial City and Dubai Investment Park, have been launched on areas that will stretch along the man-made creek.

While most of the approximately 19 developments in New Dubai that were launched at the beginning of the millennium are due to be completed by 2010, the high-rise conglomerate Jumeirah Lake Towers and Jumeirah Village will probably follow by 2014. Furthermore, the recently announced Nakheel Harbour & Tower project is expected to need more than 10 years to be completely developed and in the context of the current financial crisis, to what extent the original project will in reality be executed remains in question, as is the case for most of the developments that have been recently launched. In contrast to the developments in New Dubai, most projects within Old Dubai are in the first phases of their development and thus their overall completion is expected by 2015. While Nakheel's International City and about ten projects within Dubailand will be completed by 2010, most projects situated inland will follow in stages until 2020. Like the inland development areas, there will be several completed projects toward the south in Jebel Ali, however, the main part of the developments is expected to follow in various stages over a longer period of time beyond 2025. Nakheel's Waterfront for instance will be developed in more than eleven phases designed as individual projects. As the Waterfront's first phase is expected to be complete by 2015, the first project within DWC will be completed years earlier than its overall completion. Although two industrial and business-park developments in addition to the new CBD project Downtown Jebel Ali are expected to be completed by 2015, almost 65% of the current development area in Jebel Ali of around 420 sq km will be developed within the next twenty years. Consequently, almost 23% of the total area

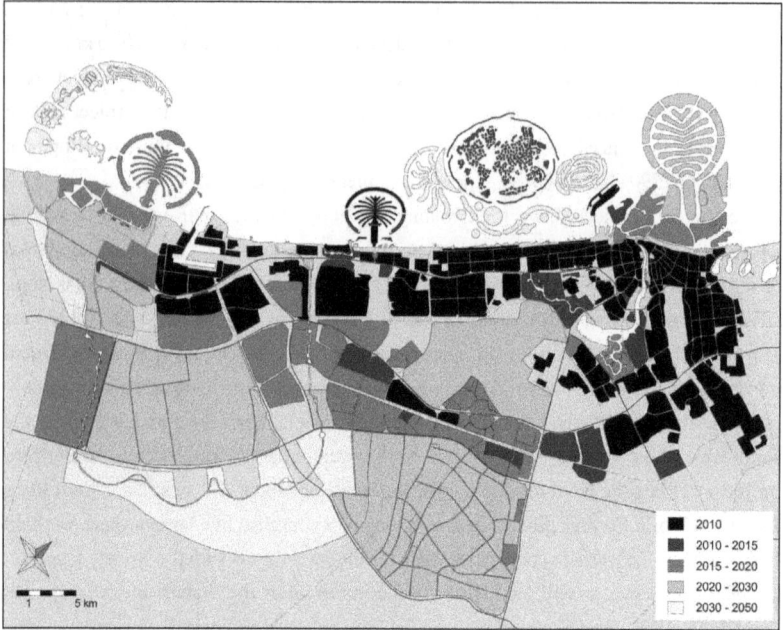

Fig. 15: The phasing of projects according to developers (2008).

of projects being carried out in Dubai at the present time will be completed after the year 2020. Around 53%, including the first phases of mega-projects such as Tatweer's Dubailand and Nakheel's Waterfront, will be completed by 2015. Around 18%, almost 90 sq km, has been either recently developed or is due to be completed by 2010. Thus, together with the new airport in Jebel Ali and the remaining project areas within certain developments, the total size of Dubai's development area is 880 sq km, which will add about 255 sq km to its former urban area within a period of about 25 years.

Consequently, all the current developments together will triple the former urban area within only two or three decades upon condition that all the projects are completed, which however has become doubtful since the beginning of the international financial crisis in 2008. The large impact of the financial crisis on the speculation-driven real estate market has slowed down the construction of many projects and furthermore has caused whole developments or their later phases to be cancelled. This particularly concerns projects that have just begun to be developed and that were scheduled to be completed within the next ten years as well as several large-scale projects, for example, the Palm Deira, which are facing a possible reduction of their original master plans. Nevertheless, Dubai has kept its record as one of the world's fastest building sites with its development of more than 80 sq km of land within the last ten years.

While about 33% of the current project area will be occupied by exclusively residential developments, the remaining 67% will be developed as mixed-use projects comprising a major share of housing developments. Typical suburban low-rise residential developments

78

such as Emirates Living are currently being developed in the form of around 19 projects on an overall area of more than 100 sq km. These surburban developments have been designed for the upper real estate market, offering security as gated communities in addition to leisure areas such as, for example, integrated golf clubs. Most of these residential low-rise developments are located in New Dubai and Dubailand in the outskirts of the former urban area. In recent years, developers have started several large-scale, mid-rise residential projects in order to provide the lower real estate market with affordable dwellings. Almost 700,000 future residents are expected to live in multi-storey apartment blocks on an overall area of about 35 sq km by 2015. Like the low-rise residential projects, the mid-rise residential projects have been designed around a certain theme that defines the architecture and landscaping. Nakheel has launched residential mid-rise developments such as the two Gardens projects in New Dubai in addition to its International City in Al Warsan. The third category of residential projects has been the high-rise developments, which are currently represented by about nine projects on an area of almost 25 sq km along the coast. More than 800,000 people are expected to live in these high-rise conglomerates, which are designed with public areas in the form of landscaped parks or waterfront promenades. Most of the residential high-rise projects are located in New Dubai, where five projects, including Emaar's Dubai Marina, have been launched. All 36 residential projects that are currently under development will cover an area of more than 160 sq km and will accommodate more than 1.7 million people by 2020.

More than 37 of the projects currently being carried out in Dubai can be categorised as mixed-use developments integrating residential and commercial areas. Around 11 of these mixed-use projects have been established as free economic zones covering an area of about 60 sq km. Furthermore, more than 12 small-scale free zones are expected to follow in the coming years. All free-zone projects usually include residential and commercial areas apart from their specific facilities, which can take the form of, for example, technology parks, research centres or industrial zones. Apart from Dubai Holding and its subsidiary TECOM, which is currently in charge of more than eight free zones, the holding Dubai World, the owner of the very first FTZ in Jebel Ali, has begun to develop several future free-zone projects. Moreover, free economic zones such as Dubai Silicon Oasis and the Dubai International Financial Centre have been established as independent authorities without any company or holding in the background. About 26 of the current developments are designed as mixed business districts with often more than 60% residential use of their GFA. These mixed-use developments often integrate office towers, residential towers, mid- to low-rise residential projects, hotels and retail districts in the form of shopping malls, for example, Downtown Burj Dubai, Business Bay, Downtown Jebel Ali, Waterfront City and the Nakheel Harbour & Tower project. Several mixed-use developments are located within Dubailand and thus they often integrate theme parks or leisure areas in addition to their residential and commercial districts. All the current mixed-use developments together have been launched on an area of about 334 sq km and

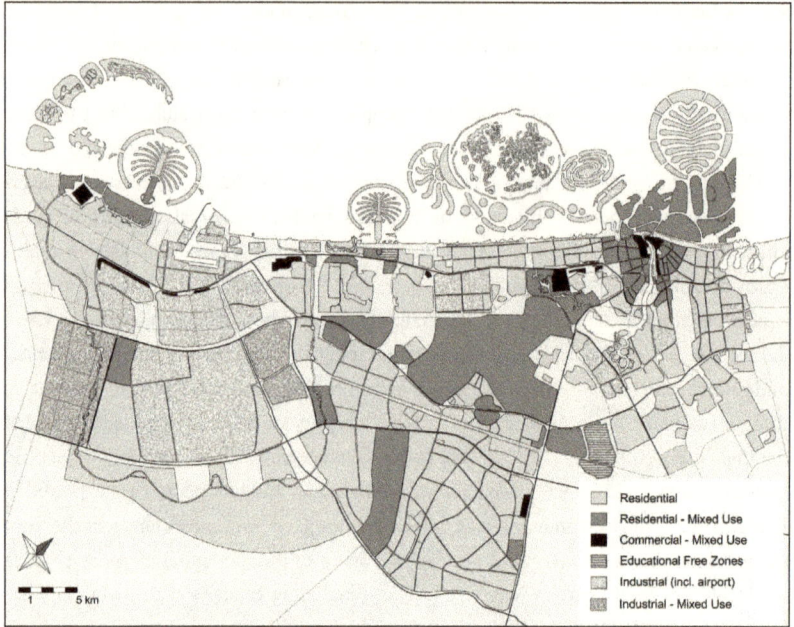

Fig. 16: The expected land-use distribution.

will provide residential units for more than 4.7 million future residents and, according to their current expectations, more than 4 million future jobs.

Along with Dubai's main axis, Sheikh Zayed Road, many developments have been developed in recent years, particularly in the area south-west of Jumeirah, which is known as New Dubai. This new district of Dubai will not only include exclusive residential developments such as Emirates Living and Dubai Marina for about 500,000 future residents but also one of the biggest business districts, which will include the cluster of TECOM's free economic zones along Sheikh Zayed Road. In addition, New Dubai will be one of the biggest tourist destinations with the almost completed Palm Jumeirah and over 40 hotel developments on its waterfront. The main CBD at the beginning of Sheikh Zayed Road will be expanded by mainly three projects, namely, the DIFC, Downtown Burj Dubai and Business Bay. This new high-rise cluster with over 300 towers will transform the former linear CBD into a new downtown district that will be connected to Dubai Creek by a canal project. Although the old Dubai centre has more and more lost its key function as the main business district since the 80s to the new development areas in the south-west, it has been established as an urban centre by the development of several new projects along the Creek, for example, Culture Village, the Lagoons or Festival City. Furthermore, the DHCC project in the south of Bur Dubai, the Dubai Maritime City at the Port Rashid and the new CBD of the Palm Deira will revitalise the city centre and turn it into one of Dubai's biggest future business districts. All in all, about 90 sq km will be added along and within the borders of the former urban area, adding housing units for almost 2 million people.

Inland, the neighbouring projects Dubai Silicon Oasis and Dubai Academic City will form a new educational, research and production conglomerate on an area of about 19 sq km. Next to it, what is currently the biggest project being carried out in Dubai, Tatweer's Dubailand with over 40 project areas comprising entertainment projects and its own downtown, will be one of Dubai's future urban centres. In addition, Nakheel has started its Waterfront development as a completely new city on the border with Abu Dhabi which will be connected to current and future developments by the 70-km long Arabian Canal. On an area of 140 sq km that will be surrounded by the canal project, there will be a fourth future urban centre in the form of the DWC development integrating residential and commercial areas in addition to free zones and the Al Maktoum International Airport as its centrepiece. Between the DWC, the Waterfront and the harbour in Jebel Ali there has been the launch of a new CBD project called Downtown Jebel Ali stretching along Sheikh Zayed Road. While in the past Dubai's CBD has been mainly located along the beginning of Sheikh Zayed Road, the new developments have diversified and decentralised the former urban structure into many 'cities' and business districts in various locations. Nevertheless, the majority of the early developments were launched in areas along the major urban growth corridor following Sheikh Zayed Road and several projects have been initiated along Dubai Creek in order to restore the old city as the future urban centre. The large new development areas of the Dubailand, Waterfront and DWC projects, which together cover an area of around 550 sq km, will add new main centres to Dubai within the next 10 to 15 years. Furthermore, many smaller projects in the form of free zones and industrial developments such as Dubai Industrial City and Dubai Investment Park have been established as cities within the city on the basis of the integration of all their major needs within their project areas.

Most exclusively residential projects are currently being developed in New Dubai or as part of Dubailand toward the inland of the emirate. In addition to the CBD, a large high-rise conglomerate has been established in New Dubai, where five projects on an area of around 7 sq km form a cluster of about 345 towers with an average height of about 140 metres. In the future, Nakheel's Harbour & Tower development will add about 40 high-rise buildings to the most dense residential areas of Dubai. In comparison, the CBD and its extension will include more than 400 office and residential towers on an area of around 10 sq km. Apart from these two main high-rise clusters, there will be a third high-rise conglomerate along Sheikh Zayed Road in Jebel Ali. Downtown Jebel Ali will consist of more than 100 high-rise buildings on an area of about 2 sq km. Future high-rise clusters will be developed as part of the Madinat Al Arab project and Waterfront City, which are both part of the Waterfront development. Apart from the fact that high-rise developments are still mainly centred along the main development axis along the coast, there are several attempts to spread this typology in different locations, for example, along inland waterfronts such as the Dubai Creek or man-made canals as well as within future centres such as Dubailand and the DWC project, which will both include a high-rise conglomerate

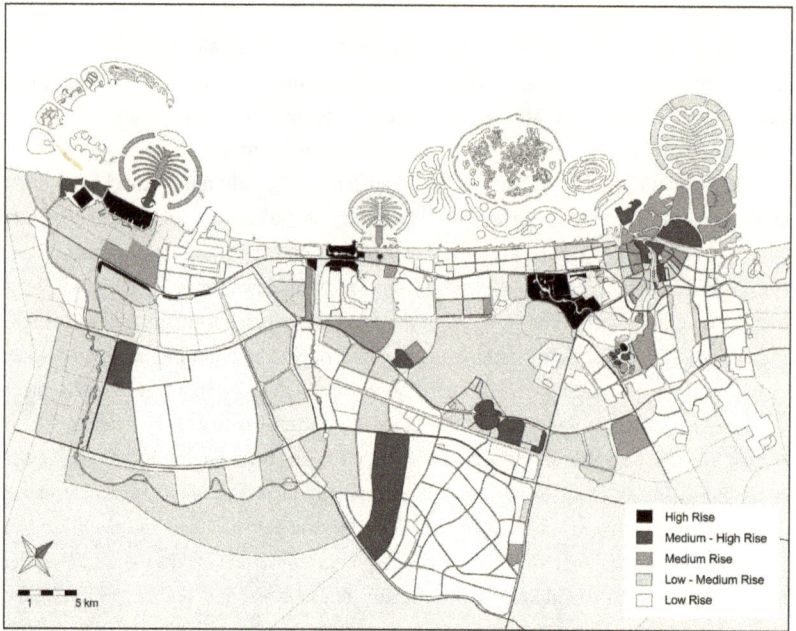

Fig. 17: The distribution of high- and low-rise projects.

as a downtown development. All high-rise projects will cover an area of more than 50 sq km and provide housing units for about 1.7 million future residents. The largest part of the current overall project area is covered by mixed-use developments with a mixture of low- to mid-rise developments in addition to a few high-rise buildings covering an area of more than 324 sq km. While it is currently expected that these medium-dense projects will offer dwellings for about 4 million people, the suburban low-rise developments will on the other hand only house about 240,000 people on an area of around 112 sq km.

Nevertheless, the recent developments have led to an increase in the average urban density to more than 6,000 people living on one square kilometre in comparison to less than 5,000 sq km at the beginning of the building boom in the late 90s. By 2020, density is expected to reach about 8,000 sq km due to an anticipated population increase from about 2 million to approximately 5 million people living on an urban area of more than 620 sq km. According to the current volume of real estate projects, almost 10 million people would be able to live on an area of approximately 900 sq km by 2030, which would mean a density increase to 11,000 sq km. In past decades, the most dense populated area has been the Dubai city centre with about 9,000 inhabitants per sq km. Due to the new mid-rise to high-rise developments along Dubai Creek and at the coast of the city centre, density is expected to grow to more than 10,000 inhabitants per sq km by 2015. The development cluster in New Dubai is expected to have a lower density of about 7,000 inhabitants per sq km, mainly because of low-rise residential projects and their occupation of almost 60% of the 70-sq km development area. By contrast, the cluster of

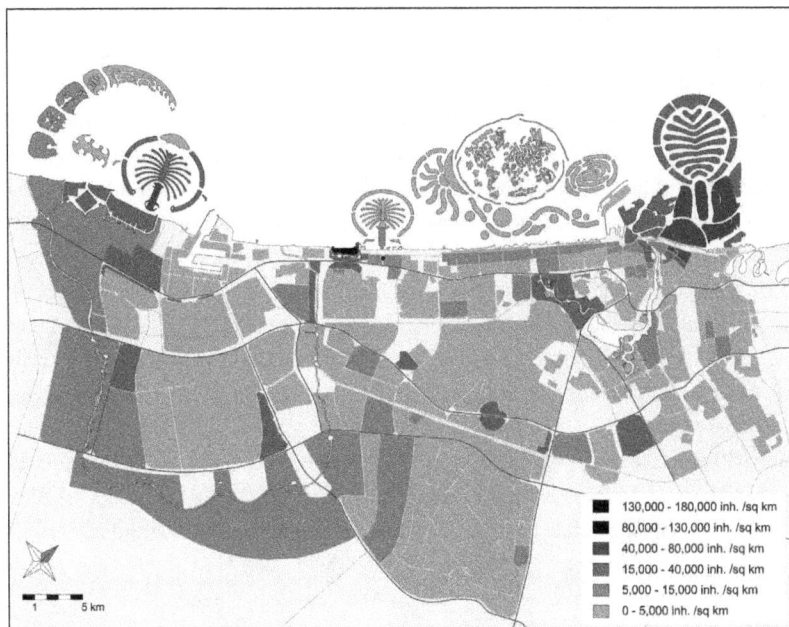

Fig. 18: The expected distribution of urban densities.

high-rise projects along the coast will house almost 3 million people on about only 10 sq km, which will lead to one of the biggest urban densities of around 30,000 inhabitants per sq km. The inland including Dubailand is expected to reach an average density of around 7,000 inhabitants per sq km after the completion of all 25 current and 24 future projects on an area of more than 300 sq km. According to its current master plan, the Waterfront development on Dubai's south-western border with Abu Dhabi will be developed as one of the densest future urban areas with more than 11,500 inhabitants per sq km. The high-rise projects in the centre of this development will reach maximum densities of more than 30,000 inhabitants per sq km like the already built high-rise clusters in New Dubai. In addition to the main CBD on Sheikh Zayed Road, where 300,000 people will live on an area of around 10 sq km, the new CBD project in Jebel Ali will have a similar density of around 30,000 inhabitants per sq km. Furthermore, the DWC development with a density of about 14,000 inhabitants per sq km and Dubai Industrial City with almost 10,000 inhabitants per sq km will contribute significantly to the major increase in density of the whole metro area of Dubai within the next five to ten years.

The rapid urban growth of recent years and the expected completion of projects within the coming years have led to a new demand for infrastructure developments, particularly with regard to the exponentially growing traffic. Thus, the government reacted in 2005 when the Road and Transport Authority launched the first large-scale, public transportation project in the Gulf in the form of the rapid transit Dubai Metro connecting the most important present and future centres of Dubai. The first phase of the project includes two different

83

lines, the Red Line and the Green Line, which together will have a length of over 70 km and about 47 stations after their completion by 2010. The Red Line will comprise 35 stations leading from Jebel Ali Port along Sheikh Zayed Road to the International Aiport. The line will be 47 km long and will connect large residential development clusters in New Dubai with the CBD area and the old centre. In addition, there has been the parallel development of the 20-km long Green Line, which will lead from the Jaddaf area at the lower Creek through Bur Dubai to Deira and the Airport Free Zone. Its 22 stations within the downtown area are expected to release the traffic and connect the new developments at the lower Creek with the former urban area. While the Red Line will be completed by the end of 2009 and only 4.7 kilometres of it will be underground at the Dubai city centre, the Green Line will follow at the beginning of 2010, of which about 14 kilometres will run underground and only 6 kilometres will run on an elevated viaduct. After completion, the Dubai Metro is expected to carry about 355 million passengers a year until two future lines will expand capacity by 2020[200]. The 47-km long Blue Line will follow the Emirates Road and connect the current inland developments with the DXB airport. Like the Blue Line, the 49-km long Purple Line will lead from the airport toward the inland but instead of following Emirates Road, the Purple Line will be built along Al Khail Road connecting the CBD area at the lower Creek with the new Al Maktoum International Airport in Jebel Ali[201].

In addition to the expansion of the Dubai Metro to an overall length of 318 km by 2020, there will be 270 km of tram lines[202]. At the beginning of 2008, construction began on the first of about seven tramlines, the Al Sufouh Tram. It will run for more than 14 kilometres from the Dubai Marina to the Burj Al Arab with stops that will include Dubai Media City and the Mall of the Emirates[203]. Furthermore, Emaar has published plans to develop a tram line called Downtown Burj Dubai Tram as part of its project along Sheikh Zayed Road[204]. In addition to the Dubai Metro and Dubai Tram, the construction of the Dubai Monorail began in 2006. While the first monorail system at the Palm Jumeirah is due to be completed in 2009, most of the remaining seven lines that are planned to be built have not been launched yet. Like the Palm Jumeirah Monorail, all further monorail lines have been designed to connect large developments such as the City of Arabia, Dubai Festival City and Waterfront City with the Dubai Metro network.

Apart from an expansion of the railway networks, there are efforts to develop several canals in order to use them as part of Dubai's transport system. In 2007, the Water Bus System was introduced in Dubai Creek, which will be expanded by current developments such as the Business Bay, whose canal will connect the lower Creek with the coast in Jumeirah. Similarly, there will be a canal system within Mohammed Rashid Garden City and the Waterfront in Jebel Ali, where the Arabian Canal is expected to connect several developments along its distance of about 70 kilometres. Consequently, there will be many new marinas within reclamation projects, for example, the Palms, or along the coast, for example, Dubai Marina. Parallel to the development of new transport systems, the Road and Transport Authority is planning to expand the road system with an additional 500 km of road plus 120 multi-level interchanges by 2020. Furthermore, there are plans to introduce more than 3,000 buses and to install separate bus lanes and zones for pedestrians and cyclists in order to enhance the diversity of Dubai's traffic. Thus, by 2020 the RTA is expecting to increase the use of public transport to 30% of the total traffic[205].

2.4 The Phenomenon of Dubaification on the Arabian Peninsula

2.4.1 The Imitation of Dubai's Model of Post-oil Urbanism

The biggest cause of the current spread of Dubai's development strategy within the Gulf has been the rapid growth and diversification of the emirate's post-oil economy. Its fortunate geopolitical location between the markets of Asia, Europe and Africa in addition to the remaining oil wealth of the region has formed the fundamental basis for the urbanisation of recent decades. Although Dubai's fossil resources have been significantly less than that of many of its neighbours, it used its chance to develop as a trade and financial centre through its liberal policies. The installation of free trade zones and investments in new economic fields such as tourism, which has also become an important town-marketing factor, have been the main reasons for rapid economic diversification. Finally, the liberalisation of the real estate market has transformed Dubai into one of the biggest investment hubs of the region and worldwide. Along with the increasing demand for freehold properties, the urban governance has become more and more privatised in order enhance the speed of development. Consequently, the introduction of self-governing free economic zones and the founding of major real estate developers have decentralised the urban administration. In order to imitate Dubai's success of exponential growth in recent years, several neighbouring emirates and countries have started to adopt its strategy of open markets, particularly regarding the real estate business and the installation of free economic zones. The same geopolitical location, comparable economic conditions and a similar political structure have led to almost the same development basis in most Gulf states and thus an opportunity to follow Dubai's development strategy. While Dubai has gained the indisputable leading position within many post-oil economic sectors, it has simultaneously challenged its neighbours to enter a new kind of regional competition in which speculation-driven development has begun to play a major role. The main future development goal of all Gulf states has been to establish an oil-independent economy and because of Dubai's rapid growth in recent years, the danger of being left behind has increased the pressure to follow the strategy of aiming for rapid urban growth by opening the market and privatising governance.

One of the first consequences of the new strategy has been the transformation of the former built environment from the form of low densities and monotone typologies into a more dense and diverse urban landscape. The new competition in the real estate market has led to the widespread development of multi-storey buildings within all urban areas including the suburban outskirts, where a mixture of typologies had been rare due to privacy concerns. Parallel to the increase of densities, the urban structure has started to become more and more decentralised due to new commercial and business districts outside of the old CBD areas in the form of large shopping malls, free economic zones or mixed-use developments. The tendency of developing self-contained cities within the city has led to an increasing mixture of land use and the introduction of a broad variety of building typologies from high- to low-rise. Furthermore, the competition between developers and investor groups has established a new urbanity in the form of, for example, retail districts, waterfront promenades or landmark architecture in order to market the individual development. The resulting need for media attention has become one of the main reasons for the development of iconic projects testing the borders of engineering, for example, the highest towers or reclaimed islands in the shape of certain themes. The high volume of real estate investments in recent years has transformed the Gulf into a laboratory of architectural extremes, which has profoundly contributed to the changing perception of the Middle East and its development possibilities. Last but not least, the recent integration of new public transport systems is part of the evolutionary process of the current urbanisation, which has freed the oil cities from their previous limitations.

2.4.2 The Risks of Implementing Dubai's Development Strategies
While the current reasons why Dubai's development strategy is being imitated are mainly due to its rapid economic success in recent years, the risks of following the new model lie mainly in the speculation-driven aspect of Dubai's current urban development. A growing dependency on the financial market has made Dubai increasingly vulnerable to breakdowns such as the current international financial crisis, which began in summer 2008. While the creation of a future market by introducing almost no restrictions with regard to speculation has proven to be effective in times of liquidity and faith in future growth, outer factors such as the current financial breakdown or other possible threats such as wars or terrorist attacks constitute major risks for a development that is mainly investor-driven. Although the vitality of Dubai's economy can today neither be evaluated nor predicted, the crisis has slowed down urban growth tremendously in recent months. Any long break within the construction of the projects poses a major threat for future investments. Thus, both outer and inner circumstances can have a large impact on attracting investors, who are naturally interested in short-term profits instead of long-term commitments. Furthermore, the growing competition and lack of cooperation within the Gulf have enhanced the pressure on each country to release its restrictions and liberalise its market. Consequently, the economies have become more and more dependent on private investors, who have used this new vacuum of restrictions to develop their speculative projects.
In order to attract investor groups and cope with the large scale of many projects, urban governance has become more and more decentralised due to the decision-making authority shifting from a central administration to the decentralised planning of several master developers, who usually prepare land use and infrastructure plans in addition to

general building guidelines within their developments. Consequently, the macro-urban planning of the overall urban area, which has generally remained the responsibility of the urban planning department, has often not integrated the new projects of the private sector and thus it has become completely outdated. The ever changing and growing urban area due to large-scale projects has made it almost impossible to develop a comprehensive master plan that would have a realistic chance of being implemented in spite of the growing freedom of investors. Thus, the result so far has been a patchwork-like urban landscape in which projects have been planned and developed in isolation from each other and so hardly connect or interact with each other. A current major concern has been the development of infrastructure, which has proved to be very difficult to plan ahead due to the unpredictability of future developments. The passive and reactive infrastructural planning has led to a major lack of supplies and thus to a serious reduction in liveability within the new urban areas. In addition to planning difficulties, the speculation-driven development has led to a rapid increase in living costs, particularly regarding rent. Consequently, a large number of Dubai's guest workers have moved to neighbouring emirates where the cheaper rent has resulted in the construction of dormitory towns and thus an increasing traffic of commuters toward Dubai. Moreover, the large number of apartments or villas for the upper real estate market has further enhanced this exodus of low- and medium-income groups searching for affordable dwellings.

While in Dubai the local population has been a shrinking minority that has largely benefited from the increasing land prices because of growing income through the lease of property, the nationals of neighbouring Gulf states have been in a less fortunate situation and thus the risk of unemployment and dependency on public subsidies has become a major future threat. Furthermore, the increasing number of guest workers has caused a large economic loss due to the general lack of re-investment of salaries within the local economies. The risk of long-term dependency on foreign labour due to the exponential urban growth and lack of a competitive local workforce can be seen as one of the consequences of the rapid urbanisation, which has been based on importing labour instead of a long-term process of supporting and integrating the local population within the new economic sectors. In addition to the social risk of a decreasing equity, the strategy of an exponential urban growth in a desert region has proven to be increasingly unecological. The growing need for energy and water has become a major threat for any future development because of the increasing costs of maintaining ecological stability. This particularly concerns water production through desalination, which leads to a growing degree of salt content in the sea and thus an exponential need for energy for future desalination. In addition to the rapid rise in the price of land and building materials, missing laws and guidelines have led to a general lack of investment in a climate-appropriate built environment. Apart from the environmental and social risks of adapting a mainly private, investment-driven development, the quality of the new built environment has suffered from short-term profit orientation. The result has been a large amount of monotone repetition of certain typologies and architectural designs in order to reduce costs and produce a high quantity of buildings in a short period of time. Furthermore, the current development of architectural and engineering superlatives has led to an unpredictable increase in future maintenance costs, particularly regarding island reclamations, inland canals and high-rise projects with glass fronts.

2.4.3 Outlook

When Dubai started to transform its governance structure and open up to new development ideas during the 90s, most neighbouring emirates and countries were still maintaining their old development patterns, which had been established at the time of the beginning of oil urbanisation. After Dubai demonstrated the rapid economic success of its strategy, particularly with regard to the introduction of an open real estate market and its laissez faire attitude to private investment, many neighbours started to follow the idea of building up a post-oil economy by reducing restrictions on a speculation-driven development. In particular, countries or emirates with a comparable location and conditions began to introduce a similar legal environment for foreign investment. Due to different socio-economic, political and cultural realities, no other country has been able to import Dubai's development model one-to-one. Particularly in cases where the majority of the population consists of locals, the fast implementation of a speculation-driven development strategy has been seen as an increasing risk for future social problems in the form of unemployment, segregation and thus political instability. Furthermore, the remaining wealth of fossil resources has put some countries under less pressure to transform their economies as rapidly as countries where the oil peak has already been left behind. The presence of Islamic conservatism and strong ties to Islamic law in certain countries have led to a growing controversy between the consumer-oriented development and still existing traditions. In particular, the introduction of tourism and along with it the establishment of a growing entertainment industry have been seen as a major controversy within the regional culture and thus rejected in places, for example, Saudi Arabia.

In addition to several neighbouring emirates within the UAE, Bahrain and Qatar have started to follow the development steps of Dubai in order to join the growing investment market in the region and thus to establish future economic prosperity and growth. As the smallest country of the GCC, Bahrain began to develop its oil-independent economy in the 70s and 80s when it became a global centre for offshore banking. Due to social and political circumstances, Bahrain lost its role as one of Dubai's biggest competitors at the end of the 20th century when the whole region was outshone by Dubai's exponential growth. In recent years, Bahrain, Qatar and Abu Dhabi have taken on the challenge of competing with Dubai by taking the approach of an individual interpretation instead of a pure imitation of the same development strategy. While Kuwait has been limited in joining the growing competition during recent years because of its geographical proximity to the on-going conflict in Iraq, the rulers of Oman and Saudi Arabia have remained hesitant to adopt the development strategies due to their different social, economic and cultural circumstances as the largest territorial states in the Gulf. Nevertheless, both countries have initiated the implementation of new economic strategies in combination with large-scale projects driven by private investments. Today, all the GCC countries have been affected to differing extents and in different ways by Dubai's development of a future global market in the Gulf. Thus, the transformation of the built environment has not been limited to Dubai, where developers such as Emaar have started to export their projects beyond the emirate's borders and even beyond the region itself. Within a very short period of time, a large number of developers have been founded all over the region by various investor groups in order to build a rapidly increasing amount of real estate projects. The new market-oriented urban development has rapidly transformed the structure of the regional built environment, which had previously been dominated by a repetitive urban pattern caused by sluggish administrative urban governance in previous decades.

---

1 Schmid, 2008, p. 83.
2 Davidson, 2008, p. 10.
3 Very little evidence can be found about the origin of the name Dubai, which can either be ascribed back to the word daba, which is a type of regional locust, the word dhub, which is the local name of a certain lizard, or the combination of the Hindi word doh for two and the Arabic word bayt for house, which refers to two white houses that supposedly stood side by side at the creek (Davidson, 2008, p. 10).
4 Davidson, 2008, p. 13.
5 Scholz, 1999, p. 211.
6 Davidson, 2008 p. 22.
7 Davidson, 2008, p. 9.
8 Heck, 2004, p. 92.
9 Schmid, 2008, p. 85.
10 Pacione, 2005, p. 257.
11 Pacione, 2005, p. 259.
12 Wirth, 1988, p. 29.
13 Pacione, 2005, p. 256.
14 Pacione, 2005, p. 257.
15 Heck and Wöbcke, 2005, p. 131.
16 Schmid, 2008, p. 86.
17 Pacione, 2005, p. 256.
18 http://www.dubaiairport.com/DIA/English/TopMenu/About+DIA/DIA+and+History/, 25.10.2008.
19 Wirth, 1988, p. 37.
20 Wirth, 1988, p. 39.
21 Pacione, 2005, p. 260.
22 Wirth, 1988, p. 43.
23 Wirth, 1988, p. 47.
24 Pacione, 2005, p. 261.
25 Wirth, 1988, p. 45.

26 Wirth, 1988, p. 49.

27 Schmid, 2008, p. 90.

28 Davidson, 2008. p. 114.

29 Schmid, 2008, p. 91.

30 Pacione, 2005, p. 260.

31 Pacione, 2005, p. 260.

32 Pacione, 2005, p. 257–258.

33 Pacione, 2005, p. 262.

34 Heck, 2004, p. 132.

35 Schmid, 2008, p. 94.

36 Amir Mohammed bin Rashid Al Maktoum became the official ruler of Dubai on 4 January 2006 (http://www.sheikhhmohammed.co.ae, 28.10.2008).

37 Pacione, 2005, p. 257.

38 http://www.dsc.gov.ae/DSC/Pages/Statistics%20Data.aspx?Category_Id=0226, 28.10.2008.

39 http://realestate.theemiratesnetwork.com/articles/freehold_property.php, 28.10.2008.

40 Schmid, 2008, p. 114.

41 Davidson, 2008, p. 129.

42 Schmid, 2008, p. 114.

43 http://www.ameinfo.com/man_made_islands_article/, 28.10.2008.

44 http://www.ameinfo.com/67119.html, 28.10.2008.

45 ICT: information and communications technology

46 http://www.madinatjumeirah.com/, 28.10.2008.

47 http://www.burjdubaiskyscraper.com/, 28.10.2008.

48 The author Saskia Sassen introduced the term 'global city' to denote the growing network of global centres of finance and trade in the form of cities caused by increasing globalisation.

49 Dubai Strategic Plan (2015) 29.10.2008, p. 18

50 Dubai Strategic Plan (2015) 29.10.2008, p. 19

51 Dubai Strategic Plan (2015) 29.10.2008, p. 21

52 Dubai Strategic Plan (2015) 29.10.2008, p. 22

53 Dubai Strategic Plan (2015) 29.10.2008, p. 26

54 Davidson 2008, p. 206

55 http://www.migrationinformation.org/dataHub/GCMM/Dubaidatasheet.pdf, 30.10.2008.

56 http://www.ameinfo.com/140234.html, 30.10.2008.

57 http://www.demographia.com/db-worldua2015.pdf, 30.10.2008.

58 http://www.ameinfo.com/113976.html, 30.10.2008.

59 Schmid, 2008, p. 113.

60 http://www.sheikhmohammed.co.ae, 31.10.2008.

61 Sheikh Hamdan was involved in the government as president of the Dubai Municipality and as Minister of Finance in the UAE.

62 About 570 million euros.

63 http://www.ameinfo.com/143945.html, 31.10.2008.

64 Schmid, 2008, p. 114.

65 http://www.gulfbase.com/site/interface/NewsArchiveDetails.aspx?n=60483, 31.10.2008.

66 Schmid, 2008, p. 115.

67 Schmid, 2008, p. 115.

68 Schmid, 2008, p. 117.

69 Dubailand was previously known as Dubai Tourism Projects Development Company. Dubai Tourism Projects Development Company was put in charge of developing the world's biggest project for leisure and tourism called Dubailand on an area of about 200 million square metres.

70 The DMC is a multi-purpose maritime zone. The DMC consists of an Academic Quarter, Marine District, Harbour Residences and Industrial Precinct.

71 Schmid, 2008, p. 118.

72 http://www.arabdecision.net/show_func_3_12_5_0_3_1345.htm, 02.11.2008.

73 The RTA was founded in 2005 and, like the DEWA (Dubai Electricity and Water Authority), was established in order to supply infrastructure more effectively by being governed as a corporation within the public administration.

74 Property Guide Dubai, 2007, p. 174.

75 http://de.wikipedia.org/wiki/Emirates_Hills, 04.01.2009.

76 http://www.dubaiinternetcity.com/, 04.01.2009.

77 http://thepurplejournal.wordpress.com/2008/08/03/dubai-internet-city-the-middle-easts-biggest-it-infrastructure/, 02.01.2009.

78 http://dubaiholding.com/en/our-companies/tecom-investments/, 04.01.2009.

79 http://en.wikipedia.org/wiki/Dubai_Media_City, 04.01.2009.

80 http://www.dubaipearl.com/About.aspx, 02.01.2009.

81 http://www.palmjumeirah.ae/the-palm-story.php, 02.01.2009.

82 http://thepalmdubai.com/palm-island-dubai/about-the-palm-islands/news-updates/palm-jumeirah/06.05.09-Palm-Jumeirah-work-on-schedule.html, 04.01.2009.

83 http://www.trumpdubai.com/, 02.01.2009.

84 http://www.ameinfo.com/76184.html, 04.01.2009.

85 http://www.telegraph.co.uk/property/3344183/Palm-before-a-storm.html, 02.01.2009.

86 http://www.dubai-marina.com/, 04.01.2009.

87 http://www.dubaimarinarealty.com/CityProjects.html, 04.01.2009.

88 http://www.dubai-marina.com/skyscrapers.html, 04.01.2009.

89 http://www.pentominium.com/, 04.01.2009.

90 http://www.dubai-properties.ae/en/Projects/JumeirahBeach/Index.html, 04.01.2009.

91 http://www.middleeastelectricity.com/upl_images/news/Jumeirah%20Lake%20Towers%20becomes%20free%20zone%20%20July%2019%202006%20Gulf%20News.pdf, 05.01.2009.

92 http://www.jumeirahlaketowers.ae/jlt_launch/about_jlt.aspx, 04.01.2009.

93 http://en.wikipedia.org/wiki/Jumeirah_Lake_Towers, 04.01.2009.

94 http://www.emporis.com/en/wm/bu/?id=210141, 04.01.2009.

95 http://www.jumeirahheights.com/en/master-plan/, 03.01.2009.

96 http://www.dubaipromenade.ae/en/masterplan/an-inspired-masterplan.html, 11.01.2009.

97 http://www.dubaipromenade.ae/en/facts-and-figures/facts-and-figures.html, 11.01.2009.

98 http://www.jumeirahislands.com/en/overview/jumeirah-islands-fact-sheet.html, 04.01.2009.

99 http://www.arabianbusiness.com/property/unit/523-jumeirah-park, 04.01.2009.

100 http://www.jumeirahpark.com/, 04.01.2009.

101 http://www.gulfnews.com/articles/04/09/25/133296.html, 04.01.2009.

102 http://www.jumeirahvillage.com/facts.php, 04.01.2009.

103 http://www.jumeirahgolfestates.com/en/section/selling-now/faq/general, 13.012009.

104 http://www.dubaiproperty4u.co.uk/impz.php, 15.01.2009.

105 http://www.realestate-dubai-property.com/property/dubai/impz/development/51.cntns, 13.01.2009.

106 http://www.uaepropertytrends.com/ptrends/mvnforum/viewthread?thread=2135, 13.01.2009.

107 http://www.alfurjan.com/en/media-centre/press-release/nakheel-launches-al-furjan.html, 04.01.2009.

108 http://www.thegardens.ae/thegardens.html, 04.01.2009.

109 http://theburjalalam.com/?paged=8, 04.01.2009.

110   http://www.nakheelharbour.com/#/faq, 13.01.2009.

111   http://www.difc.ae/district/index.html, 06.01.2009.

112   http://www.worldarchitecturenews.com/index.php?fuseaction=wanappln. projectview&upload_id=1034, 11.03.2009.

113   Property Guide Dubai 2007, p. 117.

114   http://www.dubaicityguide.com/geninfo/news_dtls.asp?newsid=11198, 04.01.2009.

115   http://www.gulfnews.com/business/General/10256900.html, 04.01.2009.

116   http://www.emaar.com/index.aspx?page=emaaruae-downtownburj-burjdubaiboulevard, 05.01.2009.

117   http://www.dubai-properties.ae/pdf/Business-Bay-bro.pdf, 05.01.2009.

118   http://archive.gulfnews.com/indepth/creekextension/more_stories/10123188.html, 05.01.2009.

119   http://www.dubai-properties.ae/en/Projects/BusinessBay/ExecutiveTowers/Index.html#, 05.01.2009.

120   Property Guide Dubai 2007, p. 245.

121   http://www.dubaitowersdubai.com/, 05.01.2009.

122   http://www.khaleejtimes.com/DisplayArticle.asp?xfile=data/business/2005/December/ business_December176.xml&section=business&col=, 05.01.2009.

123   http://www.futtaim.com/content/companyprofile.asp?profileid=1325, 05.01.2009.

124   http://www.arabianbusiness.com/500339-putting-the-culture-back-into-the-village, 06.01.2009.

125   http://www.dhcc.ae/EN/AboutDHCC/Pages/Location.aspx, 06.01.2009.

126   http://www.dubai-search-and-find.com/healthcare-city.html, 06.01.2007.

127   Dubai International Real Estate

128   http://www.kieferle-partner.com/news_5423_20080808489c0f1ab285e-Jewel-of-the-Creek.html, 06.01.2009.

129   http://www.dubaimaritimecity.com/, 07.01.2009.

130   http://www.arabianbusiness.com/542751-dubai-maritime-city-on-schedule-for-2012-completion, 07.01.09.

131   http://www.nakheel.com/en/developments, 07.01.2009.

132   http://www.estatesdubai.com/labels/Palm%20Deira.html, 07.01.2009.

133   Property Guide Dubai 2007, p. 249.

134   http://www.dubai.de/artikel/230-Palm-Deira-waechst-kontinuierlich-vor-der-Kueste-Dubais.html, 07.01.2009.

135   http://www.nakheel.com/en/developments, 07.01.2009.

136   http://www.theworld.ae/mp_infrastructure.html, 07.01.2009.

137   http://www.theworld.ae/mp_islands.html, 07.01.2009.

138   Property Guide Dubai 2007, p. 269.

139   http://www.nakheel.com/en/developments, 07.01.2009.

140   http://www.internationalcity.ae/en/the-city/facts-and-figures.html, 08.01.2009.

141   http://www.ameinfo.com/166451.html, 08.01.2009.

142   http://www.dso.ae/en/about-dsoa/vision.html, 08.01.2009.

143   http://www.nytimes.com/global/dubai/four.html, 08.01.2009.

144   http://www.dso.ae/en/about-dsoa/master-plan.html, 08.01.2009.

145   http://www.nytimes.com/global/dubai/four.html, 08.01.2009.

146   http://www.dubaifaqs.com/dubai-academic-city.php, 08.01.2009.

147   http://www.dubaicityguide.com/GENINFO/news_dtls.asp?newsid=20243, 08.01.2009.

148   http://realestate.theemiratesnetwork.com/developments/dubai/dubailand.php, 08.01.2009.

149   http://www.tatweerdubai.com/En/cd-2-1, 08.01.2009.

150 http://realestate.theemiratesnetwork.com/developments/dubai/dubailand.php, 08.01.2009.
151 http://www.dubailand.ae/project_details.html, 08.01.2009.
152 Property Guide 2007 Dubai, p. 82.
153 http://www.arabianbusiness.com/index.php?option=com_companylist&view=list&compa nyid=5805, 16.01.2009.
154 Property Guide 2007 Dubai, p. 167.
155 Property Guide 2007 Dubai, p. 230.
156 http://www.dubailand.ae/project_details.html, 10.01.2009.
157 http://www.globalvillage.ae/AboutUs/Facts.aspx, 10.01.2009.
158 http://www.falconcity.com/faq.asp, 10.01.2009.
159 http://www.propertyportal.ae/alkaheel.php, 11.01.2009.
160 http://www.dubai-properties.ae/the-villa-overview.html, 11.01.2009.
161 http://dubai-properties.ae/en/Flexible_Images/pdf/mudon_factsheet_en.pdf, 11.01.2009.
162 http://www.ameinfo.com/31019.html, 15.01.2009.
163 http://dubai-properties.ae/en/Projects/Tijara_Town/master_plan.html, 18.01.2009.
164 http://www.gowealthy.com/gowealthy/wcms/en/home/real-estate/uae/dubai/dubailand/al- barari/index.html, 18.01.2009.
165 http://www.cityofarabiame.com/about-us/about-city-of-arabia.html, 11.01.2009.
166 http://www.cityofarabiame.com/our-projects/mall-of-arabia/overview-mall-of-arabia.html, 11.01.2009.
167 http://www.cityofarabiame.com/about-us/about-city-of-arabia.html, 11.01.2009.
168 http://www.mizin.ae/arjan.html, 13.01.2009.
169 http://dubaiholding.com/en/media-centre/news/2008/October/mizin-to-unveil-dubais-first- boutique-villa-community-at-cityscape-08/, 13.01.2009.
170 http://www.tatweerdubai.com/En/cd-4-6#, 13.01.2009.
171 http://www.dubailand.ae/facts_figures.html, 11.01.2009.
172 http://www.ameinfo.com/161240.html, 28.01.2009.
173 http://www.civicarts.com/mohammed-bin-rashid-gardens.php, 28.01.2009.
174 Property Guide 2007 Dubai, p. 253.
175 Translation of the poem: 'Take wisdom from the wise. It takes a man of vision to write on water. Not everyone who rides a horse is a jockey. Great men rise to greater challenges.' (http:// www.dahproperty.com/ThePalmJabelAli.asp, 11.01.2009)
176 http://thepalmdubai.com/palm-island-dubai/about-the-palm-islands/news-updates/palm- jebel-ali/03.05.05-Remarkable-features-unveiled-on-Palm-Jebel-Ali.html, 11.01.2009.
177 http://www.waterfront.ae/, 12.01.2009.
178 http://www.waterfront.ae/, 12.01.2009.
179 http://www.waterfront.ae/, 12.01.2009.
180 http://www.cowshedproperties.com/about-dubai/the-waterfront/, 12.01.2009.
181 http://www.waterfront.ae/, 12.01.2009.
182 Gulf Construction April 2008, p. 108–112.
183 http://dubaiholding.com/en/media-centre/news/2008/October/dubai-industrial-city-com- mences-leasing-2-8-million-sq-ft-of-open-storage-yards/, 17.01.2009.
184 http://www.dubaiindustrialcity.ae/Pages/home/faq.aspx, 13.01.2009.
185 http://www.realtyna.com/dubai_real_estate/dubai-investment-park.html, 13.01.2009.
186 http://www.dipark.com/index.php?lang=en, 13.01.2009.
187 http://www.zawya.com/story.cfm/sidZAWYA20080307082009, 18.01.2009.
188 http://www.ameinfo.com/102262.html, 18.01.2009.
189 http://www.downtownjebelali.com/#, 18.01.2009.
190 http://www.dwc.ae/dwc.html, 14.01.2009.

191 http://www.arabianbusiness.com/543142-first-flight-out-of-al-maktoum-airport-by-end-of-2009, 14.01.2009.
192 http://www.dwc.ae/dwc_dubai_logistics_city.html, 14.01.2009.
193 http://www.dwc.ae/dwc_dubai_logistics_city.html, 14.01.2009.
194 http://www.internationalfreightweek.com/upl_images/news/GCClogisticsmarkettodouble-growthonbackofenergyboom26May08UAE.pdf, 14.01.2009.
195 http://www.dwc.ae/dwc_real_estate_division.html, 14.01.2009.
196 http://www.dwc.ae/images/pdf/28%20nov%2007%20DUBAI%20WORLD%20CEN-TRAL.pdf, 14.01.2009.
197 http://www.dwc.ae/dwc_real_estate_division_commercial_city.html, 14.01.2009.
198 http://www.dwc.ae/images/pdf/28%20nov%2007%20DUBAI%20WORLD%20CEN-TRAL.pdf, 14.01.2009.
199 Gulf Construction, April 2008, p. 96.
200 http://www.railway-technology.com/projects/dubai-metro/, 22.01.2009.
201 http://archive.gulfnews.com/articles/06/10/06/10072630.html, 22.01.2009.
202 http://www.dubaifaqs.com/al-sufouh-tram.php, 28.01.2009.
203 http://www.dubai-online.com/news/plans-for-dubai-tram-system-unveiled/, 28.01.2009.
204 http://www.arabianbusiness.com/537297-emaar-denies-problems-on-dubai-tram-project, 28.01.2009.
205 http://www.arabianbusiness.com/539418-bus-lanes-cycle-zones-part-of-12bn-traffic-mas-ter-plan, 29.01.2009.

II

Case Study: Kingdom of Bahrain

3 The Urbanisation of Bahrain

3.1 The History of Urbanism in Bahrain

3.1.1 The Pre-oil Settlements in Bahrain

Evidence has been found that Bahrain is the most probable location of the legendary land Dilmun, which has been mentioned by the ancient Mesopotamian civilisations. Thus, it is estimated that the first human settlement of the islands goes back to at least the year 5,000 BC. Furthermore, archaeological findings indicate that Bahrain has been a regional trade centre for several millennia[1]. In 626, the tribes of Bahrain embraced Islam and in the following centuries the islands were ruled by governors of the caliphate until the Qarmatians, a movement within Shia Islam, took over power and transformed Bahrain into one of their strongholds. In 1058, the Persians invaded Bahrain, which led to several conflicts and shifts of power between Persians and Arabs until the 16th century when Portugal, the first colonial power in the Gulf, conquered Bahrain in 1521 on account of its wealth of pearls. The Portuguese forts resisted Ottoman attacks over the following decades until they were vanquished when the Persians once again invaded Bahrain in 1602. After being defeated by Oman, the Persian Empire bought Bahrain in 1720 but lost it only about sixty years later when the Al Khalifa family took power[2].

In 1763 the Al Khalifa family, which is part of the tribal confederation Utub, left its former residence in Kuwait in order to settle in Zubairah on the north-western coast of Qatar[3]. After several conflicts the Al Khalifa family took over control of Bahrain in 1783 and decided to move there in 1796. For the first several years the Sunni rulers settled in Jaw in the eastern part of the main island until they decided to move to Muharraq and Riffa-Ash-Sharqi at the beginning of the 19th century. Apart from the period of time between 1799 and 1811 the Al Khalifa family has been governing Bahrain and its development until the present day[4]. For centuries, the majority of the island's population has been Arabs of the Shia section of Islam, who differ culturally and religiously from the Sunni rulers as well as in terms of their tribal origins. Most of the Shia population lived from fishing and agriculture, including large date plantations in the north and west of the main island. Along with the Al Khalifa family, other Sunni clans moved to Bahrain, particularly merchants, who were interested in the island's wealth of pearls. Due to the military and economic advantage of the Sunni minority, it was possible for them to take control of the whole population by introducing taxes and taking away land from families who refused to be loyal. Furthermore, the rulers were interested in replacing Bahrain's former political structure of municipal self-governance, which was established by the leading Shia scholars of the time and their interpretation of Islamic law, with a new political hierarchy in which the most powerful clan controlled the unbuilt land and was thus entitled to rule. Before the Al Khalifa family changed the political structure, the approximately 52 settlements of Bahrain had mainly been governed by the local communities themselves and a religious council consisting of Shia scholars who were responsible for decisions related to the allocation of land and adherence to the religious laws. All unbuilt land was generally seen as free to use and not under the ownership of any individual or clan. If a request for land was made, it was common to pay a certain fee to a ruling council, which usually consisted of about three religious scholars[5].

One important factor that enabled the Al Khalifa clan to sustain its ruling position was the fact that it bought the loyalty of the Sunni families that immigrated to Bahrain by relieving them of any taxes, with the exception of customs on imported or exported goods, and providing them with the right to keep their individual tribal laws and armed forces[6]. This favouritism discriminated against the Shia population and led them to become increasingly dependent economically on the Sunni families who leased to the Shia the agricultural land that had formerly belonged to them and forced them, as their tenants, to pay duties and provide services[7].

With agriculture being dependent on an adequate supply of sweet water, the cultivation of date trees was mainly concentrated in the northern and western parts of the main island where sweet water resources were located as well as on the islands of Muharraq, Sitra and Jiddah. In addition to working as farmers on date plantations, a large part of the population was forced to work as fishermen and pearl divers in order to earn their livelihood. Particularly in the 19th century, the pearl economy flourished in Bahrain in addition to the construction of boats, which formed another income basis for the population and was the second largest export economy after pearl fishery[8]. While many smaller settlements were located close to the agricultural fields inland and along the coast, two cities started to grow and prosper in the form of Manama on the northern coast of the main island and Muharraq in the south of the island Al Muharraq. The main reasons for the development of these two urban centres were their location at relatively deep waters that made it possible to build harbours, their proximity to the settlements of the ruling clans and their close proximity to the north-western coast of Qatar, from where the Al Khalifa family moved and thus trading contacts had been sustained. Although both cities were founded simultaneously, Muharraq took the lead as the capital because of its defensive location and because it was the residence of the ruling family. Manama on the other hand developed into a subordinate settlement but was the biggest gateway to the main island due to its harbour and souq[9].

Apart from the export of pearls, boats and dates, Bahrain developed into a growing trading hub of the region because of its establishment as a trans-shipment centre for pirated goods, which led to a growing British interest in the islands[10]. The first contract between the British Empire and the rulers of Bahrain dates back to 1820 when the British East India Company started to develop contractual relations with ruling tribes along the coast of the Persian Gulf. The initial goal of these contracts was to protect their trade routes against interfering piracy. In the following decades, the contractual partnership between the British Empire and the Al Khalifa family developed more and more due to mutual interests. Although the leaders of Bahrain lost a certain part of their autonomy, they benefited from the British, who helped them to preserve and widen their ruling position against the Shia population. In a contract from 1861, the British agreed to protect the ruling families, who for their part had to agree to use no armed forces against the citizens of Bahrain. Bahrain's new dependency on a colonial power was displayed in 1869 when the British representative decided to intervene in a conflict between the Al Khalifa family and announced Isa bin Ali as the new ruler. Later on, contracts made between 1880 and 1892 practically made Bahrain a British protectorate on the basis of agreements that all external affairs were to be decided by the British and that the selling and leasing of land to a third party was no longer possible[11]. Based on the British principle of indirect rule, the Al Khalifa clan officially remained in charge although the British Empire and its representative won the contractual right to change the succession, trading customs and harbour policies. Thus, the rulers of Bahrain became more and more dependent on British support and less attached to their old alliances with certain clans and tribes[12].

In terms of its economy, Bahrain became increasingly dependent on the export of pearls during the 19th century. In 1910, about 70,000 people lived in Bahrain, of which about 20,000 worked in the pearl industry[13]. The city of Manama, which had been a small town with about 8,000 residents in 1818, grew into a constantly expanding harbour city with a population of more than 25,000 people at the beginning of the 20th century. Because of its role as a port, many different nationalities settled in Manama and created its early multinational social fabric. The centre of Manama was the souq close to the harbour, which was surrounded by various residential areas called fareej in which the inhabitants lived segregated according to ethnicity, tribal background and religion. The Shia population for instance was divided into several distinct groups including the local Arabic speaking people called Baharna, the Ajam, the Persians, the Hasawis and the Qatifs. The Sunni population consisted of various small groups in addition to the majority of Arab bedouin clans who arrived along with the ruling family in the 18th century. Also, several Arabic-speaking merchant families from the south of Persia called Hawala settled in Bahrain and kept their ties to their origins across the Gulf, and together with the ruling families they formed the most wealthy group within Bahrain's society. Thus, the city of Manama was divided into a number of different areas that provided for the basic needs of the people in the form of mosques, small markets and so-called majlis that functioned as meeting points for the community. In 1919, Manama consisted of fourteen such areas, which formed the basic urban structure for the following decades. The names of the areas originated from either the residents' ethnicities or the commercial activities that took place there[14].

According to the principles of Islamic oasis towns, the residential areas of towns enjoyed an autonomous internal life in which each community solved problems of the built environment according to their own specific rules. Thus, internal disputes were generally solved among neighbours in the presence of other community members. In more complicated cases the argument had to be brought to sheikhs who were chosen to act as judges and make decisions according to Islamic law and local customs. For this purpose, communities appointed two supreme judges of their own, one Sunni and one Shia, in addition to many others of lower rank. Typical disputes that were taken to court involved boundary disagreements, unclear water rights and houses overlooking one another. The absence of a central administration and general building laws created a diverse built environment based on a collective effort, and thus the architecture expressed the local customs and know-how of the masons, craftsmen and laymen of each community[15].

The urban structure of Islamic oasis towns was often determined by the gradual development of irrigation systems, land subdivision and tenancy systems. Springs and shallow wells provided the local population with sufficient water for drinking and irrigation[16]. Due to an abundance of springs, the most common irrigation system was the method known as seeb, in which water was lifted by animals or a counterpoise and ran along surface channels through the fields to the sea because of a smooth slope. These parallel streams formed the property boundaries of fields and the basic layout of what would later develop into the network of streets of cities such as Manama, which initially began as a market place for farmers[17].

Apart from the irrigation system, the street layout was determined by the need to provide shade and shelter from the hot and humid climate. The north-south oriented streets of Manama enhanced cross ventilation by letting in a strong north-western wind, known as shamal, which reduced the effect of the long and humid summers in addition to removing disturbing smells from the souq. The narrowness of the winding side streets provided inhabitants with shade in their residential areas while in wider main streets light fabrics were hung up to create shade. The architecture of courtyard houses with attached wind towers was another adjustment to the extreme climatic circumstances. Wealthier families could afford to live in houses of stone but a big part of the population lived in huts with sloping roofs and walls that were woven from palm fronds[18]. This low-cost type of housing, locally known as barasti, benefited from a rather good system of cross-ventilation. There were many barasti huts at the borders of cities such as Manama as well as in villages, where they were erected on plots that were surrounded by fences in order to preserve the privacy of the residents[19]. Consequently, the general urban structure of the pre-oil settlements of Bahrain was very similar to the structure of other oasis towns in the region, namely, two main urban centres with ports and markets were surrounded by smaller villages close to agricultural areas or the coast. Both the urban structure and architecture were defined by the rules of the Islamic culture and the necessity to adjust to the desert climate.

3.1.2 The Urbanisation of Bahrain during the Oil Era

The Transition Period until Bahrain's Independence:
Before the first oil was found in 1932, Bahrain's pearl fishery had begun to collapse after the invention of cultured pearls in Japan during the 20s. While at the beginning of the 20th century about 2,000 boats were engaged every year in the pearl business, this number shrank to 436 boats by 1933. Apart from the thousands of employees directly working in the pearl industry, the breakdown of pearl fishery affected many merchants and craftsmen who were involved in the pearl trade. The boat building industry in particular decreased rapidly from about 120 boats produced every year at the beginning of the 20th century to only 8 per year by 1936. The consequence was the migration of many families to Kuwait or Saudi Arabia and the decay of several settlements[20]. In spite of the economic downturn, the relationship between Bahrain and the British Empire took its next step when a 'political agent' was installed by the British in order to advise the rulers on the development of governmental institutions and the establishment of new laws concerning the regulation of taxes and properties. Due to the strategic location of Bahrain in the Gulf, the British imperial administration chose to use it as a regional centre in Arabia, thus establishing it as a model for a new way of governance based on a standardised legal system and a central administration. This is particularly relevant to Manama as it became the new residence of the Political Agent in 1919 and was an important port city in the Gulf[21].

During the economic problems in the 20s, the British political agent set up new administrative institutions in order to establish a system of government agencies and courts that were independent of any specific tribe, including the ruling family. The political goal was to unify the fragmented and decentralised governance that was formed by the wide range of different groups and their approach of following their own specific rules and laws. Thus, the political agent introduced a special court in 1919 where disputes between the foreign and local population could be settled. Only one year later, the municipality of Manama was established in order to deal with all basic public services including health, transport, water and electricity. A small police department was also under its control. While the municipality in Muharraq followed in 1927, further municipalities were founded in several other small cities such as Riffa, Al Hidd and Sitra during the 50s. In 1923, a trading board was established in order to solve problems within the mercantile and industrial sectors and at the end of the 20s the Bahrain Court was established in addition to the installation of a governmental department consisting of influential Shia merchants in order to administer the land properties of the Shia population[22].

In addition to the new institutions, the political agent was engaged to develop a new system of taxes and property rules in order to increase the economic effectiveness of the country and make a more strict and central control within governance possible. Therefore, it was legally agreed that agricultural land could be individually owned instead of remaining the private property of the Al Khalifa family when it was either cultivated for over 10 years or allocated through a certain certificate by the ruler himself[23]. Furthermore, compulsory labour of farmers on date plantations was prohibited and certain taxes such as the water tax regarding the Shia population were abolished. Apart from the standardisation of taxes concerning date production and fishery, the former amount of possible credit that could be given to pearl divers and boat owners was limited. Due to these relatively small changes, the first form of a modern administration came into existence. On the one hand, it restricted the ruling family as a decision-making force but on the other, it strengthened its position in relation to influential tribes and clans whose privileges were profoundly diminished by the new laws. Consequently, the new individual land owners, almost all of whom were members of the ruling family, started to become more and more dependent on subsidies because of the higher costs of cultivating their date plantations, the abolishment of compulsory labour and their lack of qualifications with regard to the kind of work to be done in the newly established public sector. Thus, the introduction of the new legal and administrative framework meant a severe cut in the income sources of the leading Sunni families, who finally became dependent on subsidies. Because of these subsidies the government had a much smaller budget available for developing the country[24].

Although the Shia population were relieved from being forced to work or pay additional taxes, the new land laws did not affect their living conditions. However, the new political rights of the Shia population and the economic degradation of Sunni families led to a violent conflict that began in 1921 and reached its peak in 1923 when the ruling sheikh, Isa Al Khalifa, was replaced by his son Hamad bin Isa Al Khalifa due to pressure from the political agent[25]. The British had an interest in increased social stability in Bahrain because they were convinced that Bahrain had rich oil resources and had already concluded concessional contracts as early as 1914 and 1925. However, at the end of the 20s, it was the American oil company Standard Oil of California that finally won the contract to produce oil. In 1931, its subsidiary, Bahrain Petroleum Company (BAPCO), found oil and began oil production in the Awali field in 1934. One year later, the oil company Texas Corporation bought 50% of BAPCO's shares and thus ownership of Bahrain's oil was equally split between the companies California and Texas, which together were called Caltex. When the first oil refinery was built in 1936, Bahrain became the first oil-producing country in the Gulf. Although only about 17.5% of the oil profits belonged to Bahrain, the new economic sector already provided more than 50% of the state's income in 1940 in addition to thousands of jobs. Furthermore, the new oil wealth led to major investments in the infrastructural development of the country[26]. While the first schools had already opened at the end of the 19th century and beginning of 20th century in the form of two schools of the American Mission, one Persian minority school and two Arab schools, many schools including schools of higher education followed in the time of oil production[27]. Consequently, the number of students rapidly increased from about 600 in the year 1931 to around 1,900 in 1940 and more than 20,400 in 1960. At the same time, the public health sector was developed and many infrastructural projects were carried out, including electrification which had already begun in 1928. Only four years later, the installation of the first telephone network followed and the construction of the first bridge between Manama and Muharraq, which was built between 1929 and 1942.[28] Due to a higher life expectancy and the immigration of thousands of guest workers, the population of Bahrain grew rapidly from about 89,970 inhabitants with a foreign population of around 16,000 foreigners in 1941 to 182,203 people with a foreign population of 38,389 people in 1965[29]. Since the 20s, land registration and the verification of land ownership had been an increasing priority of the new bureaucratic governance and thus a Land Department was introduced that started to survey all properties and legal issues. Between the mid-30s and 1970, many aerial photographs and maps were produced in order to draw property lines and make it possible to plan the modern infrastructure. All unclaimed land was automatically under the ownership of the government, which deprived some families of land they had acquired by inheritance. Furthermore, the traditional right in Islamic societies to develop unowned land, known as ihiya, was completely abolished by the administrative system. The new registration procedures and resulting shift in land ownership had a large socio-political impact as agricultural land lost its value due to the decay of the date trade and was sold at low prices to wealthy groups including Persian merchant families[30]. Because of growing political tension, in 1939 Persian migrants were forced to either take on Bahraini citizenship or sell their properties. However, the trend of plots being sold by the ruling family to these merchants did not stop, which led to a high number of private properties, especially in cities such as Manama. Former agricultural areas turned into private housing, investment sites or private gardens at the outskirts of cities, where they often prevented the execution of urban extension plans by public authorities[31]. This in turn nurtured the early idea of reclaiming land along the coasts, which was relatively easy due to the shallowness of the sea.

In 1965, a planning study was done in order to show the areas where land reclamation was possible, which added up to about 274 sq km, an area equivalent to more than 40% of the main island. Because of springs and tidal flows, many swamps covered the coastal areas, causing a serious risk of the outbreak of epidemics such as cholera and thus there was a second reason apart from the need for development areas for reclaiming land on a large scale.

One of the first reclamation areas was the pier in Manama, which was created in order to improve the landing and shipping of cargo, which was made difficult by the shallowness of the coastal water except during the high-water seasons[32]. In 1923, a narrow line between the sea and the city was reclaimed and added to the urban area of northern Manama to become a sea road along which most new administrative and commercial buildings were built. At the end of the 60s, an area of about 8,400 sq m was added along this sea road in order to create a plot for the new Customs House. Consequently, the former coastline of northern Manama was pushed back in order to provide space for further government buildings. In addition to the new business centre along the sea road, many suburban areas were developed on reclaimed land in Manama along its eastern and southern coasts. In several cases, thousands of cubic metres of town rubbish were collected and discharged into the marshes along the coast and then covered with earth and sand[33]. All reclaimed land was considered government property and thus a large area was developed for public projects, which followed the design of Western architects in the form of massive cement structures within an orthogonal layout of streets and plots[34]. Although Muharraq was the capital until 1971, the British planners focused early on on the development of Manama, which was growing rapidly due to its advantage of being located on the main island and its function as the biggest port city.

As in many Middle Eastern countries, urban modernisation was complicated by the development of a modern street network because of the dense pedestrian-based urban fabric. The automobile, which was first introduced in 1914, was predominantly used by British officers, members of the ruling family and rich merchants. When the number of vehicles rapidly grew from about 295 cars in 1944 to 3,379 in 1954 and to 18,372 in 1970, the need for an effective road network became one of the priorities of the first urban planners. Consequently, the traditional irregular form of a system of narrow winding streets was replaced by straight arteries and a grid layout, which led to the opening and transformation of old urban areas. In Manama, two types of road were used in order to develop the accessibility of the urban area by car – ring roads were used to extend the urban area in belt-like forms and to connect the new areas with the centre in the north by following the oval form of the old core. In addition, parallel roads in finger-like patterns were built to connect the ring roads from north to south and to form the main grid. Furthermore, King Faisal Road was developed on a length of 3 kilometres along the northern coast, where a large area of around 1.7 sq km was added by land reclamation. Because of its accessibility and connection to Muharraq via causeway, King Faisal Road became the main CBD of Bahrain[35]. In 1941, the Sheikh Hamad Causeway replaced the former causeway from 1929, which consisted of an old wooden bridge, because of the rapidly increasing traffic between the two main cities after the first airport was built in the north-eastern part of the island of Muharraq during the 30s. Further developments such as new palaces in Qudaibiya and a British military base with its port in Juffair caused the expansion of the road network southward and thus encouraged urban growth in these areas[36].

As early as 1919, the development of business and administration was mainly concentrated in Manama, leading to many settlements outside of Manama becoming dormitory towns. Consequently, Bahrain developed from a structure of many small self-sufficient settlements and two equally important port cities to a structure wherein Manama became the entire country's main commercial and service centre. The construction of a new road to Budaiya at the north-western coast of the main island connected many smaller agricultural settlements with the growing urban centres, which caused a number of compounds to be developed for guest workers and their families. Although the first municipality was set up in the 20s, urban development was not guided by any structural plan but by a combination of major public projects and small private ones. The first institutions governing central town planning and the legal framework for building regulations were not established until the 60s when the Physical Planning Unit was set up in 1968. Until the first comprehensive zoning plans were implemented, any kind of zoning or planning was only carried out on the level of isolated areas. Otherwise, the urban development was passively determined by pre-existing factors such as what basic infrastructural supply had already been built or the way in which land had already been subdivided into plots, among others. Thus, while certain areas were strictly planned, for example, Qudaibiya during the 30s, which was zoned according to a rigid orthogonal grid pattern, other settlements grew in a fragmented manner without any planning[37].

In addition to the growing road network, the introduction of new building methods began to change the former built environment during the first phase of oil urbanisation. This transformation was relatively slow as a part of the population still had to live in barasti huts until the mid-60s, when 2,464 of these huts were counted in the outskirts of Manama. Nevertheless, efforts had been undertaken to diminish their numbers due to the risk of fire, which destroyed about 440 huts in 1961[38]. At the same time, the government began the first public housing programs in order to supply the growing population with a sufficient number of dwellings. Along with the establishment of welfare policies, the government became a major provider of housing units and social facilities in order to increase the living standards of the population. In order to relieve congestion in Manama, the local population was encouraged to move to recently developed suburbs such as Isa New Town, which was designed by British planners and served as a new reference for modern housing in Bahrain in the decades following its construction in 1968. Furthermore, the first apartment buildings were built in order to accommodate the large numbers of foreign labour and the old buildings were often replaced. In 1963, Manama already comprised 959 flats in addition to 8,341 ordinary stone houses and 1,456 shops along the new road network[39].

Consequently, the early oil urbanisation in Bahrain led to an increasing urban sprawl with low-density housing areas at the outskirts that were developed by a modern street network. This new trend in the urban development was intensified due to a growing percentage of unbuilt land, which was a consequence of the new regulations regarding individual property rights which inevitably encouraged some speculation. The introduction of legal and administrative institutions by the British authorities formed the basis for a new form of urbanism, which step by step became more centralised and administered. Because Bahrain had started to produce oil relatively early on, it was one of the first countries in the Gulf to enter the oil urbanisation process and thus one of the first to undergo a transformation of the previous built environment into a new urban structure including all three main elements of the oil city, namely, a dense mixed-use urban centre as main the residential area of foreign labour, a new business district including public institutions, the first banks and office buildings along the main road towards the airport and, thirdly, suburban housing areas and satellite dormitory towns.

Urban Development after Gaining Independence:

When in the middle of the 20th century the idea of a united Arab world grew all over the Middle East, Bahrain was affected by a budding political movement in the form of the foundation of political clubs in the 50s and an increasing demand for more political participation. Although the division between the Sunni and Shia populations had a negative impact on the attempt to create national unity, the High Executive Committee (HEC) was founded in 1954 following the growing demand for a people's voice after a strike was carried out by bus and taxi drivers following a decrease in their salaries. This committee consisted of an equal number of members from both communities and began to express their demands to the political agent and ruling family. The most important demands included the creation of an elected parliament, the installation of a civil and criminal code in addition to the introduction of a court of appeal and the foundation of labour unions. After the rulers ignored all these demands, a conflict began to arise and the HEC, which had changed to the Committee of National Union (CNU), initiated a general strike. Although the strike enabled the CNU to start negotiations with the ruler in 1956, there was no significant result and the conflict continued. Consequently, the movement was stopped by force and the imprisonment of the leading members meant the final end of the CNU, which can be considered to be one of the first democratic movements in the Gulf[40]. In the background, the British authorities supported the Al Khalifa family in order to preserve the status quo and their ruling position, which gained growing importance when the USA, a new competitor in the Gulf region, increased its interests in Bahrain[41].

In order to sustain the old political hierarchy, social stability had to be rebuilt and thus a work department for the regulation of state pensions was established in addition to a department for agriculture in order to increase agricultural productivity. Furthermore, a new institution for the administration of 56 villages and a department for general social concerns were set up in the 50s in addition to a department for water services and the first press department. Apart from the new attempt to increase the public control of social concerns, the participation of the population was acknowledged in the form of various elected committees dealing with administration and the removal of the general rule that all high administrative positions had to be held by members of the Al Khalifa family[42]. Since the conflict of 1956, a state of emergency hindered any further strikes until 1965, when a strike by BABCO employees broke out after the dismissal of many hundreds of employees. The demands included the re-employment of all workers dismissed since 1961 and the acknowledgement of the right to found labour unions in addition to freedom of press and speech. Furthermore, they comprised the right of gathering, the release of all political prisoners and last but not least the removal of all the foreign employees of the company. After a second violent breakup of the protests with the involvement of the British military, the rulers merely decided to release a few laws in order to improve working conditions and to allow one Arabic newspaper. But more importantly these growing conflicts with the population contributed to the British decision in 1968 to release Bahrain into independence. In the following years, several infrastructural projects

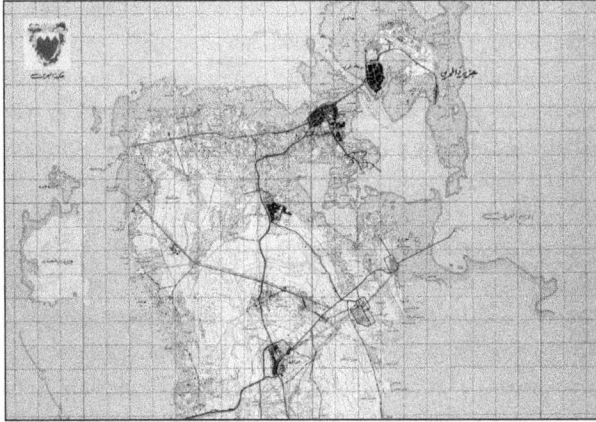

Fig. 1: The islands of Bahrain in 1969.

were built, the first national army was founded and the administrative organisation was expanded in order to prepare the small country for its new political status. After the failure of a federation of Bahrain, Qatar and the UAE under the umbrella of one united government, the state of Bahrain declared its independence on 14 August 1971[43].

While the nation's independence was being established, oil production at the Awali field reached its peak of 76,600 barrels per day in 1970. In the following decades, production decreased to about 40,000 b/d by 1993 due to the depletion of the Awali field and consequently Bahrain, which was once the centre of the first oil boom, became the lowest oil-producing country in the Gulf. Apart from the Awali field with its limited resources of about 70 million barrels, 50% of the production of the Abu Safaa offshore field along the border with Saudi Arabia belonged to Bahrain due to a border treaty concluded with Saudi Arabia in 1958. In 1996, the rulers of Saudi Arabia decided to assign the entire production of oil from the offshore field to Bahrain in order to strengthen the Al Khalifa family, who maintained their ties to the neighbouring kingdom. In 1973, the rulers of Bahrain managed to buy 25% of Bahraini oil, which had almost completely belonged to the American oil company CALTEX. Only one year later these national shares in the country's oil production were increased to 60% and in 1979 the remaining 40% was bought by the Bahraini government[44].

One of the characteristics of Bahrain's oil economy is the fact that only refined oil has ever been exported and because Bahrain's refinery has the capacity to treat more oil than the country itself produces, it has been able to treat oil from Saudi Arabia, which has been supplied by an underwater pipeline since 1945. In 1971, almost two thirds of the oil refined by Bahrain came from Saudi Arabia and thus the processing of oil became one of Bahrain's most important economic sectors. In order to reduce costs, the number of employees in the oil sector decreased from 8,455 in 1958 to 2,692 in 1992. This involved almost two thirds of the Bahrainis among these employees losing their jobs, although there was a rise in their percentage in the oil-sector workforce from 67% to 86% in this same period of time. In the mid-70s, the production of natural gasoline began in Bahrain, which was mainly used for the production of energy[45].

Due to a limited amount of oil and gas resources, and thus the limited export of oil and gas, the agenda of Bahrain's economic development strategy was focused early on on industrialisation and the construction of a modern infrastructure[46]. Since 1971, major attempts and investments have been undertaken to develop industrial capacity and infrastructure in order to transform Bahrain into a service sector-based economy. As a consequence, neighbouring countries and members of the GCC became interested in the economic development of Bahrain and joined in the new investment possibilities. Although economic diversification and the development of major infrastructural projects were launched in Bahrain after gaining independence, the first steps of this development were already completed in the 60s when the new harbour Mina Salman was developed in the south of Manama as well as an industrial area next to it that was Bahrain's first free zone in order to attract export companies. Until 1975, this harbour was the only deep-sea harbour in the Gulf, which led to its growing importance as a regional trans-shipment centre. In addition to the new harbour, the airport in Muharraq was expanded in 1971, becoming one of the most important hubs in the region. Furthermore, the first phase of urban modernisation led to the decision to develop a new satellite town in the south of Manama called Isa Town, which was completed in 1968 and provided housing units for about 34,500 people[47]. At the end of the 60s, plans were made to develop an aluminium industry and a dry dock, particularly for oil tankers. Both projects were built during the 70s when the small country was becoming more and more globally connected, thus making it possible to establish a new economic role in the Gulf[48]. One example of this was the founding of the airline Gulf Air through public shares of Bahrain, Oman, the UAE and Qatar in 1974, the headquarters of which were set up in Bahrain. A further factor in Bahrain's rapid economic development was the introduction of the Bahrain Off-shore Market in 1975, which led many foreign banks to move to Bahrain and participate in its national financial market in spite of certain restrictions[49].

Bahrain's successful entry into international financial business and several infrastructural projects led to an early attempt to develop further service sectors. In order to make Bahrain more attractive, the infrastructure had to be expanded and the urban areas modernised, which led to several projects, including a desalination plant and a power station in addition to the expansion of the harbour and its road connection with the island of Sitra. At the end of the 70s, Arabian Gulf University was founded, which was made possible by public investments from Kuwait, the UAE, Qatar, Bahrain and Oman. It was built for about 5,000 students and reflected a growing attempt to create a knowledge-based economy. Between 1981 and 1986, the 25-kilometre long causeway to Saudi Arabia was constructed by means of Saudi Arabian investments and marked a new era of cooperation between Saudi Arabia and Bahrain. General cooperation between all the Gulf states reached a new stage after the Gulf Organisation for Industrial Consultancy (GOIC) was founded in 1977 and the Gulf Cooperation Council (GCC) in 1981, which were both set up in order to improve economic cooperation and the coordination of industrialisation in the Gulf[50]. During the 80s, several industries were developed in the growing industrial areas at the harbour Mina Salman, on the island of Sitra, at the dry dock in Muharraq and close to the Awali field in the central eastern part of the main island. The new industries comprised the Gulf Petrochemical Industries Company, the Arab Iron and Steel Company, the Gulf Aluminium Rolling Mill Company and the Heavy Oil Conversion Company[51]. While all of these new industries were export-oriented and financed through public investments from Bahrain itself or from its neighbours, there was a general lack of industries for the home market. Thus, Bahrain's economy remained dependent on the import of goods, particularly in regard to commodities for consumption and production[52].

While in 1971 only about 17% of the total population of approximately 216,000 people consisted of non-Bahrainis, the percentage of guest workers grew during the process of industrialisation and modernisation to more than 32% in 1981 and to about 36% in 1991. At the beginning of the 90s, the population grew to more than half a million people, which was more than five times the population in 1941 and more than twice the population in 1971[53]. This rapid growth led to a fast urbanisation involving not only the two port cities Manama and Muharraq in the north but also the development of almost all other settlements in addition to the founding of new housing areas. The population of Manama, which became the new capital in 1971, grew from about 89,000 inhabitants to almost 180,000 at the end of the 90s and thus doubled its size within less than three decades. With a very similar growth rate, Muharraq's population increased from 41,000 inhabitants in the 60s to more than 93,000 by 1998[54]. While these two cities were growing rapidly due to low-income foreign labour moving to them, the suburbs and satellite towns expanded because of much of the local population deciding to live in the outskirts while middle- to high-income guest workers moved with their families into newly built compounds. While old settlements such as Jidd Hafs, Sitra and Riffa were modernised and expanded, two new satellite towns called Isa Town and Hamad Town were founded during the 60s and 80s in order to provide sufficient housing units for the growing local population.

In 1968 Bahrain's urbanism, which up until that point had been carried out according to project-based planning, became institutionalised and centralised in the form of the Physical Planning Unit (PPU), which was established in order to plan the subdivision of land, plot layout and design of the basic grid structure. It was also intended to integrate new social and economic development programmes in Bahrain's physical planning. This new planning institution replaced the former decentralised structure of various agencies and individuals that had worked in relative isolation from each other, causing a growing need for a centralised urban governance in order to cope with the rapid urbanisation[55]. The PPU was founded under an advisory planning board, which included several ministers that were responsible for land, legal, social and economic affairs under the leadership of the minister of the Ministry of Municipalities and Agriculture who acted as the board's president. One of the first actions of the PPU was a study in 1968 of physical planning requirements related to an estimated population increase from about 180,000 people to more than 340,000 by 1988. The planning study based on this projected population included an assessment of future need for housing, employment, education, recreation, water, electricity, communication and transport projects. In this way, the first master plans, which were mainly limited to basic land use decisions, the subdivision of new urban areas and the expansion of the road network, were designed for Manama, Muharraq and Riffa[56]. After Bahrain gained independence, an increase in public investments, predominantly caused by increasing oil prices at the beginning of the 70s, accelerated the urbanisation process. These petro-dollars were used in an ambitious investment programme to strengthen the infrastructure and modernise the country[57]. In 1973, the Monetary Agency was founded in order to control and coordinate the vast amounts of public investments being made in various development projects[58]. Most of the funds were used to build power, water and sewage facilities as well as to expand the airport, harbour and road network. Furthermore, the social infrastructure in the form of schools, health centres and public housing projects was gradually expanded and the costs of social services such as education and health as well as general living expenses were subsidised to a large extent by the government[59]. Thus, the oil urbanisation of the 70s led to a typical social structure of a local population that was becoming more and more financially dependent on public subsidies and a growing number of immigrants who were overtaking the private service sector.

The master plan for the capital Manama, which was designed by the British architect A. Monroe, covered an area of about 2,340 hectares, which was about seven times the urban area in the 50s. According to this plan, substantial areas were allocated to land that had to be reclaimed in the following decades[60]. All in all, more than 1,100 hectares were gradually reclaimed, which came to 47% of the total master plan area. In addition to general land use and the road grid, which included the Manama ring road, the first master plan covered the water infrastructure, electric network and proposals for the location of parks, mosques, health centres and schools[61]. Many of the new ministries, which were founded after 1971, were located in the north and north-east of Manama, where a new area was developed on reclaimed land. This new area was called the Diplomatic Quarter and contained most of Bahrain's ministries, courts and embassies, which were built as modern multi-story blocks. The northern business district was formed by the construction of several hotels and banks, which made Manama the centre of Bahrain's hotel industry with over 5,700 beds at the end of the 80s as well as a growing financial centre with about 173 banks and financial companies in 1986, including the 27-storey tower of the National Bank of Bahrain which was the tallest landmark at this time[62]. During the development of the new urban areas in the north and east of Manama, the old city centre witnessed a process of degradation because no preservation strategies were applied to it. Although its urban fabric of narrow and irregular streets remained to a large extent, most buildings were replaced by modern cement constructions. Due to the rapid increase in size of Manama's urban area, the old core of Manama constituted only about 10% of Manama's total urban area.

Fig. 2: The satellite city Hamad Town.

Due to a lack of accessibility by car and small-sized plots, most Bahraini families that lived in the centre of Manama moved from their old houses to the suburbs in order to rent their properties to companies or directly to guest workers who were searching for cheap accommodation. In a study made in 1987, it was observed that about 42% of the centre's population was Bahraini, mostly Shia. A lack of maintenance led to traditional buildings rapidly decaying and being replaced, thus causing a large part of the traditional urban fabric to be lost. Apart from the Shia mosques and ma'atams still located there, the old core lost a large part of its function as a cultural and social centre[63]. Furthermore, the immigration of foreign labour into the old districts meant that the population and its density was continuously changing and growing with as many as 952 people living on one hectare, which was three times the average density of Manama in the late 80s[64]. Although the old souq was partly modernised, it lost its traditional function as Bahrain's largest market place because of the arrival of new shopping malls and supermarkets. While the master plan of 1968 estimated that about 200,000 people would live in the capital at the end of the 90s, less than 180,000 inhabitants were counted in 1998, which meant that there was a large number of unbuilt plots. The main reason for this was the development of many new housing areas to the south of the main island that provided the local population with bigger plots as well as the development of new roads that made it possible to travel quickly to the business district in Manama. Apart from the expansion of old settlements such as the town of Riffa where the population increased from about 12,000 people in 1965 to about 63,423 inhabitants in 1998, the two new satellite towns Isa Town and Hamad Town became preferred housing areas[65].

While Isa Town was already established at the end of the 60s and gradually expanded during the 70s, Hamad Town was designed and developed by the Ministry of Housing after a proposal of the PPD in 1979[66]. The Ministry of Housing (MoH) was established in 1975 in order to solve the increasing problem of a lack of housing units for the local population.

Consequently, the Housing and Ownership Directorate at the Ministry of Labour and Social Affairs, which had been responsible for the development of housing projectc between 1968 and 1974, was dissolved[67]. The goal of the new ministry was to provide adequate housing for every Bahraini family that was unable to finance its own home. In this regard, as well as constructing dwelling units in the form of houses and apartments, the ministry began to establish policies related to housing loans and the allocation of plots[68]. In 1975, the Physical Planning Directorate moved from the Ministry of Agriculture and Municipalities to the Ministry of Housing, thus restructuring the urban planning and making the Ministry of Housing in charge of all planning directorates concerning urban development. Apart from the Physical Planning Directorate, the Survey Directorate, the Technical Affairs Directorate, the Loans & Ownership Directorate and the Administration and Finance Directorate were all under the umbrella of the Ministry of Housing, which replaced the Planning Coordination Board as the main decision making force in urban planning[69]. Between 1975 and 1996, the MoH accomplished the development of 15,690 housing units, mainly in the form of detached villas. In addition to the new towns Isa Town and Hamad Town, several housing projects were launched in various areas all over Bahrain including Umm Al Hassan, Muharraq, Arad, Hidd, A'a'li and Sanabis. While Isa Town was developed for about 35,000 inhabitants, Hamad Town was designed for 60,000 people[70]. Thus, in 1991, more than 12.5% of the overall population of Bahrain lived in these two towns[71].

In addition to its function as a dormitory town of Manama, Isa Town developed into a new urban centre after several private schools, a campus of the University of Bahrain and public institutions such as the Ministry of Education and the Ministry of Information moved there. Hamad Town however essentially remained a suburban dormitory town divided into four districts stretching 9 km from north to south with a width of 2.5 km in the western part of the main island. It covered a built area of almost 1,300 hectares and thus had capacity for about 12,000 dwelling units[72]. All in all, three main urban areas can be distinguished in Bahrain at the beginning of the 90s, namely, the old city centres in Manama and Muharraq with their existing road networks, the new urban areas that followed the general land use and subdivision plans of the Physical Planning Department and the new masterplanned towns and settlements in the outskirts, which were mainly developed through public investments. In addition, the small villages within the agricultural areas began to expand, particularly because of the construction of compounds on former agricultural land. The result was a merge of all urban areas and the disappearance of any distinct borders to settlements, although many empty plots and remaining agricultural areas interrupted what was in other respects a coalescing urban landscape. While multi-storey blocks and high-rise buildings were introduced in the business district of Manama, it was not allowed to mix these typologies with the low-rise residential areas in the outskirts due to the privacy concerns of the local families. Nevertheless, several apartment blocks were developed as part of public housing projects, which however in most cases were not higher than three to five floors. Thus, the overall urban density of about 1,700 people living on 1 sq km was considerably low and, as in many Gulf countries, oil urbanisation was causing an increasingly uneconomic and unecological urban structure of urban sprawl enabled by the exponential spread of automobile use.

3.1.3 The Impact of Post-oil Urbanism in Bahrain

Although the economic diversification of Bahrain had started earlier than in many other Gulf states, the price decline of oil in 1986 had a major impact on the drop of the country's GDP, thus proving that it was still dependent on the fossil resource. Furthermore, most of Bahrain's producing industries, including the aluminium industry, depended on cheap energy and continue to do so today. Therefore, they are considered to be less future-oriented than a knowledge-based service sector. Apart from the fall in the price of oil, Bahrain's economy was seriously affected by the Iraq-Kuwait war, which led to major costs in the military sector and thus to a huge loss of public investment in infrastructural and industrial developments. In particular, joint industrial and infrastructural projects between Gulf states were either reduced or cancelled so that instead of a project volume worth about USD 2.6 billion being carried out, only about 27% of this volume (about USD 700 million) was developed during the 90s. Apart from the attempt to diversify the industry of the country through public investments, economic development driven by the private sector began to become more and more important. One example of the growing privatisation is the case of the company BALEXCO[73] founded in 1977 by public investments, over 60% of which was sold to private investors during the early 90s after it had been transformed into a stock corporation[74]. While in the mid-90s the privatisation of Gulf Air did not succeed due to increasing financial problems and dependency on public subsidies, the private sector did carry out essential infrastructural projects such as a new energy plant. Furthermore, new privately owned companies such as Bahrain Leisure Facilities Company, Bahrain Gulf Course Company and Al Jazira Tourism Company were founded in order to establish the tourism industry in Bahrain, which was mainly kick started by the construction of the new King Fahad Causeway connecting Bahrain to Saudi Arabia. In 1994, more than 5.2 million people crossed the causeway, including an increasing majority of shopping and leisure tourists[75].

As well as thanks to a large number of shopping malls and hotels, the Bahraini tourism industry benefited from less restrictions, for example, on alcohol consumption, and a wider variety of leisure possibilities than in Saudi Arabia. Since 2006, the Bahrain International Circuit and the annual Formula 1 has led to a wider regional and global awareness about Bahrain as a travel destination. In addition to the development of its tourism sector, Bahrain's role as an offshore banking centre has become more and more important for its economic diversification. At the end of the 90s, about 47 offshore banks with an overall capital of about USD 67.9 billion existed in Bahrain. Furthermore, the number of local commercial banks grew rapidly and more than 32 companies sold their shares on the local stock market, which was introduced in 1989[76]. In order to attract more regional and international investors, the government decided to found Bahrain Development Bank in 1991, which provided low-priced loans and venture capital. In addition, the Bahrain Marketing and Promotion Bureau was introduced to support companies opening their headquarters in Bahrain in addition to supporting a public programme since 1993 involving tax concessions, a reduction of costs with regard to rent and electricity and subsidies in the case of the employment of Bahrainis. A further major change in Bahrain's economic development was the permission to found companies 100% owned by foreign capital[77]. This economic liberalism, which allowed more and more private investment to be made in the local market because of significant reductions in protectionism, has become a major keystone of Bahrain's current economy.

While Bahrain's economic development was focused on increasing business possibilities for foreign and regional investors, the local population became more and more excluded due to a lack of education for certain jobs and the more competitive foreign workforce. For instance, only about 368 students were matriculated at the Arabian Gulf University at the end of the 90s, which was originally intended to educate around 5,000 students[78]. Apart from a lack of public investment in the education of the local population, it was mainly because of the speed of the new development, which involved an immediate need for highly educated employees and masses of cheap labour, that education was left by the wayside as foreigners were brought in to fill jobs. In order to reduce the negative impact on the Bahraini population, several government programmes were launched in order to integrate the local workforce into the new economic sectors, and in comparison to other Gulf countries, the Bahraini workforce has in many ways proven to be an economic asset, particularly in regard to its willingness to work in the private service sector. Unlike in other Gulf countries, the local population of Bahrain has received less subsidies and land, thus many people, particularly from the Shia majority, have been forced to accept low-paying jobs. This adverse economic situation, in addition to the general demand for more political rights and participation, led to many riots and violence during the 90s when a new regional demand for democracy was fueled by the end of the second Gulf war and its consequences in Kuwait, where an elected parliament, the National Assembly, was re-established in 1992.

In Bahrain, around 300 well-known persons started a petition for the reinstallment of the parliament, which had been introduced in 1973 according to the constitution but removed only two years later in 1975 due to its interference in government operations. The reaction of the ruler in 1992 was the establishment of a council with a purely advisory function called Majlis Al Shoura. All its 30 members were directly appointed by the Amir himself and thus less respected by the population, which did not achieve its aim of changing the political structure[79]. In 1994, more than 22,000 people signed a new petition, which was historic because of its dimension and impact in the following years. Around 30% of the subscribers of the petition belonged to the Sunni community, a proportion that was approximately the same as that in relation to the overall population. As such, the new democratic movement could not be seen as a further continuation of the old conflict

Year	2002	2003	2004	2005	2006	2007
Total population	672,124	689,418	707,160	724,645	742,561	1,046,814
National population	417,940	427,955	438,209	448,491	459,012	529,368
Male	210,814	215,848	221,019	226,187	231,493	-
Female	207,126	212,107	217,190	222,304	227,519	-
Expatriate population	254,184	261,463	268,951	276,154	283,549	517,446
Male	175,407	180,430	185,598	190,568	195,671	-
Female	78,777	81,033	83,353	85,586	87,878	-
Total growth rate %	2.7	2.6	2.6	2.5	2.5	5.8
National population %	2.4	2.4	2.4	2.3	2.3	-
Expatriate population %	3.1	2.9	2.9	2.7	2.7	-
Age under 14	187,105	190,108	194,828	197,800	202,565	-
Age over 65	16,457	17,375	17,798	18,321	18,756	-

Fig. 3: The population development between 2002 and 2007.

between the Shia majority and the ruling Sunni minority. When Sheikh Isa rejected the petition again, tensions escalated between the population and the government, leading to many riots being carried out by the disadvantaged Shia community and the imprisonment of about 2,000 people[80].

After the death of the Amir Sheikh Isa in 1999, his son Sheikh Hamad became ruler of Bahrain and began to initiate a major change in its political system[81]. In 2002, the state of Bahrain was proclaimed a constitutional monarchy and the re-establishment of an elected parliament led to a certain degree of public participation. This was however limited to a predominantly advisory role. In the following five years the population grew from about 672,000 people to more than 1,046,000 in 2007, mainly caused by the immigration of around 250,000 guest workers, whose share in the overall population increased from about 38% to almost 50%. During this major societal change the small kingdom underwent various reorganisations of its governance, which had been shaped over the previous decades by a mainly oil-driven development. While the oil era was dominated by a huge central administration that did not function very efficiently, the new economic situation led to an attempt to increase profits and thus the outdated style of the bureaucracy was made more corporate and flexible through the introduction of new agencies designed for this purpose. This particularly affected the urban planning sector, which was dominated by the Ministry of Housing from 1975 to 2002. In addition to being responsible for housing developments, the Physical Planning Directorate and the Survey Directorate had been under the authority of the MoH until it was restructured in 2002. While the Physical Planning Directorate, which remained in charge of the overall physical planning, was moved to the Ministry of Municipalities and Agricultural Affairs, the Survey Directorate became an independent authority. Furthermore, the Ministry of Works was added to the MoH to establish a joint Ministry of Works and Housing, which however was split again into two separate ministries five years later in 2007. In recent years the first steps have been made to transform the various ministries into authorities, as is the case, for example, of the Ministry of Electricity and Water, which was changed into the Electricity and Water Authority in 2007[82].

Parallel to the reorganisation of the ministries under the king's cabinet, the crown prince

Sheikh Salman initiated the founding of the Economic Development Board (EDB) in 2002. This new agency became a driving force within urban governance by establishing several new committees such as the Urban Development and Housing Committee, which was put in charge of a large housing development called Northern New Town. The main goal of the EDB however is to diversify Bahrain's economy quickly by developing a comprehensive strategy and creating a climate that will attract direct investments from abroad. Consequently, several initiatives such as the Formula 1 Grand Prix, the liberalisation of Bahrain's telecommunication industry or bringing about the privatisation law in addition to a Free Trade Agreement with the United States were established. Moreover, the EDB has functioned as a marketing agency of Bahrain as future investment hub in the Gulf by promoting business opportunities in order to enforce foreign investments and the establishment of public private partnerships.[83] In the following years almost the entire territory of Bahrain turned into a free zone offering a low tax environment and thus one of the most cost effective locations in the GCC. In addition, there was no law that forced companies to assign a certain amount of shares or to engage a mediating agency.[84] This liberalisation has led to a growing interest of the private sector in Bahrain, particularly regarding the growing financial and real estate market. Since 2002 the development of the Bahrain Financial Harbour has been the embodiment of the new approach to reestablish Bahrain as financial capital in the Gulf.[85] While the real estate market has been already opened for GCC nationals in 1999, a new law in 2003 has furthermore allowed every non-Bahraini to buy freehold properties and thus to receive a self-sponsored residence permit[86].

The new legal situation and the general investment climate have led to a building boom in form of various projects, mainly on reclaimed land in the north and close to the urban centres. One of the first freehold property-developments has been Amwaj Islands in the north-east of Muharraq, which was already announced in 2002. In the following years several further projects on reclaimed islands have been launched by the private sector, like e. g. Reef Island and Bahrain Bay along the waterfront of Manama and the development Durrat Al Bahrain in the south of the main island. Beside the coastal developments the freehold projects Al Areen and Riffa Views have been built in the inland. All these developments have in common to be designed by a master developer of a private investor group or a public private partnership in form of mixed-use projects with integrated commercial districts, leisure facilities and hotels in addition to large residential areas. Beside the common two-storey villas the majority of projects comprise a number of multi-storey apartment buildings and even high rise. Parallel to the masterplanned projects several further areas, as e. g. the Seef District, the Al Fateh District and the Seafront District in Manama, were enabled to be built by freehold property-developments. Due to the former subdivision and allocation of land the development followed a general zoning plan, which has been updated according to the new regulations. Consequently many multi-storey buildings were built in order to use the plots as profitable as possible. The beginning transformation of the built environment has been an indicator of the profound change of urbanism in Bahrain, which has been more and more dominated by initiatives of the private sector. As result the World Trade Centre Bahrain and the Bahrain Financial Harbour became the new landmarks of the beginning post-oil era illustrating new economic opportunities and the approach to enter the global market arena.

3.2 The Structure of Urban Governance in Bahrain

3.2.1 The Public Sector

Bahrain is a monarchy where decision making is mainly governed by the king and ruling Al Khalifa family. King Hamad bin Isa Al Khalifa has been the ruler of Bahrain since 1999. While Bahrain was a pure monarchy when he first came to power, it has since become a constitutional monarchy with the introduction of a parliament in 2002. Apart from the king himself his son, the crown prince Sheikh Hamad, is another important political figure who was involved in the development of the Bahrain Defence Force (BDF) and the establishment of the Bahrain Centre for Studies and Research. A third highly influential figure is Sheikh Khalifa bin Salman Al Khalifa, the uncle of King Hamad and brother of the previous king. Since 1971 Sheikh Khalifa has been the Prime Minister of Bahrain and thus in charge of the cabinet. He is the head of the Supreme Defence Council and chairman of several other councils and committees such as the Monetary Council, the Higher Council for Civil Aviation, the Higher Committee for Projects, the Council of Petroleum, the Water Resources Council and the Supreme Council for Civil Service. The Prime Minister is the most politically influential figure beside the king, whose attempt to balance power by introducing reforms has had less impact on the Prime Minister's control of a large part of the government apparatus, the functionaries of which are his own appointees[87]. Since the beginning of the millennium the Crown Prince Sheikh Salman bin Hamad bin Isa Al Khalifa has been gaining more and more political influence in his new roles as chairman of the Committee for Implementation of the National Charter in 2001, chairman of the Economic Development Board since 2002 and as Deputy Supreme Commander of the BDF since 2006. The government itself consists of a cabinet of 23 cabinet ministers who are royally appointed. The main legislative body, the Bahraini parliament, called the National Assembly, consists of two chambers – the Consultative Council (Majlis Al-Shoura) is the upper house and has 40 royally appointed members while the Council of Representatives (Majlis An-Nuwab) is the lower house and has 40 publicly elected members[88].

Major political change at the beginning of the millennium led to a reduction in the number of governorates in Bahrain from twelve to five. These five are the Capital Governorate, the Central Governorate, the Muharraq Governorate, the Northern Governorate and the Southern Governorate. Consequently, five local municipalities were established, these being the country's smallest administrative entities, under the leadership of the Ministry of Municipalities and Agriculture Affairs (MoMA). This ministry is organised into three

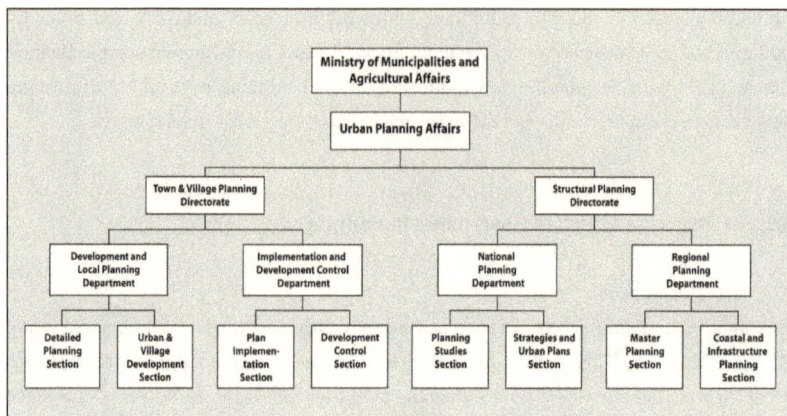

Fig. 4: The structure of the Urban Planning Affairs.

main departments which are controlled respectively by the Minister Office, the Under Secretary of Municipal Affairs and the Under Secretary Assistant of Municipality Services. Due to its role as the head of all municipalities and its authority to issue building permits, the Urban Planning Directorate was moved from the Ministry of Housing to MoMA in 2002 and re-established as Urban Planning Affairs. Today it consists of two separate directorates that deal with planning requirements on the local and national levels. These are the Town & Village Planning Directorate (TVPD) and the Structural Planning Directorate (SPD). Each of these two directorates is divided into two departments, each consisting of two sections. Thus, the SPD consists of the National Planning Department, which is sub-divided into the Planning Studies Section and Strategies and Urban Plans Section, and the Regional Planning Department, which is sub-divided into the Master Planning Section and Coastal and Infrastructure Planning Section. While the SPD with its two directorates is responsible for planning issues concerning overall urban development on a national and regional scale, the TVPD is mainly responsible for developing and implementing detailed zoning plans for every urban area. Therefore, it is structured into the Development and Local Planning Department (DLPD) and the Implementation and Development Control Department (IDCD). Thus, the DLPD is in charge of developing zoning plans, which are designed by its Detailed Planning Section, and creating studies about development concerns in certain areas, which are conducted by its Urban & Village Development Section. Moreover, the IDCD, which is divided into the Plan Implementation Section and the Development Control Section, is responsible for observing the implementation of zoning plans that have already been approved and for controlling building permits that have been issued regarding the current zoning[89].

While physical planning is executed by the urban planning department of the MoMA, infrastructure planning is the responsibility of the Ministry of Works (MoW). This consists of the planning, design, construction and maintenance of the public road network and drainage system. The MoW is also responsible for the development of public buildings,

King Hamad bin Isa Al Khalifa

Crown Prince
Sheikh Hamad

Premier Minister
Sheikh Khalifa bin Salman Al Khalifa

Government Cabinet
(23 royally appointed ministers)

Economic Development Board:

Economic Vision 2030
+
National Planning
Development Strategy
(NPDS)

Further
Ministries

Ministry
of Municipality
and Agricultural
Affairs

Ministry
of
Housing

Ministry
of
Works

Electricity
and
Water
Authority

Ministry
of
Transport

Housing and
Urban Development
Committee:

Projects:
Northern New Town
(2008 back to MoH)

Urban Planning
Directorate:

Zoning Plans
+
Building
Permission

Planning
Department

Social Housing
Projects and
Programs

Central Planning
Unit (CPU)

Infrastructure
Planning
Coordination

Planning
Department

Infrastructure
Projects

Planning
Department

Infrastructure
Projects

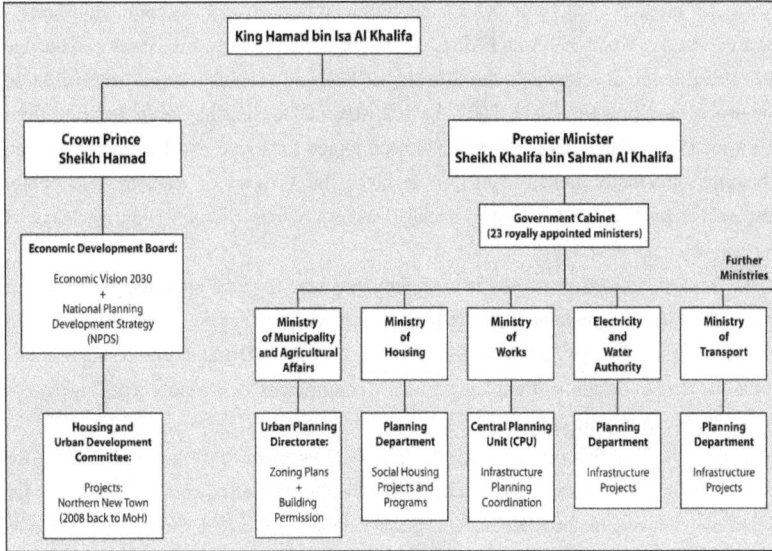

Fig. 5: The main institutions within urban governance.

including their design, construction, project management and maintenance[90]. The MoW consists of 13 directorates, which are divided into five main sections, namely, Technical Services, Human & Financial Resources, Construction Projects & Maintenance, Sanitary Engineering and Roads[91]. In addition, a Central Planning Unit (CPU) is in charge of the planning, implementation and coordination of all main infrastructure projects[92]. The CPU cooperates jointly with the Survey and Land Registration Bureau (SLRB), which is a separate authority, in order to improve planning based on current data[93]. The SLRB itself is divided into three general directorates – the General Directorate of Surveying, the General Directorate of Land Registration and the General Directorate of Resources & Information System[94]. The main goals of the SLRB are to build and to maintain a spatial information infrastructure in order to support the fundamental data requirements of the public and private sectors[95]. While the road network and drainage system are the planning responsibilities of the MoW, the Electricity & Water Authority and the Ministry of Transport are two separate public institutions that also work on the development of Bahrain's infrastructure. They must, however, cooperate with the MoW which acts as the main coordinator.

The Electricity & Water Authority (EWA) is structured into four main departments. While the department of Electricity & Water Production & Transmission is, as its name suggests, in charge of producing and transmitting electricity and water, the department of Planning & Projects plans and develops new projects as well as carrying out research work. There are also departments for distribution and customer services and administrative and financial affairs[96]. The Ministry of Transport (MoT) is responsible for the development of public transport, postal services, telecommunications and civil aviation. In recent years the ministry has been privatising public transport and telecommunications[97]. Apart from the MoW, EWA and MoT, which are the three main ministries responsible for infrastructure development, the Ministry of Housing is another important institution for urban development. Since 2007 the Ministry of Housing has been separate from the Ministry of Works, with which it had been joined for five years. Because the Urban Planning Directorate moved to MoMA in 2002, the Ministry of Housing is currently responsible for the development of social housing projects without being in charge of general physical planning.

Apart from the ministries several new committees were founded upon the instigation of the Economic Development Board (EDB), which was established under the leadership of Crown Prince Sheikh Salman in 2002. The main goal of the EDB and its subordinate committees is to design and implement a new comprehensive development strategy for Bahrain concerning the economy, society and environment (Economic Vision 2030). In order to implement this new strategy, the EDB took over much of the responsibility for urban planning from MoMA, particularly regarding the creation of a new strategic plan for Bahrain. In this regard, the EDB engaged private consultants such as Skidmore, Owings & Merril (SOM) to design the new plan. Furthermore, the Housing and Urban Development Committee was created to develop a large-scale project on the north-western coast which was intended to solve Bahrain's growing social housing problem. However, in 2008 this project was put under the authority of the Ministry of Housing. This incident is one of several examples where areas of responsibility have overlapped between the ministries, which are under the leadership of Sheikh Khalifa, and the EDB, which is led by Sheikh Salman. Due to the restructuring currently being undergone by the public sector, which includes a partial privatisation, it is expected that the assignment of responsibility will continue to change. Major efforts are currently being made to develop an integrated structure of governance in order to improve the coordination of urban development.

3.2.2 The Private Sector

The private sector in Bahrain consists of two general types of investor groups – individual investors, who either build a joint venture with other investors or carry out small-scale projects on their own, and investment banks in the Bahraini financial market. Apart from Arcapita Bank, which was the first Islamic investment bank in 1997, several additional financial institutions have been attracted by the real estate market in Bahrain. In particular, banks from Kuwait such as Kuwait Finance House and Gulf Investment House are interested in carrying out real estate projects in the small kingdom due to its growth expectations. The largest local banks that are majorly involved in investing in property development are Gulf Finance House and Ithmaar Bank in addition to Arcapita Bank mentioned previously. In contrast with Dubai, there are no large-scale real estate developers in Bahrain directly owned by the ruler himself because he owns most of the country's unbuilt areas. However, due to the fact that most of the current development areas in Bahrain are located on reclaimed land, which is owned by the ruling Al Khalifa family, the government has become a shareholder in most projects in alliance with investment banks. In most cases of major developments the leading investor group, which consists of banks or individual investors, has founded a real estate company that acts as the main developer in charge of the planning and implementation of the project's master plan. It is also this so-called 'master developer' that is responsible for the management of the whole project including the allocation of plots to sub-developers and the sale of ready-for-use real estate.

The majority of the real estate companies described above are named after the project they are developing. For example, Durrat Khaleej Al Bahrain Company is the owner of the project Durrat Al Bahrain. This real estate developer was founded by the government of Bahrain and Kuwait Finance House, both of which hold a 50 percent stake[98]. Other examples are the Bahrain Bay Development company in charge of the project Bahrain Bay, which was founded by Arcapita, and the Bahrain Financial Harbour Holding Company in charge of the Bahrain Financial Harbour, which was initiated by Gulf Finance House (GFH). Another company owned by GFH and the Bahraini government is the Al Areen Holding Company, which was created in order to develop Bahrain's largest inland development, Al Areen Resort. In some cases the developer has been established with a name different from that of its project such as the company Ossis, which was founded by a group of eight Bahraini investors in order to develop the project Amwaj Islands. In addition, real estate companies such as Tameer, which was established as a closed shareholding company and is owned by Inovest B. S. C.[99], have been created in order to function as sub-developers working on small-scale projects for the major projects such as Al Areen, Amwaj Islands and Durrat Al Bahrain. Many smaller real estate companies have been founded or moved from other Gulf countries to Bahrain in order to participate in the up-coming property market, which has become promising due to the country's close proximity to Saudi Arabia and its improved business environment with low restrictions.

3.2.3 The Decision-making Process

While a comprehensive master plan has not yet been implemented to act as the basis for urban development, individual zoning plans are designed for each area of Bahrain by the Detailed Planning Section within MoMA's Urban Planning Affairs. After the preparation of new zoning plans for subdivided areas, the plans are submitted for examination to the head of the Urban Affairs department and the minister of MoMA, who then discuss the new plans in a planning committee. This planning committee, which was only recently established, is intended to improve the coordination between physical planning and infrastructure planning, which is mainly done by the Ministry of Works. Thus, the head and chairman of this committee is the minister of MoMA while the under secretary of the Ministry of Works is vice-chairman. After discussion and further adjustments, the zoning plans are approved by the cabinet ministers and sent back to the Town & Village Planning

Directorate, whose Plan Implementation Section and Development Control Section are responsible for controlling the implementation of the approved plans. Subsequently, each developer has to apply for a building permit at the appropriate department of MoMA. After a building permit has been issued, it must be sent to the Development Control Section in order to ensure that the developer's plans are in accordance with the zoning plan. After amendments are made, which may be necessary depending on regulations and current zoning plans, construction can begin. While each developer is responsible for connecting his building to the infrastructural network on the development site, the network itself is entirely developed by the public sector, mainly by the Ministry of Works and the Electricity & Water Authority[100].

With regard to the zoning plans of master-planned developments, each individual developer is responsible for the physical planning of his own project. After the completion of a master plan, each developer has to apply for approval from the state. This application must include a traffic impact study and an environmental impact study, both of which must be conducted by the developer himself before they are checked by the Ministry of Works and its Central Planning Unit in addition to a public agency called Environment & Wildlife Affairs. After research has been carried out on the impact of a master-planned development and any necessary amendments have been made, the plans are discussed by the planning committee and subsequently approved. While in some cases such as, for example, the project Amwaj Islands, a zoning plan is designed by the Town & Village Planning Directorate according to the approved master plan, there have been other cases, for instance the Bahrain Bay project, where the project area has remained zoned as a special project area without a detailed public zoning plan. This is mainly due to current regulations, which allow investors to change their projects according to their future needs. However, each future change would nevertheless have to be approved separately by the public sector. In addition to the independent design of the master plans, a new system of granting building permits has been introduced in order to speed up the process of construction. Instead of sending each application for a building permit to MoMA, sub-developers submit their applications directly to the master developer, who is in charge of controlling their compliance with the zoning of the master plan. After the master developer has approved the applications, he sends it to MoMA, where the permits are finally issued. With regard to infrastructural supply, the master developer and the public sector often cooperate in order to provide the required infrastructure as early as possible. While the master developer is responsible for the planning and development of the main infrastructure, each sub-developer is only responsible for the connection of their own projects to the main network[101]. All in all, a general tendency can be observed that the decision making process is becoming more and more decentralised due to less bureaucratic restrictions and more planning responsibilities being placed on the private sector.

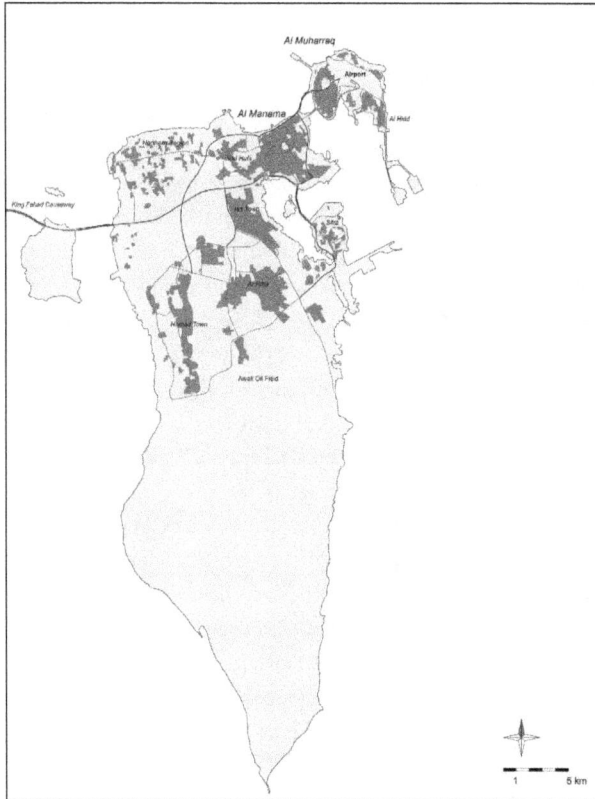

Fig. 6: The settlement structure in 2000.

3.3 The Transformation of Bahrain's Urban Structure

3.3.1 The Urban Structure at the Turn of the Century

At the end of the 20th century Bahrain's 578-square kilometre main island occupied almost 87% of its overall land area of approximately 665 sq km. Apart from the islands Al Muharraq (17.9 sq km) and Sitra (11.4 sq km) to the north and east of the main island, many smaller islands in addition to the island archipelago Hawar were populated with settlements[102]. The least populated areas have always been the central and southern parts of the main island, where Bahrain's highest point – a rocky area called Jebel Dukhan reaching 134 metres of absolute altitude – is located.

Land reclamation for urban purposes in Muharraq, Manama, Sitra and Budaiya has caused the shape of the islands to constantly change since the middle of the 20th century. Thus, the total area of Bahrain has been growing, particularly since the beginning of its industrialisation in the 50s. Between 1959 and 1991 a reclaimed area of about 23 sq km led to an increase in the total area of around 3.5%. Most reclamation was done in the north, east and south of the capital Manama, where expansion areas were needed for the

121

growing business centre and the new harbour. In the 80s a large piece of land of about 5 sq km was reclaimed in the north of the main island, later known as the Seef District, in order to allocate plots to people who had lost their land due to the construction of new highways leading from Manama to the southern part of the island and to the causeway to Saudi Arabia. While a large part of the Seef District was reclaimed in one go, other areas were reclaimed in steps of rather small areas, as for example in the case of the expansion of Juffair in the south-east of Manama. In the 90s the total area of Bahrain was about 665 sq km, of which about 307 sq km was still being used for agriculture and around 200 sq km was occupied by urban areas spreading all over the northern half of the main island and Muharraq. Around 45% of the total urban area was occupied by suburban residential areas, which had been rapidly growing from about 37 sq km in the late 70s to more than 91 sq km in the late 80s, particularly in the outskirts, where two new dormitory settlements – Isa Town and Hamad Town – had been developed. Consequently, the largest suburban housing areas were built further south of the traditional settlements and along the newly developed road network.

In 2001 the three biggest settlement areas apart from Manama with 24% of the total population and Muharraq with 14.1% were Riffa with about 12.3%, Jidd Hafs and Hamad Town, both having about 8%. Riffa, as the home town of the ruling Al Khalifa family, who mainly settled in West Riffa, has been dominated by a large Sunni population. In 2001 almost 80,000 people lived in Riffa while in the governerate of Jidd Hafs the population reached around 52,000, of which a large part was from the Shia population. The new suburban settlement Hamad Town grew from only about 29,000 residents in 1991 to more than 52,700 in 2001, thus becoming Bahrain's biggest newly developed housing area in addition to Isa Town, where with about 36,800 inhabitants around 6% of the population settled. The town Sitra had about 43,900 residents, making it Bahrain's seventh largest settlement, and similarly to Al Hidd on the island of Muharraq, which housed over 11,600 residents, a large part of the inhabitants were the employees of the industries based there. Apart from the small number of towns, a large part of the population lived in small villages and compounds in the northern, central and western areas of the main island. All in all, about 120,000 people, more than 18% of Bahrain's total population, lived in the rural areas and on the outskirts of towns, which occupied about 229 sq km. All six major towns together covered an area of around 110 sq km and housed about 467,320 people, almost 72% of the total population. Thus, the average urban density was about 4,250 residents per sq km in contrast to a density of about 524 residents per sq km in the rural areas. The areas with the highest urban densities were the capital Manama with about 5,681 inhabitants per sq km and Muharraq with around 4,378 inhabitants per sq km. Densities were lowest in Hamad Town and Isa Town, which had about 4,055 and 3,069 inhabitants per sq km respectively[103].

In 2001 the modern Central Business District was exclusively located along King Faisal Highway on reclaimed land on the northern shoreline of Manama. In addition to commercial buildings such as offices, banks and shopping malls many government offices were located within the CBD. Shopping districts such as the Central Market were established at the western end of the CBD and along its southern borders towards the old centre. The old urban centres of Manama and Muharraq developed into mixed-use areas containing office buildings, shops and apartment buildings for foreign labour. However, some residential areas of the local population have still remained in the old centres, particularly in the case of Muharraq. In addition to the city centres of the two largest towns, many shopping streets and commercial districts developed along main roads such as, for example, Budaiya Road leading from Jidd Hafs to the western coast, and along the central roads of smaller towns such as Riffa and Isa Town. Most social infrastructure such as hospitals and schools was built within the expanding urban areas of Manama and Muharraq in addition to local schools in the villages and new suburban areas. In Isa Town a cluster of private schools was developed beside the campus of Gulf University and a campus of Bahrain University, the main campus of which was built on an area south of Hamad Town. Apart from the main governmental institutions in the centre of Manama and the Diplomatic Quarter, a few ministries and public administrative buildings were developed in other areas such as Isa Town, where two ministries were built, and Juffair and Umm Al Hassan in the south-east and south of Manama.

Due to the direction of the wind coming from the north and north-west, industrial areas were developed along the southern and eastern coastline of Muharraq, the main island and Sitra. While in Muharraq the construction of a dry dock in Al Hidd led to the development of an industrial zone, in the south of Manama the harbour Mina Salman was the main reason for the local industrial development. Because of its close location to the inland Awali oil field, the oil processing industries and the petrochemical industries settled in Sitra, which consequently became the centre of heavy industry in Bahrain. Similarly, the aluminium industry in the western part of the main island was established close to the oil and gas producing areas inland in order to enable energy resources to be supplied quickly. Due to the location of the Awali oil field and the military areas of the BDF the southern half of the main island became the least populated area of Bahrain. In 1935 the British established a navy base in the south of Juffair. This area of Manama has remained a foreign military zone because the US Navy moved into the former British base after Bahrain gained independence based on an agreement in 1977[104]. Because of industrialisation and modernisation, many of Bahrain's agricultural areas were used as sites for the development of compounds and thus the total area of land used for agriculture, which was once an important contributor to the income of the Bahraini population, was reduced along the green belt in the north and west of the main island from 60 sq km in the 70s to around 15 sq km in the late 90s. Apart from a number of date plantations, particularly in the west, several smaller gardens and fields were used to plant vegetables and fruits in addition to alfalfa which was used as a forage crop[105].

During the oil urbanisation in the second half of the 20th century three main infrastructure projects in the form of the international airport in Muharraq, the harbour Mina Salman in the south of Manama and the King Fahad Causeway to Saudi Arabia had a decisive impact on the further development of the road network as well as the location of housing areas, business districts and industrial zones. The King Faisal Highway in the north of Manama was built to connect the capital with Muharraq and the airport by two causeways, including the Sheikh Hamad causeway, which was already built in 1929. The Al Fateh Highway along the eastern coast was developed to link the centre with the harbour area and new residential areas in the south and to form together with the King Faisal Highway in the north and the Sheikh Sulman Highway in the south the outer ring road of Manama. The construction of a new causeway in the south of Manama called Sheikh Khalifa bin Salman Causeway was completed in 2003 to connect the industrial areas in Muharraq and Manama. Furthermore, this new causeway led to the connection of the ring road in Muharraq and thus to the establishment of a new ring road around both cities. In the north-west of Manama at the Pearl Roundabout, the King Faisal Highway was extended to the south, thus connecting it directly to the inner ring road of Manama leading around the old city centre. The King Faisal Highway itself, which ends at the Dilmun Roundabout, became one of the central junctures between the centre of Manama in the east, the CBD in the north, the residential areas in the Jidd Hafs region in the west and the Sheikh Sulman Highway in the south. Furthermore, the Budaiya Highway, which ends at the King Faisal Highway, was built through Jidd Hafs passing by the villages and many compounds in the northern region to Budaiya at the western coast.

During the 80s and 90s the highway network was extended in order to connect the outer ring road of Manama with the King Fahad Causeway, which was completed in 1986, and the newly developed southern settlements Hamad Town and Isa Town. The Sheikh Sulman Highway in the south of Manama was directly connected to the new causeway and two big interchanges, which joined the traffic route with the highway leading from Isa Town and Riffa to the north and the new highway extension leading from the south to Manama by crossing the Budaiya Road and ending at the King Faisal Highway. In addition to these two main highways the southern part of the main island was connected to Manama by the Sheikh Jaber Bin Ahmed Alsubah Highway leading from Riffa to Sitra and then, via another causeway, to the south of Manama. Along the western coast the Janabiya Highway was built to connect Budaiya with the Causeway Approach Road and the roads leading along the date plantations towards the villages along the coast and Hamad Town in the south.

The urban fabric of Bahrain at the beginning of the millennium was divided into four different types – the traditional urban fabric of the pre-oil settlements, the suburban perpendicular net fabric in the suburbs, the urban areas of mixed typologies in the transition zone and the block structure of the new business districts. Although the modernisation of the city centres in Manama and Muharraq led to the replacement of most buildings and the development of several thoroughfares, the traditional structure of the non-orthogonal road network based on the ownership of plots has remained. Furthermore, the replacement of old houses did not lead to a remarkable change in the plot ratio because the average building height of two to three storeys of these houses was retained. Because of a high density of buildings, small courtyards continued to be used in order to supply houses with air and light.

While Bahrain's traditional urban structure has been the product of the interaction of the members of its population in their urban and rural activities, the suburban fabric of two-storey villas situated on standardised square plots is the result of the subdivision of land that has been carried out by the state since the 40s. The surrounding walls and the general rule of homogeneous building heights in order to protect the privacy of the traditional Arab family led to an urban landscape with a very low density, a high car dependency and an ambivalent residential typology, where villas in various architectural styles with showy facades are blocked from the view of the outside world by high walls. In the cases of some new settlements such as Isa Town and the new urban areas in the south of Manama, the typologies and land uses were mixed. In addition to blocks of residential houses, multi-storey apartment buildings and walled compounds, commercial buildings and hotels were built along the main streets. Similar to all new urban areas the structure of the central business district was based on the general physical planning and its zoning of the subdivisioned areas. Many plots remained undeveloped due to a lack of investment by the individual owner and thus they were available to be used as car parks. On the plots that were developed stand multi-storey blocks and high-rise buildings used for commercial and government administrative purposes, which became important landmarks and identification points in the otherwise rather monotone looking urban landscape.

3.3.2 Current Types of Urban Development

Developments Following Public Zoning Plans:
As in previous decades, many new development areas are subdivided by the state and built according to official zoning plans, which are prepared in order to determine land use and building height. Due to increasing pressure from developers, the state's existing zoning plans were recently adjusted to accommodate developers' preference for high-rise buildings. The current development zones are divided into seven different kinds of land use and 22 different building types.

The residential areas owned by Bahraini citizens are of five different types, namely, the 'special residence' type A and type B, the 'linked residence' type A and type B and the 'garden's residence'. Based on Resolution No. 27, which was approved by the Council of Ministers in 2005, a clear regulation was introduced defining what percentage of the total plot area can be constructed, the size of the plain surface of the floors, the building height, the number of parking lots needed and the front, rear and side retraction in addition to special stipulations. In the case of the 'special residence', type A for instance is defined as separate or linked residential villas, the construction of which should not exceed 180% of the total plot area and the maximum height of which should be no higher than

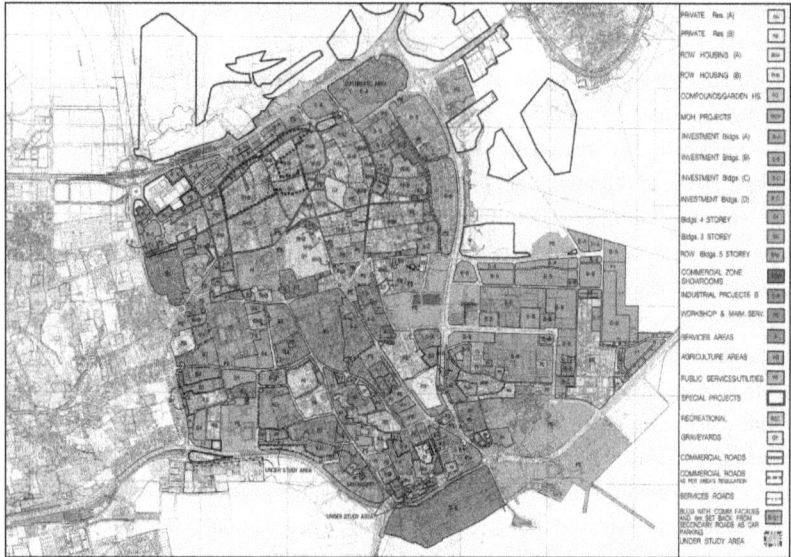

Fig. 7: The zoning plan of Manama (2008).

15 metres with a maximum number of three storeys. 'Special residence' type B differs from type A in that more residential units can be developed on one plot, including flats. 'Linked residence' type A allows for buildings with a size of up to 210% of the total plot area, while the number of floors is limited to three. 'Linked residence' type B allows for buildings, whose construction area is 300% of the total plot area in spite of the limitation to three floors. The regulations governing 'garden`s residences' concern the development of compounds consisting of separate or linked residential villas in addition to garden apartments, the construction ratio of which should not be more than 55% of the land area of all constructions. Furthermore, the building height is limited to two floors with a maximum height of 10 metres[106].

Multi-storey buildings for commercial or residential use are divided into seven types according to their maximum height and/or maximum construction ratio. All of these buildings are classified as 'investment buildings' and thus are allowed to be sold as freehold properties. The highest densities are permitted in areas with the type 'Investment Building B-A', the construction ratio of which can be as much as 1,200% of the total plot area. The plain surface area of any floor is limited to 60% of the plot area. All 'investment buildings' can be developed as commercial, residential or administrative buildings or a combination of these three different uses. The main difference between the 'investment building' types is the construction ratio of their given total plot area, which is restricted to 750% with regard to type B-B, 500% with regard to type B-C, which furthermore is limited to a maximum of 10 floors, and 300% in the case of type B-D, which can have a maximum of six floors. With regard to four-storey and three-storey buildings, the construction ratio has been limited to 240% and 180% of the total plot area. In the case of

126

Type	Abbr.	Use	Max. floor area ratio	Max. site occupancy index	Min. retractions of the land boundary	Max. height
PRIVATE Res. A	RA	Residential	1.8	0.6	3 m (front) and 2 m (side and rear)	3 floors (15 m)
PRIVATE Res. B	RB	Residential	1.8	0.6	3 m (front) and 2 m (side and rear)	3 floors (15 m)
ROW HOUSING A	RHA	Residential	2.1	0.7	3 m (front) and 2 m (rear)	3 floors (13 m)
ROW HOUSING B	RHB	Residential	3	1	-	3 floors (13 m)
COMPOUND	RG	Residential	0.55	-	5 m (front) and 3 m (side and rear)	2 floors (10 m)
Investment Bldgs. A	B-A	Residential/ Comm.	12	0.6	No front retraction, 6 m (side and rear)	-
Investment Bldgs. B	B-B	Residential/ Comm.	7.5	0.6	No front retraction, 4.5 m (side), 6 m (rear)	-
Investment Bldgs. C	B-C	Residential/ Comm.	5	0.6	No front retraction, 3.5 m (side), 5 m (rear)	10 floors (50 m)
Investment Bldgs. D	B-D	Residential/ Comm.	3	0.6	No front retraction, 3 m (side and rear)	6 floors (30 m)
Bldgs. 4 floors	B4	Residential/ Comm.	2.4	0.6	5 m (front) and 2 m (side and rear)	4 floors (20 m)
Bldgs. 3 floors	B3	Residential/ Comm.	1.8	0.6	5 m (front) and 2 m (side and rear)	3 floors (15 m)
ROW Bldgs. 5 floors	BRS	Residential/ Comm.	5	1	-	5 floors (22 m)
Commercial Zone	COM	Commercial	3	0.6	6 m (front) and 3 m (side and rear)	5 floors (25 m)
Stores/ warehouse	ST	Industrial/ Comm.	2.4	0.6	6 m (front) and 3 m (side and rear)	3 floors (18 m)
Industrial projects A	D-A	Industrial			restrictions per individual case	
Industrial projects B	D-B	Industrial	2.4	0.6	6 m (front) and 4 m (side and rear)	4 floors (24 m)
Light industries	LD	Industrial	2.4	0.6	6 m (front) and 4 m (side and rear)	4 floors (24 m)
Workshops	WS	Industrial	2.4	0.6	6 m (front) and 2 m (side and rear)	4 floors (24 m)
Service Areas	S	Industrial/ Comm.	1.8	0.6	5 m (front) and 2 m (rear)	3 floors (18 m)

Fig. 8: The building regulations of each zoning area.

the areas containing five-storey buildings, there is a distinction between areas containing 'linked multi-storey buildings', which have a maximum construction ratio of 500% of the plot area, and so-called 'commercial exhibition' areas, which have a maximum construction ratio of 300%. In addition to the regulation of residential and commercial areas, industrial areas have been divided into six different categories. These are areas of 'industrial production projects' type A, 'industrial production projects' type B, 'areas of stores and warehouses', 'areas of light industries', 'areas of workshops and maintenance services' and, finally, general 'services areas' for light industrial use and services such as repair workshops. Furthermore, agricultural areas have been defined in order to prevent them from being transformed into building land. Certain areas such as the areas of the housing developments of the Ministry of Housing or new development areas of the private sector, have not been initially zoned or defined by the urban planning directorate itself due to their own masterplans, which have usually been designed separately[107].

There are currently 21 approved zoning maps covering all five governerates including the Capital Governerate, which comprises Manama, the Seef District and the western part of Manama including Jidd Hafs, Sanabis and Belad Al Qadim. The biggest development areas are located on reclaimed land along the northern and eastern shoreline, where in addition to several commercial developments, many residential multi-storey buildings have been constructed due to the growing real estate market in Bahrain. In addition to the development of many former empty plots within the main CBD along the King Faisal Highway in the north and within the Diplomatic Quarter in the east, the new centres of construction have been the Seef District in the north-west and Juffair in the south-east. Because of the very fortunate infrastructural supply provided by the main highways leading from the Manama centre to the causeway to Saudi Arabia, several shopping malls have been constructed in the Seef District, which has become an expansion of the former commercial district. Thus, many banks and office buildings have relocated to the district, the land of which was reclaimed in the 80s in order to create residential plots that however were only partly developed. In Juffair, because of the large number of US soldiers at the US Navy base located there and the immigration of many foreign guest workers in recent years, there has been a fast development of multi-storey apartment buildings and compounds. In the case of Muharraq, four zoning maps have been approved for this governerate, one of which includes the zoning for a privately funded development called Amwaj Islands, located in the north-east of the island. While the master plan of this project was designed by its master developer, an official land use map was prepared and published by the public sector for the allocation of building permits. In the north-west of Muharraq a large new area in the form of a crescent has been reclaimed but not yet completely subdivided and has thus been earmarked as a 'freeze zone'.

In recent years most of Bahrain's urban development has been centred in and around the two main cities in the north in addition to a few scattered development areas in the south and north-west of the islands, which are mainly private-sector developments or public-sector housing projects. While in previous decades most aspects of urban planning had been the responsibility of a single ministry, the decision to move the Planning Directorate from the Ministry of Housing to MoMA in 2002 and an increase in master-planned projects by the private sector led to urban planning in Bahrain becoming decentralised. This in turn has led to various kinds of developments, mainly divided into urban expansion based on zoning plans by the public sector and individual master-planned projects by both the private and public sectors.

Master-planned Developments:
1) The New Manama Waterfront
Because of its attractive location on Bahrain's northern coast at Manama and in front of the urban centre, the old harbour was chosen to be one of the first large-scale developments of the private sector in Bahrain. At the beginning of the millennium the investment bank Gulf Finance House initiated the development of the Bahrain Financial Harbour (BFH) with the primary aim of establishing Bahrain as a regional and global financial centre. It is a mixed-use project also comprising residential, retail and leisure areas and hotels. The development of the project is divided into several construction phases beginning at the former shoreline and going up to the sea on a reclaimed area of about 380,000 sq m. All in all the whole project area will cover 28 individual development units as part of ten main projects, including the already built Financial Centre, which comprises the 53-storey twin towers called Harbour Towers, the Harbour Mall and the Harbour House. Additional project areas are the Commercial East and Commercial West in addition to the Harbour Row along the marina in the centre of the development and the Dhow Harbour at the western shoreline. The last phases of the development will consist of residential,

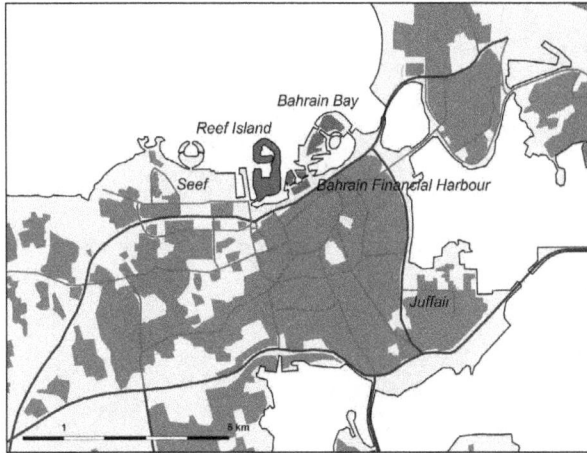

Fig. 9: The master-planned projects in Manama.

leisure and hotel projects in the north in the form of the BFH Hotel, the Diamond Tower and, in addition to the Residential North and the Residential South, which will include a cluster of three residential towers called Villamar, there are plans to develop the Bahrain Performance Centre on the northern island. After completion of the BFH by 2012 it is expected that about 7,000 people will reside and around 8,000 people will work within the development[108].

In the north-west of the BFH another project has been launched in the form of a 579,000 sq m large island called Reef Island. The development is a joint venture of the Bahraini government and the Bahrain based Mouawad Group for Real Estate Development Company. The developer in charge of the project is Lulu Tourism Company, whose goal has to establish a luxury residential island in the centre of Manama. In addition to 39 multi-storey apartment buildings providing 1,217 housing units, the project will include the development of 114 villas along the outer shorelines and along the two inner bays with their beach fronts. In addition to the residential projects there will be a shopping mall in the southern centre of the island close to a large-scale healthcare facility called the Medical Centre. Furthermore, there will be several leisure and tourist projects in the form of a marina with a yacht club, a 5-star hotel with a spa and public promenades with cafes, restaurants and small parks along the southern shoreline of the island. While the first phase of the project is scheduled to be completed in 2009, the second will be completed in 2010[109]. Only one bridge in the west will connect the project with the mainland in the Seef District and there will be no direct connection to the Bahrain Financial Harbour in the south-east. Thus, the project can be seen as an extension of the Seef District rather than of Manama's old harbour area where the land for it has been reclaimed. While the southern part of the island with its shopping mall and promenades will be open to the public, the northern part will be a gated community with limited access to non-residents.

The third development forming Manama's new waterfront is the project Bahrain Bay,

129

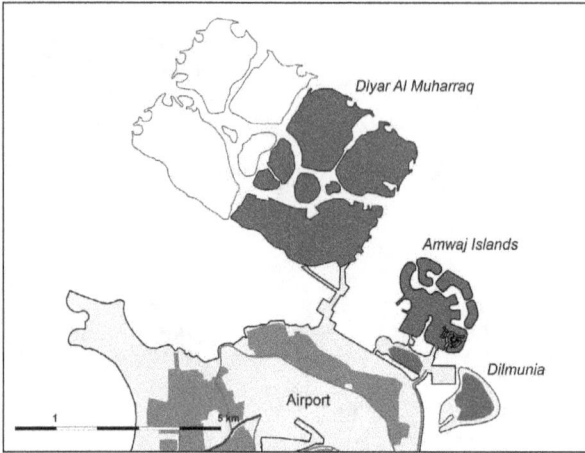

Fig. 10: The island projects in Muharraq.

which was launched as a joint venture of the investment bank Arcapita Bank B. S. C. and a Bahrain-based investment group. The first of three phases is the reclamation of an area of around 430,000 sq m along the northern shoreline of Manama in front of the Diplomatic Quarter. Bahrain Bay is structured into two main axes which will be defined by, in the case of the north-south axis, a canal going through Zone 5, and, in the case of the east-west axis, a length of park within Zone 2 and the building of Arcapita Headquarters. The project is divided into seven distinct zones as well as containing three other projects – Raffles City Bahrain, Arcapita Headquarters and the Four Seasons Hotel Bahrain as the centrepiece of the whole development on a detached island. While most districts will be mainly occupied by residential projects in the form of apartment buildings in addition to leisure areas, Zone 3, Zone 5 and Zone 7 will comprise commercial towers located along the outer access road in order to prevent traffic congestion. Two radial main roads will connect the inner zones, while the outer ring road will link the project with the BFH in the west and the intersection of King Faisal Highway and Al Fateh Highway in the east. In Zone 2, as well as the park previously mentioned, there will be a marina and promenades along the shoreline in addition to several retail districts, three mosques and one school in order to establish and market the project as self-contained. The total built area of the project will be more than 1.45 million sq m, of which about 60% will be for residential use, and it is expected that around 25,000 people will live within the development by 2015[110].

Together with the BFH and Reef Island, the Bahrain Bay project will add a reclaimed area of about 1.4 sq km and residential units for around 36,000 people to the capital. Thus, the new waterfront would reach a density of more than 25,700 people living on one sq km, which will be more than four times the current average density of Manama. According to current plans a large island will be developed in the future about 2 kilometres from the coast forming a bay with the three waterfront projects.

Fig. 11: The Amwaj Islands project.

2) The Island Developments on the Eastern Coast of Al Muharraq

Together with the Bahrain Financial Harbour in Manama, the project Amwaj Islands, launched in 2002, was one of Bahrain's first freehold property developments. But in comparison to the BFH, which was initiated to extend the business district and establish a new financial centre, the Amwaj Islands project is intended to be a mainly residential, tourism and leisure development in order to serve the up-and-coming real estate market. Messrs Saud Kanoo, Khalid Alsharif and Jameel Al Matrook, together with five other Bahraini entrepreneurs, founded the master developer Ossis. This new real estate company was put in charge of developing six man-made islands on an offshore area of around 7 sq km bought from the Bahraini government in front of the north-eastern coast of the island Al Muharraq. As master developer, Ossis has been responsible for developing the land use plan and the main infrastructure for this project in cooperation with consultants while sub-developers such as Tameer have started to build real estate projects in accordance with the guidelines of the approved master plan. There are six islands, each constituting an individual development zone. Najmah is the main island with an area of more than 1.3 sq km. Dalphene has an area of about 480,000 sq m. The three northern islands Wardeh with a size of 239,225 sq m, Tala with 204,500 sq m and Hamama with 124,492 sq m are connected to each other and the main island by one main access road and four bridges. The square island Farasha with a size of about 200,000 sq m has been developed together with the island Dalphene between the main island and the coast of Al Muharraq due to their public functions. While Farasha's project area will include an international school, a university and a business park, Dalphene will have theme parks in addition to several recreational facilities. All in all, an area of about 2.8 sq km has been reclaimed by dredging the local sand from the sea and using the geotube containment system over a one-year period from 2002 to 2003[111].

After the land reclamation the second phase of the project was the development of the main infrastructure, including roads, electricity, water, sewage and telecommunications in addition to the installation of an irrigation system and the construction of a breakwater and beaches along the shoreline. During the first two phases, which were completed in 2003 and 2006, Ossis carried out several studies, including hydrological studies regarding the land reclamation, which were done in cooperation with the consultant Hacrow, and traffic impact studies. While the British consultant Scott Wilson was engaged to design the macro-layout of all six islands, the Bahrain-based company Gulf Engineering was put in charge of the infrastructure planning. Several projects, for example, Tala Island, were designed by various other architects and engineers, including the Australian architect Davenport Campbell. Consequently, each project was individually designed in accordance with the building and land use regulations of the master plan. The project Lagoon in the centre of the main island will be established as the main commercial district of the whole development in addition to a marina on the western shoreline that will be built in front of three hotel developments and a shopping mall. While the buildings around the Lagoon can be built up to 20 floors high, buildings on the plots in the second row behind can only be built up to 15 floors and in the third row, up to 10 floors, except for two plots at the entrance to the Lagoon that have been reserved for the development of high-rise buildings. On the opposite side of the Lagoon will be the Amwaj Gateway project, which will include 94 town houses in addition to three 20-storey apartment buildings offering 375 apartments. In the south-east of the main island two residential projects have already been developed in the form of the Meena7 Towers, which comprises seven blocks of apartment buildings with about 240 residential units, and the Al Marsa – Floating City, which consists of 375 residences along a canal network covering a total area of 55,000 sq m[112].

In the north of the Al Marsa development the Dragon project has been launched in the form of 20 linked villas and an apartment building offering 31 apartments. Further to the north similar small-scale residential developments have been launched, for example, the Lagoona, Safeena and Mirage projects, which together offer 46 housing units along the shoreline. Most of the shorelines of all four residential islands will be developed using attached villas and three-storey apartment buildings in front of private beaches with the exception of certain parts of the northern shoreline of the main island on the opposite side of Tala Island where public beaches will be developed as well as along the marina where the shoreline has been reserved for hotel developments. In the second row behind the low-rise developments along the shoreline there will be residential units in the form of detached villas with two or three floors and five-storey apartment buildings. In the special case of about 11 plots along the shoreline, which are scattered in different locations on the four islands, permission has been granted to construct buildings up to 15 floors. All in all, the whole development will include about 1,300 villas and 350 apartment blocks offering around 12,600 residential units. In 2008 approximately 3,000 people were already living in the project, which is expected to house more than 30,000 residents after its phased completion by 2015. The whole built-up area of the development will be around 4.6 million sq m, which includes 400,000 sq m of hotel and commercial areas in addition to 2.8 million sq m of apartments and 1.4 million sq m of villas. Around 40% of the current buyers of residential units are investors from Saudi Arabia in addition to approximately 20% of buyers coming from the remaining GCC countries and 15% from Europe, Australia or the US. Thus, around 80% of all the residential units sold are currently owned by non-Bahrainis while only 20% has been bought by local investors, who usually rent their properties to expatriates[113].

Due to the close proximity of the International Airport and the capital Manama, two further island developments have been initiated along the north-eastern coast of Al Muharraq close to Amwaj Islands. In 2006 land reclamation began for Diyar Al Muharraq, a project that was previously known as Two Seas. In recent years its master plan has been significantly changed from an island in the shape of a seahorse to a number of islands of different sizes and shapes. The project will cover an area of around 6 sq km located off the coast of where the future airport extension will be. After the completion of its first stage in 2009 according to the current plans, it will be expanded to an overall area of 12 sq km[114]. The master developer of Diyar Al Muharraq is a newly established real estate company also named Diyar Al Muharraq, which was initiated by a joint venture of the Kuwait Finance House, which is the main investor, and a group of private investors[115]. It is designed to be a self-contained development integrating commercial properties, leisure and social facilities in addition to large residential areas constituting around 30,000 housing units that can provide a home for an expected population of about 100,000 people[116]. While most of these housing units will be freehold properties for the open real estate market, a number of them are earmarked to be affordable homes for Bahrainis and expatriates with lower incomes. Some of these social dwellings will be built on a specific area provided by the Ministry of Housing while the rest will be built on on smaller sized plots in other areas and sold by the developer directly to the public[117]. Around 800 of these lower priced residential units will be constructed within the first phases of the development, which will start in 2010, when the main infrastructure will be completed and the first facilities will be ready to use. A specific feature of the project will be large public areas in the form of parks, promenades and beaches along more than 22 km of the 40km-long shoreline[118]. The first stage will include 12 islands of various sizes which will be connected by a road system including 13 bridges and a canal network with several marinas[119]. While most residential areas will stretch to the north, mainly in form of low-rise villas in addition to several multi-storey apartment blocks, the south will be occupied by a commercial district including a cluster of high-rise buildings and one of the biggest shopping malls in Bahrain.

Another new development is the project Dilmunia, which was launched by the Bahrain-based investment bank Ithmaar Bank. It is located south of Amwaj Islands and will consist of two man-made islands covering an area of 1.25 sq km and as well as residential developments it will include retail, health and wellness facilities[120]. The master developer Ithmaar Development Company (IDC) has engaged the international consulting firm DP Architects from Singapore to design the master plan of the island, the reclamation of which is expected to take about 36 months. After the construction of the main infrastructure including roads, pedestrian and vehicular bridges, sewerage, irrigation and electrical installation in addition to telecommunication facilities, approximately 25 plots, each ranging from 12,000 to 46,000 sq m, will be developed by investors in accordance with the approved master plan[121]. While the commercial district, hotels and health and wellness facilities will stretch along the canal, which will divide the two islands, the residential areas will be developed on the northern island facing the open sea. One circular access road along the inner canal and one straight thoroughfare crossing the smaller island will form the main road infrastructure of the project. While one road will connect the project with Al Muharraq, three to four future causeways will connect Dilmunia with neighbouring projects such as Amwaj Islands in the north. Because of large green areas in the form of parks in the south-west of the island, an absence of high-rise buildings and the predominantly medium- to low-rise residential developments the plot ratio is rather low. In addition to multi-storey blocks in the centre and along the canal there will be a cluster of eight apartment blocks on the southern shoreline. The centrepiece of the project will be the medical cluster, covering an area of about 165,000 sq m, which will include a 216-bed hospital, a research centre and a wellness hospital in addition to four hotels[122]. The residential areas will comprise 38 lagoon-view villas, 36 garden-view villas and 44 sea-view villas in addition to about 780 residences in five- to eight-storey apartment buildings in addition to 582 apartments in three- to four-storey buildings. The commercial cluster, spreading over 25,000 sq m, will be developed in the form of two- to four-storey buildings housing supermarkets, restaurants and commercial outlets[123]. After completion in 2013 it is expected that around 4,500 people will be permanently living in the development, which has been marketed as 'health island'[124].

In addition to the residential and commercial island developments, there has been an attempt to establish a new industrial development called Bahrain Investment Wharf (BIW) in the south of Al Muharraq. Initiated by a joint venture of the Al Khaleej Development Company (Tameer) and the Bahraini government, the project will be developed on a 1.7 sq km area, of which 55% will be reclaimed, within the Hidd Industrial Area. The location was chosen for its direct access to major sea, air and road networks, namely, the new Sheikh Khalifa bin Salman Al Khalifa port, Bahrain International Airport, King Fahad Causeway and the future Bahrain-Qatar causeway[125]. In addition, the fact that there is already industrial infrastructure, for example, a power plant, was a major factor in the choice of project location. A large area of around 900,000 sq m will be occupied by an 'Industrial Park', which will accommodate various industries, including warehousing, storage, packaging and re-distribution facilities. Furthermore, an Associated Services area covering 400,000 sq m will be added to serve the industrial areas in addition to a 300,000 sq m 'Business Park', which will be located at two different sites[126]. The Business Park will consist of low-rise office blocks, including the BIW headquarters, a training centre, conference hall and further commercial facilities including a hotel. The remaining 100,000 sq m of the project will be developed as a 'Residential Park' that will accommodate about 20,000 employees, workers and Middle Management[127]. As a free zone, the development tries to attract foreign investment by offering 100% foreign ownership for most categories of business in addition to a highly favourable tax environment with no corporate, personal, value-added or withholding tax[128].

While the south of Al Muharraq will be occupied by an expanding industrial sector, the north-east will be dominated by residential, commercial and tourism developments. The three current island projects Amwaj Islands, Diyar Al Muharraq and Dilmunia will provide residential units for around 140,000 future residents on an overall reclaimed area of about 16 sq km. Thus, the average density of the new urban areas will be approximately 8,750 people living on one square kilometre, which is almost twice the average population density of Al Muharraq today. All three developments aim to become self-contained 'cities' integrating commercial districts, social facilities such as private schools and hospitals as well as large leisure areas in the form of beaches, promenades and parks. Although much of the shorelines of these developments will be occupied by private beaches, all three developments will have several kilometres of public beaches and promenades along the sea. For example, over 50% of the coastlines of Dilmunia and Diyar Al Muharraq will be open to the public[129]. Except for certain areas within Diyar Al Muharraq most of the residential developments will be sold as freehold properties and thus the percentage of Bahraini residents is expected to be rather low. According to current plans several further island projects will be carried out in the future to the south and east of Dilmunia island.

3) The Developments in the South of the Main Island

In the very south of the main island, 54 kilometres away from the International Airport, a cluster of 15 man-made islands called Durrat Al Bahrain has been developed by the Durrat Khaleej Al Bahrain Company, which was founded by a joint venture of the Kuwait Finance House and the Bahraini government. The whole development area covers 21 sq km stretching almost five kilometres along the coast and four kilometres into the sea, where a group of 12 islands is arranged in a circle in front of a crescent-shaped land expansion of the coast. The total land area, which was developed by land reclamation, will add about 4.1 sq km to the overall area of the development, which will include further projects on an area of around 4 sq km along the coast. The master plan of the project was designed by Atkins, who divided the development into five different project areas comprising the 'Islands', the 'Hotel Island' and the 'Crescent' in addition to the 'Durrat Marina' and the 'Golf Course' in the north. The 'Islands' will consist of six 'Atoll Islands' in the shape of half open atolls, each covering an area of around 400,000 sq m, and five fish-shaped 'Petal Islands', each covering an area of about 165,000 sq m. These two types of island are arranged in two circular rows. The Atoll Islands are located in the outer row facing the open sea. While all the Petal Islands are connected to the crescent on the main land by one circular access road, the Atoll Islands are linked to the main road by bridges to the Petal Islands. The housing units on the Atoll and Petal Islands will altogether come to 2,000 villas. These will have two storeys and be of four different pre-designed types. On the Petal Islands they will be located either along paved shorelines or further inland while all the villas on the Atoll Islands will have direct access to a beach. Instead of separate private beaches there will be community beaches along the inner bays of the

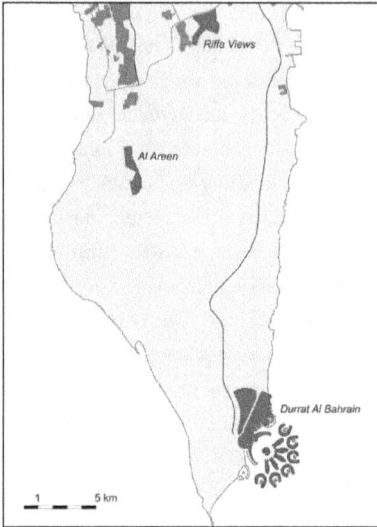
Fig. 12: The developments in the south.

Atoll Islands, at the ends of the Petal Islands and along the shoreline of the Crescent and its high-rise waterfront. All in all, there will be about 3,600 apartments in addition to offices and hotels contained in over 42 high-rise buildings along the Crescent, at the centre of which will stand a landmark tower, a marina in front of a retail district and a bridge to the Hotel Island[130].

The high-rise buildings will start at about 10 storeys on the outer edges of the Crescent and gradually build up to 33 storeys in the centre. The first two floors of each building will contain retail outlets, spas, gyms and restaurants for residents, thus forming the project's largest public promenade. While the high-rise buildings will be developed along the beach front of the Crescent, several town houses will be built in clusters around private courtyards further inland. A man-made lake will stretch along the area connecting the Crescent with the mainland, where two further residential projects called Lakeview and Gardenview, comprising low- to medium-rise apartment buildings, will be developed. The centre of the whole development will be the 'Iconic Commercial Hub' which will contain schools, a central mosque and healthcare facilities in addition to shopping and recreational areas. The 100-storey Durrat Al Bahrain Tower with a height of around 300 metres will be the centrepiece of the project and a new landmark of Bahrain. The lower floors of the mainly residential tower will be used commercially as a shopping mall, thus making it the central part of the retail district in the downtown area. Next to the centre of the development and at the bridge leading to the Hotel Island will stand the 40-storey Gateway Towers, a further landmark of the development[131]. In the north of the project several hundreds of residential units will be developed as villas within the 18-hole golf course, which will be one of the biggest tourist attractions apart from the beachfronts, retail districts and recreational facilities on the Crescent[132]. Furthermore, in a joint venture

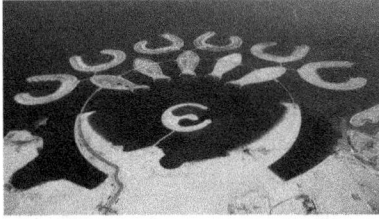

Fig. 13: The islands of Durrat Al Bahrain.

Fig. 14: Townhouses on one island.

of Tameer and Durrat Al Khaleej, the Durrat Marina is being developed in the form of three islands covering around 700,000 sq m and comprising 4,000 residential units within multi-storey towers and blocks[133].

All in all, it is expected that Durrat Al Bahrain will cater for up to 60,000 future residents in addition to around 4,500 daily visitors after the staged completion between 2010 and 2015[134]. The main goal of the project is to expand Bahrain's tourism facilities and attractions as well as develop freehold properties to satisfy the expanding real estate market. With a maximum density of about 7,500 people living on one square kilometre Durrat Al Bahrain will be much less densely populated than the island developments in the north where large areas of land are not available for development as in the south and the price of land is thus higher. In order to reduce the disadvantage of the rather long distance of more than 40 kilometres between the project and the central business district in the capital, a new two-lane highway has already been developed linking the development with Manama[135].

While Durrat Al Bahrain stretches along the south-eastern coast, a new project called Al Areen is being developed in the south-west close to Hamad Town, where the Al Areen Wildlife Sanctuary occupies a large area beside a campus of Bahrain University and the Bahrain International Circuit. Named after this wildlife park, the Al Areen project is located in the east of the territory of the Al Areen Wildlife Sanctuary on an area of around 2 sq km, thus reducing the park to about 70% of its original area. Due to highway expansion in recent years by Hamad Town, the development is a 25-minute drive away from the business district in Manama about 30 km away[136]. In 2004 the Al Areen Holding Company was established by a joint venture of the Bahraini government, who owns the Al Areen Wildlife Sanctuary, Gulf Finance House B.S.C. and a group of regional investors[137]. Apart from the construction of mainly residential freehold properties, the goal of the development is to expand Bahrain's health and family oriented tourism sector. Thus, several of the total 18 projects feature theme parks, hotels and recreational facilities. All in all, there will be five main development phases including phase one involving the construction of the main infrastructure in addition to the projects Banyan Tree Desert Spa & Resort and the theme park The Lost Paradise of Dilmun, which were already completed in 2007. Parallel to phase one, which was completed at the beginning of 2009, phase two was launched with the construction of three projects – Oryx Hills, Sunset Hills and the Domina Prestige Hotel. Furthermore, the development of the project Downtown Al Areen

137

will be started in phase two but completed in a later phase[138].

Other projects, which include the Al Waha Resort, Sarab Al Areen and the Al Areen Medical and Rehabilitation Centre, will be completed according to the individual developers' time frames by 2012 when the entire Al Areen project is expected to be completed[139]. Sarab Al Areen in the north of the project area will consist of a 116,000 sq m shopping mall in addition to a hotel and around 50,000 sq m of residential developments. The Sarab Mall will be built at the entrance of the commercial district, which stretches along the main access road of the project Downtown Al Areen. Apart from commercial use in form of retail outlets, restaurants and leisure facilities, there will be several residential projects within the 10 zones of the Downtown development. While the top three floors of the multi-storey buildings along the main boulevard will be occupied by offices or housing units, the ground floors will be reserved for retail outlets, restaurants and cafes. In the west several low-rise residential developments, a theme park and a hotel will be developed in addition to a mixed-use resort development in the north. At the southern end of the boulevard the Al Areen Medical Centre will be developed surrounded by low- to medium-rise residential projects. At the beginning of the southern half of the project is the already developed Banyan Tree Resort with its hotel, spa and low-rise dwellings. Next to it the residential project Oryx Hills will comprise 106 villas, which will be surrounded by large green areas in the form of two green corridors stretching from east to west. Further south, nine projects, for example, Sunset Hills, will form the end of the development as exclusive residential developments mainly consisting of villas in addition to town houses and medium-rise apartment buildings[140]. All in all the whole project will provide around 3,300 residential units comprising 800 villas and 2,500 apartments[141]. Thus, it is expected that a maximum of 10,000 people will live within the development, while several thousands will visit the entertainment, recreational and shopping facilities. With a future density of less than 5,000 people living on one square kilometre the project will have a rather low density in comparison to many of the current developments.

In addition to the Al Areen project, another inland development called Riffa Views has been launched to the south of Riffa. While luxurious residential developments will occupy most of the area, several leisure facilities such as golf courses will be built in order to attract upper income groups and real estate speculators. Arcapita Bank initiated the new project and founded its master developer, which is also called Riffa Views, in order to build around 1,000 villas on an area of about 2.8 sq km[142]. Almost 50% of the overall area will be occupied by two golf courses, which will be located at the centre of the development bridging the space between three projects called The Oasis, The Lagoons and The Park. Development of these began in 2007. While the Oasis district will be located in the north along the Al Muaskar Highway leading from Awali to Sitra, the Lagoons district will cover the south of the development area along three man-made lakes. In the west the Park district will be developed along an additional piece of land. It as well as the two other districts will be developed as gated communities. On a custom lot size of 850 sq m investors can build homes following the given guidelines, however most villas will be developed by Riffa Views itself. Although each district has its own individual architectural design, there will be seven basic types of dwellings of various sizes, from two-bedroom townhouses of just 193 sq m to six-bedroom villas of around 790 sq m[143]. While only 30% of the development area will be composed of built areas including homes, roads, parking and leisure facilities, the remaining 70% will be occupied by the golf courses, man-made lakes and landscaped areas. In addition to the golf courses and their facilities, for example a golf clubhouse and golf academy, there will be a tennis facility close to the Riffa Views International School[144]. With an expected population of around 3,000 residents and thus a density of only about 1,000 people per square kilometre, Riffa Views is one of Bahrain's most exclusive residential developments with its leisure areas.

Durrat al Bahrain, Al Areen and Riffa Views are all located in the south of the main island and are significant in that they are all targeted at the tourism sector and are designed to be luxurious residential projects with a high percentage of landscaped areas and thus a rather low density. The main reason for this development trend is the government's general attempt to expand Bahrain's tourism industry on its undeveloped land. The government's support of the development of the necessary infrastructure such as new highways in addition to projects such as the Bahrain International Circuit have led to a growing interest from the private sector in investing in real estate in the south despite the disadvantage of the long distance to the business districts in the north. Due to the availability of land and developers' attempts to increase the value of their developments, large landscaped areas or leisure facilities such as golf courses have been integrated within the low-rise gated communities. Nevertheless, integrated commercial districts, which will include shopping malls, in addition to leisure areas will differentiate the projects from plain suburban neighbourhoods and thus establish new attractions in the south. The current 13 sq km development area of all three projects combined will be able to house around 73,000 people. However, the number of permanent residents is expected to be lower because of the large percentage of holiday homes, mainly owned by Saudi Arabian investors, and high rental prices.

4) The Developments in the North-West of the Main Island

On the north-western coast close to Budaiya the Bahrain-based AAJ Holding Company has launched a beachfront gated community called Marina West comprising 11 high-rise towers from 20 to 32 floors. The whole development area will cover around 75,000 sq m, which is mostly reclaimed land, and the overall built-up area, including residential, retail and leisure space, will reach about 350,000 sq m[145]. While ten residential towers will offer 1,168 apartments, ranging from a 92 sq m one-bedroom simplex to a 329 sq m four-bedroom duplex, one central tower will be developed as a 286-room five-star hotel. A two-storey podium linking a row of five residential towers on each side will be occupied by retail outlets, cafes, restaurants and a covered car park for about 2,000 cars. In addition, the gated community will include a health club and sports and recreational facilities, including swimming pools, tennis courts, one private beach and a marina[146].

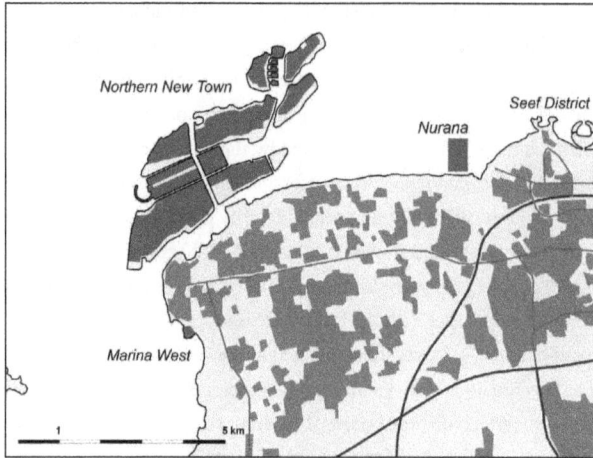

Fig. 15: The projects along the northern coast.

After the project's completion by 2010 approximately 3,000 to 4,000 future residents are expected to move in. Except for the Marina West project the north-western region has been the subject of less focus for development in recent years due to its lack of large cohesive areas of unbuilt land.

The reclamation of land from the sea in order to obtain new project areas is also occurring in the north-west of Bahrain where 14 man-made islands stretch 4 km along the coast. These islands, which were completed in 2008, belong to the largest public sector development in Bahrain, currently known as North Bahrain New Town. All in all the archipelago consists of four large and five medium sized islands in addition to five small islets covering a total area of around 7.4 sq km. The initial goal of this project, which was launched by the Crown Prince himself, was the development of around 15,000 dwelling units as part of a social housing programme for the Bahraini population. While the Housing and Urban Development Committee, which was founded by the Economic Development Board, had been in charge of the project for several years, it was moved back to its original instigating authority, the Ministry of Housing, at the end of 2008. The first phase of North Bahrain New Town will consist of the development of social housing on the two islands, known as Island 13 and Island 14, stretching along the coast. An access road through the village Diraz will connect the project to Budaiya Road leading from the western coast to Manama. In the future there are plans to develop a second access road through the village Barbar as well as a new causeway, which will connect the development to the Seef District in the east. Apart from the development of housing units in the form of villas, town houses and apartment buildings with up to 10 floors, there are plans to develop integrated social facilities such as schools, a regional university, a regional hospital and mosques in addition to recreational and leisure facilities including several public beaches and parks. Island 14, the largest island of the development, is divided into more than 120 lots, of which more than 50% will be developed as residential areas. Furthermore, large areas along the centre of the island will be used for major public facilities such as a hospital and schools in addition to shops along the roads. Several playgrounds, small parks and one beach in the north will be the major public areas. Island 13 will comprise around 40 blocks including a large area in the south-west which has been reserved for a stadium and additional sports facilities. In addition to shops within central blocks, there will be a promenade along the northern shoreline where two beaches will be developed. In the east a bridge will lead to a landscaped park on a small island in the form of a banked up hill overlooking the whole development.

While Island 11 and 12 will provide large areas for commercial and recreational use, for example, a large park, the northern islands will be mainly occupied by residential developments, most of which will be developed by the private sector. In contrast to the development's southern islands, where a major part of the social housing programme will be constructed, the northern islands will be partly developed by freehold property projects and will thus be available for sale to non-Bahrainis. The public centre of the whole development will be the Central Park with a mosque on the western coast of Island 11 and a retail district on Island 12. In addition to parks there will be 16 beaches, most of which will be public sand beaches in addition to several gravel beaches. The main infrastructure, including water, waste water and electricity will follow the main access road connecting the centres of all six main islands. A mixture of different multi-storey apartment blocks and two- to three-storey town houses and villas will create a density of around 20 dwelling units for every hectare of land. Thus, it is expected that between 75,000 and 100,000 people will be able to settle in the New Town development, which will have a large impact in the northern region, particularly regarding traffic. After the completion of the main islands by 2016, there are plans to expand the project by adding about five islands in the west. Also, due to the building of a future causeway linking the project with Seef District, there have been proposals for more island developments along the coast towards the east.

In addition to the New Town development and the Marina West project, which have both already been launched, a project called Nurana has recently been announced. It will be built on 2 sq km of reclaimed land to the north-west of Bahrain Fort. The project is being promoted as a vibrant and sustainable community inspired by the historical, geographical and cultural characteristics of its surroundings. NS Holdings Company, the owner of the Nurana project, appointed Manara Developments to be responsible for the development. In turn, Manara Developments engaged Davenport Campbell to design the master plan. Approximately 60% of the project will be earmarked for residential use and the remaining 40% will provide a mixture of land uses including areas reserved for commercial, retail and hospitality facilities[147]. Although the land reclamation for the project already began at the end of 2008, a master plan has still not been published and no estimates have been made regarding the number of future inhabitants.

According to current information, the northern region will be expanded by more than 7.5 sq km of reclaimed land providing housing units for around 100,000 people. The future urban density of the new reclaimed areas of around 13,000 residents living on one square kilometre will lead to an increase in the overall population density of the northern region from about 1,180 residents per sq km in 2001 to in the future more than 3,900 residents per sq km – an increase of 330%.

3.3.3 The Resulting Urban Structure

Changes in Bahrain's urban structure have been mainly due to areas of urban expansion that have been developed on reclaimed land, particularly along the northern coast. These areas have either been directly added to the mainland or developed as islands located close to the shoreline. In Manama the on-going reclamation in the Seef District and Juffair has pushed the coastline more and more into the sea, resulting in the further expansion of already reclaimed areas in the north-west and south-east of the capital, which has grown by about 3 sq km over recent years. In addition, a large area of around 4.4 sq km was reclaimed in the north-west of Muharraq, which except for a housing project of the Ministry of Housing has remained undeveloped. While the three master-planned projects Bahrain Financial Harbour, Reef Island and Bahrain Bay situated on the northern coast of Manama cover an area of about 1.4 sq km, there are three developments currently being built in the east of Al Muharraq, namely, Amwaj Islands, Diyar Al Muarraq and Dilmunia, which are over 10 sq km. Furthermore, the future expansion of Diyar Al Muharraq will add an additional 6 sq km of reclaimed land and further island developments in the south-east of Al Muharraq will lead to a major increase in the small island's land mass. While the area of Al Muharraq has increased by more than 66% since 2001, Manama has expanded by around only 12% of its former urban area over recent years. In this regard however it must be mentioned that since the 70s large areas have been reclaimed along the northern, eastern and southern shorelines of the capital area which to a large extent have remained undeveloped. In the south the project Durrat Al Bahrain adds more than 4 sq km of reclaimed land to the main island, while in the north-west the New Town project adds around 7.4 sq km. All in all, Bahrain will gain around 30.4 sq km of developable area due to the land reclamation to be carried out within the eight on-going master-planned developments and the areas of urban expansion on the coasts in Manama and Muharraq. Consequently, the overall land area of the Kingdom of Bahrain has increased by more than 4% since 2001.

Including the inland developments Al Areen and Riffa Views, the current master-planned projects of the private sector together cover around 24 sq km. Apart from the six island developments in Manama and Al Muharraq in addition to the Marina West project in the northern region, three projects have been launched in the south of the main island. These are Durrat Al Bahrain, Al Areen and Riffa Views. On their own they cover around 50% of the entire project area being developed by the private sector as a whole. At around 8 sq km, Durrat Al Bahrain is currently Bahrain's largest master-planned project, followed by the 7.4-sq km New Town project and Diyar Al Muharraq, the first phase of which

142

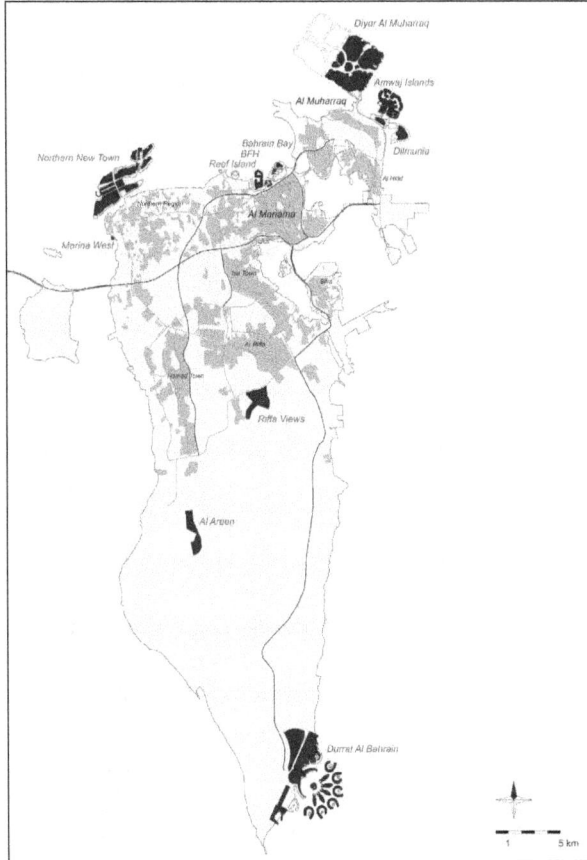

Fig. 16: The main projects in Bahrain.

constitutes around 6 sq km. In contrast, several projects such as the Bahrain Financial Harbour, Reef Island, Bahrain Bay and Marina West have been launched on areas far smaller than 1 sq km. The master-planned developments, including inland and waterfront projects by the public and private sector, together cover an area of around 31.4 sq km, of which more than 85% is being developed on reclaimed land. Due to the close proximity to the business districts of the capital in addition to the already built infrastructure, eight of the total of eleven projects have been launched in the north of Bahrain.

In order to make freehold properties more attractive and to make developments appealing to tourists most master developers integrate commercial, leisure and recreational areas into residential projects. The three developments in the south of Bahrain are particularly targeted at tourists and thus business. Consequently, hotel and resort projects have become popular elements of any master-planned development, from small-scale projects such as Marina West to large projects such as Diyar Al Muharraq. In addition to the integration of retail districts in the form of shopping malls or shops along promenades, most developers

143

of bigger projects try to include social infrastructure in the form of private schools and health centres in order to create self-contained neighbourhoods. However, Bahrain's main business districts and thus most workplaces have remained in the former urban centres and most developments usually become exclusive dormitory towns. Furthermore, social services such as schools and parks are generally not to be found in new urban areas and have remained within older districts due to the high price of land.

Based on its accessibility via main highways, the Seef District has developed more and more into the major business location of Bahrain, expanding the central business district in the north of Manama towards the west. Parallel to the development of office towers and shopping malls in the Seef District, the construction of the Bahrain Financial Harbour and the World Trade Centre has expanded the business district in northern Manama and will most likely ensure its future as a major economic centre in Bahrain. While Manama continues to be Bahrain's business district, two new industrial centres will be created. One is an industrial park called Bahrain Investment Wharf which will be located in the south of Al Muharraq. The other is Sitra Technology City, which will be Bahrain's main industrial area in the south. However, most recent developments have been residential due to the exponentially growing freehold property market in Bahrain, which has to a large extent been influenced by an increasing interest from Saudi Arabian investors and nationals from other GCC countries. As a consequence of the construction boom, there has been a massive influx of guest workers to Bahrain, the number of whom has more than doubled from 254,184 in 2002 to 517,446 in 2007. Consequently, the number of foreigners in Bahrain's population has grown to 1,046,814, which is almost 50% of its total population compared to just about 37% five years before. The majority of these new immigrants work in the lower service sector, particularly in construction, and thus live in work camps or the old districts of Manama where companies rent multi-storey apartment buildings to accommodate their employees. Middle and upper income groups generally live in either compounds, which are widely spread over the urban area from the north-west region of Bahrain to southern Manama, or to recently built apartment buildings in Al Hoora, Al Mahooz, Juffair or Seef District, where hundreds of multi-storey buildings were constructed in a short period of time. Very high income groups rent or buy dwellings within new master-planned developments such as Amwaj Islands, where as of 2008 about 3,000 people are already living.

Apart from the fact that most of the current developments will be not completed before 2015, rental prices are expected to increase and thus most companies or employees will be less interested in renting dwellings within these luxurious developments, which will consequently remain exclusive for upper income groups. However, in some cases such as the New Town project and Diyar Al Muharraq there have been attempts to integrate affordable housing projects for lower income groups. For example, there are around 15,000 dwellings in the New Town project that are part of the social housing programme of the Ministry of Housing. Another example is a housing project for 20,000 workers and their families in Bahrain Investment Wharf. All 12 current master-planned projects will in total provide residential units for about 366,000 future residents, almost 35% of the current population, of which about 293,000 will be accommodated in projects along the northern coast of the main island and on the eastern coast of Al Muharraq. While the three island developments in Manama will be able to accommodate around 35,000 residents on around just 1.4 sq km, the three projects in Al Muharraq will provide dwellings for about 135,000 people on an area of more than 16 sq km. The reason for the higher density in Manama is the development of residential high-rises for Bahrain Bay and the Bahrain Financial Harbour, where 32,000 people will be able to reside on an area of about just 800,000 sq m. However, the real future densities will depend on permanent residents who can rent or buy dwellings in prime developments and whose number should be expected to be much lower than the current estimations of developers. In the cases of the southern projects Durrat Al Bahrain, Al Areen and Riffa Views, which have been designed as exclusive properties and tourist destinations, the maximum density is already expected to be rather low at around just 6,000 people living on 1 sq km. By contrast the New Town project in the north-west is expected to reach an urban density of more than 13,000 residents per sq km. In addition to the completion of many thousands of apartments in the Seef District and Juffair, average urban densities will increase rapidly in certain urban expansion areas.

The highly concentrated urban development along Bahrain's coasts is due to four main factors. These are the availability of land through reclamation, the ownership of land by the government, proximity to the main infrastructural network and prime locations for seaside property projects. Due to increasing land prices and a drive to maximise profits, this coastal urban development is characterised by multi-storey buildings and high-rises. These are also being constructed in areas where the former zoning and infrastructure was initially only designed for low- to medium-rise developments. Low- to medium-rise typologies are to be found more inland and on the new reclaimed islands. An example of this development trend is Durrat Al Bahrain, where several high-rise buildings are being built along the mainland coast while the outer islands are reserved for villas. In the case of the Seafront District in Manama, several high-rise projects are being constructed such as the Bahrain Financial Harbour while Reef Island, further out to sea, has low-rise residential projects. This development trend has intensified due to the relatively fast development of infrastructure along the coasts in combination with the general attractiveness of waterfront locations, where beaches or sea views enhance the prices of freehold properties.

---

1   Scholz, 1999, p. 84.
2   Heck and Wöbcke, 2005, p. 102.
3   Meinel, 2003, p. 75.
4   Scholz, 1999, p. 84.
5   Scholz, 1999, p. 85.
6   Scholz, 1999, p. 85.

7   Scholz, 1999, p. 86.
8   Scholz, 1999, p. 86.
9   Hamouche, 2008, p. 186.
10 Scholz, 1999, p. 87.
11 Scholz, 1999, p. 87.
12 Scholz, 1999, p. 88.
13 Scholz, 1999, p. 86.
14 Hamouche, 2008, p. 191.
15 Hamouche, 2008, p. 192.
16 Hamouche, 2008, p. 186.
17 Hamouche, 2008, p. 187.
18 Hamouche, 2008, p. 188.
19 Hamouche, 2008, p. 189.
20 Scholz, 1999, p. 91.
21 Hamouche, 2008, p. 192.
22 Scholz, 1999, p. 88.
23 Scholz, 1999, p. 88.
24 Scholz, 1999, p. 89.
25 Scholz, 1999, p. 90.
26 Scholz, 1999, p. 91.
27 Hamouche, 2008, p. 205.
28 Scholz, 1999, p. 91.
29 Nabi, 2000, p. 4.
30 Hamouche, 2008, p. 195.
31 Hamouche, 2008, p. 196.
32 Hamouche, 2008, p. 196.
33 Hamouche, 2008, p. 197.
34 Hamouche, 2008, p. 198.
35 Hamouche, 2008, p. 198.
36 Hamouche, 2008, p. 199.
37 Hamouche, 2008, p. 201.
38 Hamouche, 2008, p. 202.
39 Hamouche, 2008, p. 203.
40 Scholz, 1999, p. 92.
41 Scholz, 1999, p. 93.
42 Scholz, 1999, p. 93.
43 Scholz, 1999, p. 94.
44 Scholz, 1999, p. 95.
45 Scholz, 1999, p. 96.
46 Scholz, 1999, p. 97.
47 Scholz, 1999, p. 97.
48 Scholz, 1999, p. 98.
49 Scholz, 1999, p. 100.
50 Scholz, 1999, p. 101.
51 Scholz, 1999, p. 102–103.
52 Scholz, 1999, p. 104.
53 Nabi 2000, p. 4.
54 Nabi, 2000, p. 5.
55 Nabi, 2000, p. 6.

56 Nabi, 2000, p. 6.
57 Nabi, 2000, p. 7.
58 Scholz, 1999, p. 101.
59 Nabi, 2000, p. 7.
60 Nabi, 2000, p. 9.
61 Nabi, 2000, p. 10.
62 Hamouche, 2008, p. 206.
63 Hamouche, 2008, p. 208.
64 Hamouche, 2008, p. 205.
65 Nabi, 2000, p. 10.
66 Nabi, 2000, p. 11.
67 Ministry of Housing, 1996, p. 14.
68 Ministry of Housing, 1996, p. 16.
69 Ministry of Housing, 1996, p. 18.
70 Ministry of Housing, 1996, p. 23.
71 Ministry of Housing, 1996, p. 47.
72 Ministry of Housing, 1996, p. 74.
73 Bahrain Aluminium Extrusion Company
74 Scholz, 1999, p. 104.
75 Scholz, 1999, p. 105.
76 Scholz, 1999, p. 106.
77 Scholz, 1999, p. 105.
78 Scholz, 1999, p. 105.
79 Meinel, 2002, p. 164.
80 Meinel, 2002, p. 165, 172.
81 http://www.theestimate.com/public/122900.html, 14.02.2009.
82 http://www.zawya.com/story.cfm/sidZAWYA20071212050038/Bahrain:%20King%20orders%20setting%20up%20of%20Electricity,%20Water%20Authority, 15.02.2009.
83 http://www.bahrainedb.com/AboutEDBIntro.aspx, 14.02.2009.
84 http://www.arabianbusiness.com/530704-bahrain-opens-door-to-kingdom, 14.02.2009.
85 http://www.bfharbour.com/html/aboutbahrain/milestones.htm, 14.02.2009.
86 http://www.bahrainfreeholdproperties.com/freeholdproperty_aboutbahrain.aspx, 14.02.2009.
87 Khalaf 2003, p. 18 http://www.iue.it/RSCAS/RestrictedPapers/conmed-2003free/200303Khalaf05.pdf, 16.02.2009.
88 http://en.wikisource.org/wiki/Constitution_of_the_Kingdom_of_Bahrain_(2002)#Section_3_The_Legislative_Authority_National_Assembly, 16.02.2009.
89 Al Ghazal, 07.05.2009.
90 http://www.works.gov.bh/default.asp?action=category&id=14, 19.02.2009.
91 http://www.works.gov.bh/default.asp?action=article&id=160, 19.02.2009.
92 http://www.tradearabia.com/NEWS/CONS_135696.html, 19.02.2009.
93 http://www.gisdevelopment.net/proceedings/mest/2007/Papers/day3/P51.pdf, 19.02.2009.
94 http://www.slrb.gov.bh/AboutSLRB/default.aspx?PageId=95&Lnk=Link1, 19.02.2009
95 http://www.slrb.gov.bh/AboutSLRB/default.aspx?PageId=98&Lnk=Link4, 19.02.2009.
96 http://www.mew.gov.bh/default.asp?action=category&id=34, 19.02.2009.
97 http://www.transportation.gov.bh/en/modules.php?name=Content&pa=showpage&pid=16, 19.02.2009.
98 Property Investment Guide 2007, p. 151.
99 An investment company regulated by the Central Bank of Bahrain.

100   Al Ghazal, 07.05.2009.

101   Al Ghazal, 07.05.2009.

102   Scholz 1999, p. 83.

103   http://www.statoids.com/ubh.html, 22.02.2009.

104   http://www.globalsecurity.org/military/facility/manama.htm, 22.02.2009.

105   http://www.nationsencyclopedia.com/economies/Asia-and-the-Pacific/Bahrain-AGRICUL-
TURE.html, 22.02.2009.

106   http://websrv.municipality.gov.bh/ppd/doc/rule_buidingregulations_en.pdf, 26.02.2009.

107   http://websrv.municipality.gov.bh/ppd/doc/rule_buidingregulations_en.pdf, 26.02.2009.

108   http://www.bfharbour.com/html/faq/faq_development.htm, 26.02.2009.

109   http://www.reef-island.com/main.asp, 27.02.2009.

110   Gulf Construction January 2007, p. 62.

111   Nouri, 25.11.2008.

112   Nouri, 25.11.2008.

113   Nouri, 25.11.2008.

114   Gulf Construction January 2009, p. 68.

115   http://www.diyar.bh/en-faq.html, 28.02.2009.

116   Gulf Construction January 2009, p. 70.

117   http://www.diyar.bh/en-faq.html, 28.02.2009.

118   Gulf Construction January 2009, p. 70.

119   Gulf Construction January 2009, p. 72.

120   Gulf Construction January 2008, p. 98

121   Gulf Construction January 2008, p. 101

122   Gulf Construction January 2008, p. 103

123   Gulf Construction January 2008, p. 105

124   Gulf Construction January 2009, p. 61

125   Gulf Construction January 2007, p. 88

126   http://www.bahiw.com/masterplan.htm, 03.03.2009

127   http://www.bahiw.com/biwprofile.htm, 03.03.2009

128   Gulf Construction January 2007, p. 90

129   Gulf Construction January 2008, p. 107.

130   http://www.durratbahrain.com/en/investors-ar/investors.html, 01.03.2009.

131   http://www.durratbahrain.com/en/explore/crescent/zones/zones.html, 01.03.2009.

132   http://www.durratbahrain.com/en/explore/golf-course/golf.html, 01.03.2009.

133   http://www.durratbahrain.com/en/explore/durrat-marina/durrat.html, 01.03.2009.

134   http://www.durratbahrain.com/en/investors-ar/investors.html, 01.03.2009.

135   http://www.durratbahrain.com/en/project/location.html, 01.03.2009.

136   http://www.alareenresort.com/alareenoverview.htm, 03.03.2009.

137   http://www.alareenresort.com/history.htm, 03.03.2009.

138   http://www.alareenresort.com/progress.htm, 03.03.2009

139   http://www.alareenresort.com/progress.htm, 03.03.2009

140   http://www.alareenresort.com/masterplan.htm, 03.03.2009

141   Saffy 2007, p. 251.

142   http://riffaviews.com/about/, 03.03.2009.

143   http://riffaviews.com/signature-estates/lagoons-estate/, 03.03.2009.

144   http://riffaviews.com/about/, 03.03.2009.

145   Gulf Construction 2008, p. 119.

146   Gulf Construction 2008, p. 120.

147   Gulf Construction 2009, p. 58.

4 The Transforming Built Environment of Manama

4.1 Manama's Previous Urban Structure

4.1.1 Historical Background of Urban Development

In 1960 the urban area of Manama was concentrated on an area of less than 4 sq km along the north-eastern shoreline of Bahrain's main island. At this time the outskirts in the south were still occupied by agricultural areas and small rural settlements. All in all, the small port city with a souq in the north and rural outskirts in the south had a total area of around 12 sq km. Although large areas of land were undeveloped in the south land was reclaimed in many areas in the north and north-east, where the first causeway was built linking Manama and Muharraq. Muharraq was Bahrain's capital city until 1971. It became an important hub in the 30s due to the establishment of the International Airport and its gradual expansion since then. Consequently, a business district grew along the road on the northern coast that links Manama with the airport in Muharraq. After Bahrain gained independence and Manama was made the new capital more and more areas were reclaimed along the northern shoreline in order to expand the country's first central business district and to develop a modern administrative centre. During the 60s and 70s, when the old harbour area was in the process of transforming into a modern CBD, the old harbour Mina Manama lost its function as Bahrain's main port due to the construction in 1962 of Mina Salman in the south, which became Bahrain's first deep water harbour. In the following decades this new harbour area was expanded on reclaimed land and became a centre of light industries following its declaration as a free economic zone. The early road network at this time consisted of a ring road leading around Manama's city centre, a coastal road connecting Manama with Muharraq and a road to the south linking Manama with the harbour Mina Salman. There was also a road connecting Manama's city centre with settlements in the west and south including Jidd Hafs and Riffa.

Due to Bahrain's early oil production the modernisation of Manama's city centre began rather early with the construction of a road network and the replacement of old buildings. For the most part, the basic layout of roads and the division of land remained true to the original old urban structure because of the ownership by many different individuals of small parcels of land. However, new urban areas were organised around an orthogonal road system. Along the northern coast line two- to six-storey blocks were developed on reclaimed land and on areas of the former souq, where modern cement buildings replaced the former urban fabric. On average the blocks were about 120 metres long and around 20 metres wide and stretched along the old main roads that run from north to south because of the old canals that lead from the agricultural areas in the south of Manama to the sea. The direction of these roads was also influenced by the direction of the wind, which provided natural ventilation for the densely built souq. The floors above shops on street level in the souq were often used as accommodation for foreign labour or as offices. The most important landmark of old Manama was the building Bab Al Bahrain

Fig. 1: Manama in the year 1951.

on Government Avenue marking the entrance of the souq opposite the old harbour. The two-storey building, which initially housed government offices, had already been built by the British during the 40s and was later refurbished in the 80s. While certain buildings in the new business district in the north were developed with up to 10 floors, the districts further south mainly consisted of buildings with two to three floors. This was due to the limitations imposed on building height by a lack of accessibility by road and a lack of space for parking. Consequently, new businesses moved into the new urban areas such as the new Central Market development in the north-west while the old districts were predominantly occupied by apartments of lower-income guest workers as well as small shops. In 1987 only around 42% of the city centre's population was Bahraini, while the immigration of foreign labour was growing by 6.1% per annum[1]. In spite of many of its inhabitants being foreign workers and the move of most Bahrainis to the outskirts, Manama's city centre has remained of cultural importance due to a large number of mosques and matams[2].

Manama's old centre consists of the Souq District, the Al Naim District and Ras Al Ruman. In the north Government Avenue marks the former coastline, which was shifted more than 300 metres by the reclamation of an area of around 2 sq km in the middle of the 20th century. Many shops, banks and hotels in addition to office and administrative buildings were developed along both sides of this important access road. In addition to Yateem Centre, one of Bahrain's first shopping centres, a shopping complex called Gold Centre was built in the souq. However, small shops continue to line the roads in the traditional style of the old souq and it remains the shopping district of choice for lower income groups. In the new business district in the north of Manama two parking lots are situated between Bab Al Bahrain and the old harbour. Around these parking lots stand the 14-storey Regency Hotel, a medium-rise office building in the west and two office blocks

150

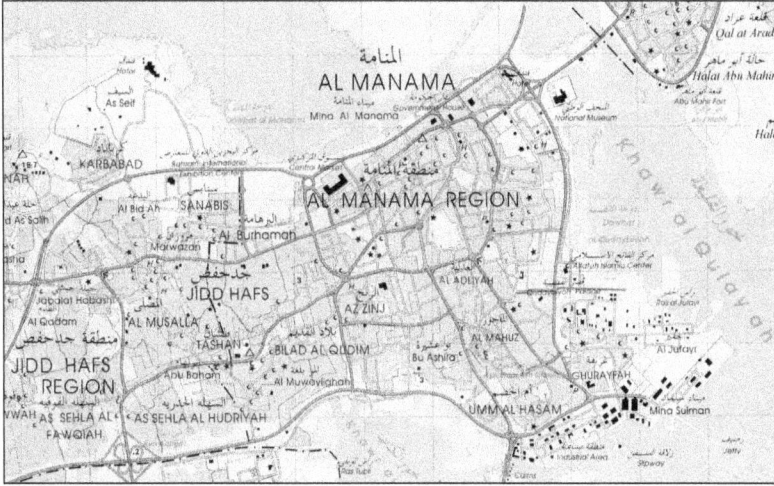

Fig. 2: Manama in the year 1996.

in the east. Next to them the 27-storey tower of the National Bank of Bahrain stands as one of Bahrain's tallest landmarks with a height of 101 metres. Except for a few high-rises such as the Bahrain Tower most buildings in the northern business district were built as multi-storey blocks with up to 10 floors, mainly occupied by administrative and commercial offices in addition to shops on the ground floors. Large parking lots cause an interrupted urban landscape and thus a rather low built density with long distances between buildings, which is in clear contrast to the urban structure of the old city centre. While in the west the Central Market at the Pearl Roundabout marked the end of Manama's northern business district, the Diplomatic Quarter was developed on its eastern edge on land that had been reclaimed. The area of the Diplomatic Quarter had already been reclaimed during the 60s and 70s but stood mostly empty until more intensive construction of buildings began in the 90s. Initially, the Diplomatic Quarter was chosen to be developed as a prime location for international embassies and governmental institutions due to its location in the north-east where two causeways link the capital to the International Airport in Muharraq. In addition to the Sheikh Hamad Causeway the Sheikh Isa Bin Salman Causeway was built on reclaimed land in the north during the 90s in order to improve the transport link between both cities. At the end of the 20th century the district housed five ministries, three hotels and several embassies. In addition, five embassies were built in the adjacent Al Hoora District in the south, where apart from a few governmental buildings including two ministries several residential and commercial blocks were developed along the main access road leading from north to south called Exhibition Avenue. In addition to shops and restaurants a medium-sized shopping centre and several hotels turned the area into an important tourist destination mainly attracting GCC nationals. Close by the Al Hoora District and Sheikh Hamad Causeway the National Museum was built on an area along the coast. In addition to Sheikh Hamad Palace and the parliament building,

151

Fig. 3: The contemporary centre of Manama.

the Ministry of Education and the Public Library are located in the Qudaibiyah district. In the east of Quadaibiyah a large area of more than 220,000 sq m along the Al Fateh Highway is occupied by the Al Qudaibiyah Palace facing the Al Fateh Mosque opposite. Further south the Gulf Hotel complex was built on the previous shoreline before further reclamation extended the area to the east, where the Juffair District developed into a growing residential area for foreign guest workers, including a large number of soldiers and employees of the US Navy Base in the south.

Near the US Navy Base the harbour Mina Salman and an industrial free zone are located on the south-eastern coast. The southern districts Umm Al Hassam and Buashirah along the Sheikh Isa Bin Salman Highway developed into medium-density mixed-use areas due to many low-rise compounds and multi-storey blocks. The beginnings of the urbanisation of the now former agricultural areas in the south were characterised by the development of low-rise dwellings, which were often built within compounds. The early road network was determined by the borderlines of the various fields of individually owned land. After a period of scattered suburban settlements southern Manama became more and more built up as well as diverse due to the construction of apartment buildings for foreign labour in addition to several commercial districts, particularly along main access roads. Furthermore, several governmental institutions and foreign embassies settled in Buashirah, Al Mahooz and Umm Al Hassam. While a housing project of the Ministry of Housing in addition to several multi-storey blocks increased the density in the south of the district Al Mahooz, a large number of plots in the central districts of Manama, which include Al Adliyah, Al Saqiyyah, Al Zinj and Al Salmaniyah, have remained undeveloped. To the east Al Adliyah developed into a residential district with a commercial centre in addition to a few governmental administrative buildings and embassies. Along Salmaniyah Avenue between the districts Al Salmaniyah in the north and Al Saqiyyah and Al Zinj a large

152

area is occupied by hospitals and the Gulf University. Two large cemeteries between the districts Al Salmaniyah and Qudaibiyah in addition to a large public park in the south of the Al Naim divide the old urban area from the urban expansion areas, which were developed during Bahrain's period of oil urbanisation.

Several villages such as Jidd Hafs, Sanabis and Al Salihiyah became districts of Manama when the capital governerate was expanded to the west in 2002 and the governerate of Jidd Hafs was dissolved. During the first period of Manama's urban expansion the former agricultural settlements grew into each other and merged, thus losing their distinct borders. Consequently, the agricultural green belt between these settlements gradually shrunk to just two major contiguous areas between the two main roads connecting Manama with the western part of the main island. Before the Sheikh Khalifa Bin Salman Highway was developed during the 90s, Budaiya Road was one of the major access roads leading from the old city centre to the west. Thus, many commercial districts, a hospital and administrative buildings were developed along this former development axis. In the north of Sanabis the new Sheikh Salman Highway linked Manama with the causeway to Saudi Arabia and with suburban settlements such as Hamad Town. Consequently, the large reclaimed area in the Seef District, which was once developed for low- to mid-rise residential developments, turned into an expanding commercial district with a cluster of five large-scale shopping centres.

4.1.2 The Urban Morphologies during the Oil Era

In 2001 around 205,800 people lived in the governerates of Manama and Jidd Hafs on an area of about 51 sq km. Thus, around 50 people lived on one hectare, which is a relatively low average urban density when compared to the maximum density of over 900 residents per hectare in the old souq district. In the outskirts of Manama the urban sprawl that grew during the oil urbanisation had a very low urban density, which was mainly due to plots remaining undeveloped because of speculation and the fact that multi-storey blocks were not allowed to be build close to suburban settlements. Consequently, at the end of the 20th century Manama was a typical oil city comprising a dense old city centre, a modern business and administrative district stretching along the highway towards the International Airport as well as low- to medium-rise suburban outskirts interrupted by vast areas of undeveloped land. The road system of Manama essentially consisted of three major highways along the coasts connected to an inner and an outer ring road running around the old city centre. The connection between the former and the latter consisted of four major roads running from north to south. Due to recent urban growth three main expansion areas have evolved known as the Seafront District, the Seef District and the Juffair District. While the Seafront District became a target of major developments because of its function as the main central business district of Bahrain, the Seef District has also become an important business district because of its attractive location close to the city centre and its accessibility by a highway extension. Juffair's development was due to the US Navy Base and the move of thousands of soldiers there. Consequently, large areas were reclaimed along the coasts of all three districts in order to provide lots for the increasing demand of investors, who have been attracted by the liberalised real estate market and the possibility of developing freehold property projects.

4.2 The Three Urban Expansion Areas

4.2.1 The Seafront District
The area of the Seafront District under examination is the portion stretching more than 3.2 kilometres along the northern coast covering a reclaimed area of around 1.8 sq km. Its current expansion areas in the north are three master planned developments, namely, Bahrain Financial Harbour, Bahrain Bay and Reef Island, which together will add an extra 1.4 sq km of reclaimed land to the Seafront District after their completion by 2015. In addition, an area of around 330,000 sq m has been reclaimed in order to develop a new access road linking the Bahrain Financial Harbour and Bahrain Bay and to extend the park area stretching along the coast. While the western part of the Seafront District is occupied by the Central Market and small shopping centres, the eastern part is occupied by the Diplomatic Quarter covering an area of around 410,000 sq m. In the centre of the Seafront District many administrative and commercial buildings have been developed in addition to several hotels. Despite the many modern buildings the old Bab Al Bahrain Building has remained an important landmark of the district marking the beginning of the old city centre, which lost its seaside location due to constant reclamation. In addition to its function as Bahrain's administrative and business centre the Seafront District has recently become attractive as a residential area. This new trend is mainly due to the liberalised real estate market, which has led to a large number of projects that are expanding and transforming the Seafront District's urban structure.

Today, the main access road of the district is the King Faisal Highway along the coast, which links Manama to Muharraq and to the western and southern part of the main island. In the south the Sheikh Hamad Causeway is linked to Government Avenue, which had been the northern coastal road before the major land reclamation during the 60s. At the western end of the Seafront District the Pearl Roundabout has remained an important juncture of main roads heading to the west and south and linking King Faisal Highway with Sheikh Khalifa Bin Salman Highway. A large part of the area to the south of the Pearl Roundabout is used for public services, including a fire station and a school. However, most of the developable area in this part of the Seafront District is still unbuilt and according to current building regulations they will be developed with commercial, administrative or residential buildings following the current zoning plan, which permits a construction ratio of 300% of the land area and a maximum building height of six floors. To the east, the buildings of Central Market and four medium-scale shopping centres occupy an area of about 58,000 sq m in addition to large parking areas of almost

Fig. 4: The Seafront District in 2008.

50,000 sq m. While there is a police station, a health centre, a school and a mosque to the south of the Central Market, a 17,000-sq m area in the north has been made free for the development of Investment Buildings Type B-A offering a possible construction ratio of 1,200% of the land area with no restrictions on building height. In accordance with this new zoning a residential tower with 21 floors has recently been completed next to three multi-storey buildings with six to ten floors. Further east several large plots together totalling more than 880,000 sq m are undeveloped or being temporarily used as parking areas by employees working in neighbouring buildings. Due to the current zoning all plots can be developed with Investment Buildings of type B-A. In accordance with these guidelines, a 23-floor residential tower has recently been completed on an area of about 880 sq m. In order to provide parking lots for each of the around 70 apartments a five-storey car park has been built at the base of the building. Six adjacent plots have already been developed with medium-rise buildings mainly comprising housing units, offices and one hotel. However, because of many undeveloped lots the average built density of this area of the Seafront District is rather low with a plot ratio of just 1.3.

Continuing east, the two-storey main building of the Ministry of Municipalities and Agricultural Affairs, which was built in the 50s, stands on a plot of about 16,000 sq m on Government Avenue. It also has many additional offices in Gold City – a building on the opposite side of Government Avenue. Next to the ministry, an eight-storey building of a bank stands in front of a low-rise commercial building. Next to the bank on its eastern side stands another eight-storey commercial building and the hotel complex of the Regency Intercontinental Hotel, which comprises the old 14-storey main building and a recently added 10-storey building. Next to the hotel, a four-storey commercial building stretches

155

Fig. 5: The Bab Al Bahrain building.

Fig. 6: The World Trade Centre.

from north to south and the old post office faces the Bab Al Bahrain building. Two large parking lots covering around 15,500 sq m are located in the centre of the Seafront District between Bab Al Bahrain and Bahrain Financial Harbour. To the east of these parking lots one six- and one seven-storey office building stand beside the tower of the National Bank of Bahrain, which, at a height of 101 metres, had been Manama's tallest building before the recent completion of the twin towers of the Bahrain Financial Harbour and the Bahrain World Trade Centre. Next to the National Bank of Bahrain tower stands another high-rise block called Manama Centre, which covers a large area of more than 20,000 sq m. The complex consists of two eight-storey office buildings with shops on the ground floors connected by a third building forming an inner courtyard with a small green area. To the east three medium-high administrative buildings, including Government House, stand on a plot of around 43,000 sq m, of which 84% consists of car parks and green areas. This government area is surrounded by 3-metre high walls and access to the public is very restricted. On the opposite side of Government Avenue four commercial high-rise buildings of banks with 15 to 22 floors form the north-eastern border of the old centre.

Further east, the Sheraton Hotel stands next to the Bahrain Commercial Complex, also known as the Sheraton Complex, which includes a shopping centre and the 18-storey Sheraton Tower. Recently, the shopping centre was expanded by the Bahrain World Trade Centre development to a total retail area of about 25,000 sq m³. The Bahrain World Trade Centre itself stands on a plot of around 24,000 sq m, which is mostly covered by the shopping centre extension and the twin towers. After a construction period of four years the 50-storey towers were completed in 2008, becoming one of Bahrain's most remarkable modern landmarks due to their unusual design and their height of 240 metres, which makes them the second tallest buildings in Bahrain. The three wind turbines attached to the three bridges connecting the two towers are the distinctive feature of the development. Opposite the Sheraton Complex on the other side of Government Avenue there is a large area which used to be the premises of the British Embassy but is now vacated and designated as a special project area. Next to it a large area of unbuilt land has been approved for the development of Investment Buildings Type B-A. Opposite the Sheraton Hotel another hotel complex stands beside a commercial building and the Electricity and Water Authority on an area in the Diplomatic Quarter.

The entire Diplomatic Quarter covers an area of more than 410,000 sq m and currently comprises a total of 52 buildings in addition to three hotel complexes. While hotels are located at the eastern and western borders of the Quarter, several administrative buildings including five ministries are located in the centre and along King Faisal Highway overlooking the coast. While around 26% of the area consists of roads and sidewalks, about 24% of the district area consists of parking sites, which are mainly located in the south-west and north-east. While 38 office towers constitute around 7% of the total area, about 6% is occupied by 11 administrative buildings and another 5% by three large hotel complexes, namely, the Holiday Inn and the Diplomat at the eastern end of the district and the Golden Tulip at the western end. Thus, approximately 22% of the total plot area of the Quarter consists of hotel complexes while around 49% consists of plots reserved for commercial, residential or administrative developments, of which only around 50% are currently built. Most of the remaining unbuilt plots, which are currently zoned for the development of Investment Buildings Type B-A, are being used for parking. Furthermore, due to the average building height of about 10.5 floors, the overall plot ratio of the Diplomatic Quarter is about 2.8, which is almost twice the average plot ratio in the the other areas of the Seafront District.

Apart from administrative government buildings and hotel complexes, most of the commercial buildings in the Seafront District are used by banks and other financial institutions. In order to establish the district as Bahrain's future centre of financial business, the Bahrain Financial Harbour has been constructed on reclaimed land in the old harbour area of Mina Al Manama in front of Bab Al Bahrain. After an area of around 380,000 sq m was reclaimed in 2003, the Financial Centre complex occupies 28,989 sq m at the entrance of the project on King Faisal Highway. In 2007 the Financial Centre, which consists of the Harbour Towers, the Harbour Mall and the Harbour House, was completed. The two 53-storey Harbour Towers are currently Bahrain's tallest buildings at a height of 260 metres. The whole Financial Centre offers a total of 105,300 sq m of office space and 15,200 sq m of retail and dining space in addition to 3,500 parking spaces[4]. The 10-storey Harbour House, which is connected via a pedestrian bridge to the main building complex, is a commercial building located at the man-made marina in the centre of the project area. While the Financial Centre has already been completed, the rest of the projects of the Bahrain Financial Harbour are still under development, for example, Commercial East, which will include the already built 22-storey BIIC Tower[5]. At the present time, the Bahrain Financial Harbour can only be accessed via King Faisal Highway until the completion of a new access road on reclaimed land, which will connect the Bahrain Financial Harbour with the highway junction at the Sheikh Isa Bin Salman Causeway. In addition to the reclamation work that is part of the Bahrain Financial Harbour project and the adjacent Bahrain Bay project, the construction of the Reef Island development is being carried out on a reclaimed area which was initially intended for the creation of a public park on two small islands. However, this idea was replaced with plans to build one single island that will be about five times bigger than the two islands originally planned.

While there are very few green areas within the business district between King Faisal Highway and Government Avenue, there is a large seaside park stretching along the coast on an area of around 196,000 sq m. It can be accessed via a pedestrian bridge close to Bab Al Bahrain Square or, in the main, by car. It is thus not a very integrated part of Manama's old city centre. Due to the expansion of King Faisal Highway over time, this major road essentially divides the large seaside park covering around 11% of the Seafront District from the urban areas on the other side of the highway to the south. Not including recently reclaimed land, the Seafront District covers around 1.77 sq km. While about 28% of this area consists of roads including King Faisal Highway, around 47.6% consists of commercial plots. However, only around 27% of the area currently consists of lots containing completed commercial developments (mainly banks and hotels). Thus, empty lots account for more than 20.6% of the area, which are usually used for parking. According to the current zoning plan, more than 82% of these empty plots can be developed with Investment Buildings Type B-A, while around 65,000 sq m in the south-west has been restricted to the development of Investment Buildings Type B-D, which have a much lower height. Although 7.5% of the total plot area of the Seafront District has been reserved for government institutions, only around one third of this is actually occupied by administrative buildings due to the large need for parking sites. While official parking lots such as that in front of Bab Al Bahrain make up only around 1.4% of the Diplomatic Quarter, more than one third of the Quarter, including unbuilt plots, is currently used as parking space. In spite of the fact that the Seafront District, as Bahrain's main business centre, has been mainly reserved for commercial and administrative use, a few residential towers have recently been developed, making up less than 1% of the district. Because of large unbuilt areas and low-rise complexes such as the Central Market, the average plot ratio of the Seafront District is just 1.75. However, in some cases such as certain residential high-rises, the plot ratio can be as much as 12.0.

In the future, the reclaimed land of the development areas of the three master-planned projects along the coast will come to around 1.4 sq km. Furthermore, it is expected that many of the currently unbuilt plots in the older parts of the district will be developed over the coming years in accordance with the guidelines of the current zoning. Thus, developers may build a high-rise with a construction ratio of up to 1,200% on an area of about 300,000 sq m and six-storey blocks with a construction ratio of up to 300% on an area of around 65,000 sq m. Furthermore, the Diplomatic Quarter will be expanded to the south, where an empty plot of around 43,500 sq m will be developed with Investment Buildings Type B-A. In addition, two so-called special project areas have been zoned. Like the three master-planned projects on the coast these areas have not yet been assigned a particular building type in order to allow future investors to have a more flexible layout of their master plan. One of these special project areas will replace the Central Market with a mixed-use development covering more than 162,000 sq m. The second special

158

Fig. 7: The zoning plan of the Seafront District (2008).

project area of around 22,000 sq m is located in the east of the Seafront District on the former plot of the British Embassy opposite the Bahrain World Trade Centre.

All in all, the size of the overall area available for development in the Seafront District is around 592,500 sq m, of which at least 58% will be developed with high-rise projects. The recent completion of two residential towers in the western part of the district marked the beginning of a new trend of constructing residential high-rises in Manama's central business district. Thus, in addition to commercial buildings such as banks most future projects can be expected to be apartment buildings due to the growing freehold property market in Bahrain. Based on its waterfront location and proximity to business and leisure areas, the Seafront District has become an attractive location for investment. Subsequently, the construction of three developments along the coast was recently begun, thus benefiting from easy access via King Faisal Highway as well as a prime seaside location in the historic harbour of Manama.

1) Bahrain Financial Harbour:

While the Financial Centre, which is the first development phase of the Bahrain Financial Harbour project, has already been completed, the next two development phases are scheduled to be completed by 2015. The whole development area is divided into 10 different projects, which are further subdivided into a total of 28 plots. On either side of the Financial Centre two commercial developments called Commercial East and Commercial West have been built on an area of around 38,000 sq m comprising around six towers with up to 30 floors, which are connected by a shared multi-level car park. While the new Dhow Harbour will occupy the south-western part of the Bahrain Financial Harbour, a marina will be built at the centre of the project. According to current plans, a 1,200-metre long promenade will stretch along this marina including a medium-rise

Fig. 8: Bahrain Financial Harbour.

development called Harbour Row that will contain shops and restaurants in addition to offices and apartments. This promenade will stretch as far as the Residential South island, where the Villamar project has been recently launched. With more than 50 floors, the three residential towers will rise up to 220 metres and their bases will be linked by a winding four-storey building, which will contain shops and restaurants. On a separate island of around 8,900 sq m in the west of the Residential South island, the 45-storey Diamond Tower will include commercial facilities and residences. While the north-western island has been reserved for a large-scale hotel complex, the northern island has been designed to be a second residential island called Residential North, where in addition to six residential high-rise projects a building called the Bahrain Performance Centre will be developed with the intention of it becoming the cultural centre of the Bahrain Financial Harbour. It will comprise an opera house and a conference hall in addition to exhibition spaces.

The basic road network of the Bahrain Financial Harbour will include an access road around the commercial projects in the south and a road running from north to south in its eastern part linking the three main islands and the adjacent project Bahrain Bay with King Faisal Highway. All in all, around 15% of the project area will be traffic area, which will mainly consist of roads due to the development of several car parks providing around 11,000 parking spaces. At the present time, one car park has already been completed in the Financial Centre, which offers about 4,000 parking spaces, and a second car park with a height of 10 floors is currently under construction as part of the Commercial East project[6]. Apart from the traffic area, over one third of the overall development area will be occupied by green areas spreading between buildings, which will partly be used as parks and public leisure areas. Only around 33% of the total GFA of more than 1,115,000 sq m will be developed with apartments. The remaining area will be mainly occupied by the commercial buildings of the Financial Centre, the Commercial East, the Commercial

West, the Harbour Row and the hotel complex. All the residential projects will comprise around 10 towers with up to 50 floors, including the three towers of the Villamar project, which will offer around 900 apartments[7]. While it is currently expected that about 7,000 people will reside in the Bahrain Financial Harbour, around 8,000 workers are expected to come and go daily[8]. The most important public centre of the development will be the promenade along the marina and the canals constituting around 5% of the total area. Furthermore, the construction of the Dhow Harbour along the western border of the Bahrain Financial Harbour is intended to compensate for the replacement of the old Mina Al Manama. While the Financial Centre has the highest construction ratio within the Bahrain Financial Harbour at around 800% of the plot area, the average plot ratio of the Bahrain Financial Harbour as a whole will be approximately 3.5, which is more than twice the density of the old urban areas of the Seafront District.

2) Bahrain Bay:

After the completion of a recently reclaimed area of more than 800,000 sq m to the east of the Bahrain Financial Harbour, the construction of Bahrain Bay, which covers an area of around 435,000 sq m, was begun. This development consists of two islands that form a man-made bay surrounding an oval island in the centre. Apart from the development area of these three islands, more than 200,000 sq m is intended for the construction of an outer ring road. The rest of the reclaimed area along the coast will mainly be occupied by an extension of the existing seaside park, which will complement the man-made bay that stretches for around 1.5 km along the coast. According to the approved master plan, seven distinct zones in addition to three special project areas have been defined. These three special projects will be the Four Seasons Hotel Bahrain, which will occupy the oval island in the centre of the bay, Arcapita Headquarters and Raffles City on the more eastern of the two islands forming the bay. While the plots along the inner rim of the bay will be developed with low- to medium-rise buildings, high-rise developments will be restricted to the plots along the main ring road in order to prevent traffic congestion in the centre. In addition to the outer ring road, one two-lane and one four-lane central circulation road, which will be linked to the outer ring road by four communication roads, will be built in order to provide sufficient accessibility by car. Including the sidewalks and promenades along the shorelines in addition to another six roads that are mainly to be used as service access roads, the overall traffic area will constitute around 40% of the total development area. In addition to a 40-metre wide green buffer zone, which will separate the development from the outer ring road, there will be a total green area of around 60,000 sq m that will contain three parks, four pedestrian corridors, one public square and a 1.4-km long promenade along the inner bay and the central canal. In addition to the South Park at the western border and the North Park on the eastern island, which will stretch in the form of green strips with a width varying from 20 to 30 metres, the Arcapita Park will surround the Arcapita Headquarters.

The over 170-metre long eight-storey building of Arcapita Headquarters will cover an

Fig. 9: Bahrain Bay.

Fig. 10: Arcapita Headquarters.

area of around 7,600 sq m and together with the South Park on the opposite shoreline it will form one of the two main design-axes of the development. Next to the Arcapita Headquarters, which is already nearing its scheduled completion for 2010, the construction of the Raffles City project is being carried out on an area of around 43,000 sq m. This project features three curved residential towers with up to 38 floors and five rows of 50 town houses that will be built on a shared base along the waterfront. Called Sky Villas they will be provided with terraces overlooking the bay. In addition to this an eight-storey apartment block will be constructed in the fourth row of town houses providing 200 serviced apartments. After completion by 2015 there will be a total of around 850 dwellings constituting 68% of the 288,000-square metre GFA. The remaining area of more than 92,000 sq m will be mainly occupied by a shopping centre, which will be located in the shared base[9]. Along the western border of the project the High Street Plaza will be established as the largest square in Bahrain Bay covering 4,350 sq m. With a plot ratio of more than 6.7 Raffles City will have almost the same average plot ratio as Bahrain Bay. Between Raffles City and Arcapita Headquarters the construction of an access road has commenced in order to link the outer ring road with the oval island. In the centre of the 54,000-sq m island the Fours Seasons Hotel complex will cover most of the area with various facilities and a tower. This tower was originally designed to have a height of 285 metres but this has recently been reduced to a much lower scale.

The remaining development area of Bahrain Bay is divided into seven zones starting at its western border, where Zone 1 comprises a residential low-rise district along the marina. The total plot area of Zone 1 is 23,163 sq m, on which can be built buildings with a total GFA of 90,592 sq m. Due to the construction of medium-rise buildings the resulting plot ratio of 3.3 will be one of the lowest of the whole development. In addition to seven four-storey Water Residences, which will be built on reclaimed islands in the sea, there will be two six-storey apartment blocks named South Marina Apartments in addition to one yacht club house at the marina[10]. The area of Zone 2, which will be linked to Bahrain Financial Harbour, will contain the over 12,000-sq metre South Green Park, the Boutique Hotel, which has a separate green area of its own, and two residential towers, which will each be built on a multi-storey base. While the total lot area of Zone 2 is 35,147 sq m, due to building regulations only an area of 13,000 sq m will be built because of the development of high-rises of as many as 33 floors. Thus, the two residential towers will

162

add 101,424 sq m of residential area to a total GFA of 141,269 sq m, which leads to a plot ratio of around 4.0. The six-storey Boutique Hotel will be located at one end of the South Green Park, which will stretch up to the promenade along the inner bay[11]. To the north of Zone 2, Zone 3 will comprise four lots covering an area of 37,337 sq m, which will be developed with one office tower with up to 45 floors and six 26-storey residential towers, which will be built on a multi-storey base with car parks and retail facilities. Based on the currently permitted GFA of 385,982 sq m, the plot ratio of Zone 3 at 10.3 is the highest one of the entire development[12]. With regard to the adjacent three lots to the south, Zone 4 comprises six- to eight-storey apartment blocks and town houses covering 68% of the 18,683-square metre zone. While the majority of the permitted GFA will be occupied by residential units, shops will be established on the ground floors. In spite of the rather low heights of the buildings, Zone 4's plot ratio is as much as 5.2 due to permission to build most of the lot area[13].

Two high-rise projects, including the 59-storey Palace Apartments, will form Zone 5 stretching over three lots on both sides of the central canal and along the outer ring road. While the 45-storey Palace Apartments will occupy an area of 8,875 sq m on the western main island, a commercial complex comprising an eight-storey base and a 26-storey tower in addition to a school will cover an area of 8,356 sq m. In addition to a four-storey retail podium the Palace Apartments will incorporate three amenity levels covering 6,300 sq m above the retail base and a car park on three basement levels. The tower will have a total GFA of about 66,700 sq m, of which more than 87% will be occupied by residential units[14]. In addition to public promenades along the canal, the High Street Plaza, which belongs to Raffles City, is expected to become an important public space of the entire Bahrain Bay development. Due to a total lot area of just 17,231 sq m, the plot ratio of Zone 5 at 8.4 is one of the highest in Bahrain Bay[15]. To the east of Zone 5, Zone 6 will be a mixed-use district with a large residential component in the form of two high-rises in addition to one commercial tower spreading across three lots along the northern border of the Raffles City development. The 180-metre high commercial tower will offer, together with its six-storey base, a GFA of 57,836 sq m, which is restricted to commercial use. Next to the commercial development there will be two 26-storey residential towers on a multi-storey base, which will include a school. All in all, the total permitted GFA is 169,601 sq m on a total lot area of 24,639 sq m, leading to an average plot ratio of 6.9[16]. To the east will be North Park in Zone 7, which will be a mixed-use district. The 50-storey North Tower, which will be built in the North Park, will be one of Bahrain Bay's tallest buildings and apart from offices it will have about 24,300 sq m of residential area and a hotel spreading over 18,900 sq m. Including another 26-storey tower and its six-storey base the total office area of Zone 7 will be 90,642 sq m and its retail area will be 1,200 sq m. Because of a total lot area of just 16,916 sq m, Zone 7 has an above-average plot ratio of 7.9[17].

All in all, Bahrain Bay's total GFA, which includes the developable GFA of all seven zones and of the three special projects, is expected to be around 1.45 million sq m spreading on a total developable lot area of around 200,000 sq m. Thus, the development's average plot ratio of 7.3 will make the project one of the most dense urban areas in Bahrain. Around 60% of the GFA will be occupied by residential units consisting of apartments in addition to a few town house developments. While Raffles City will offer around 850 dwellings, the residential projects within Zone 1, Zone 2, Zone 3, Zone 4 and Zone 6 will cover a total GFA of 718,266 sq m, which will mainly be developed with various sized housing units. According to current expectations and plans for a maximum of around 25,000 people to move to the development after its completion by 2015, each resident will be provided with an average living space of 27 sq m. Consequently, Bahrain Bay would become one of Bahrain's most populated areas with a density of more than 500 people living on each hectare of the 435,000-square metre project area. Most of the future residents will be accommodated in the high-rise-developments of Zone 2, Zone 3, Zone 5 and Zone 6 along the outer ring road, while Zone 1 and Zone 2 in addition to Raffles City will contain lower-density neighbourhoods along the inner bay. Each of the total of 30 lots will be developed with integrated car parks, which will be mostly be underground, in order to supply residents and daily commuters with sufficient parking space. All in all, there are plans to integrate 11,000 parking spaces within the seven districts in addition to car parks in the three special projects. In order to prevent heavy traffic in their centres, the zones containing commercial high-rises such as Zone 3, Zone 5 and Zone 7 will be located close to the outer ring road.

3) Reef Island:

In the west of the Seafront District the construction of the Reef Island project is being carried out on an area of around 579,000 sq m. It is expected to be completed by 2015. Apart from three high-rise buildings the project mainly consists of medium- or low-rise buildings. While the residential projects will constitute about 79% of the total developable lot area of around 498,450 sq m, the remaining 20% will be occupied by one hotel, a medical centre and a shopping centre in the middle of the development in addition to one marina and yacht club at the central bay. More than 12% of the project area will be covered by its road infrastructure in the form of one main access road running around the shopping centre and leading to the villas in the north. While at centre of Reef Island the shopping mall, medical centre and hotel at the marina will be open to the public, the northern villa-district will be a gated community of three developments, namely Beach Villas, Lagoon View Villas and Sea View Villas. The remaining residential developments of Reef Island will comprise 43 multi-storey residential buildings, of which 39 will be developed along the shoreline. In the south-west a 60-storey residential tower called the Icon Tower will be Reef Island's biggest landmark in addition to two 23-storey high-rise buildings on the lot next to the bridge connecting Reef Island to the coast. The location of high-rises and of the three commercial projects close to the only bridge linking the development to the Seafront District is intended to reduce traffic movement within Reef Island.

Fig. 11: Reef Island.

While more than 90% of the around 1,300 dwellings will be apartments within the 8- to 10-storey apartment buildings and the three high-rise projects, the villa-developments will offer just 114 residential units. However, about 43% of the total lot area of Reef Island will be occupied by these two-storey villas due to their large average lot size of 1,400 sq m. In addition to five public green areas that are each around 2,000 sq m and strategically located in order to allow public access to the sea, there will be a 500-metre long public promenade along the southern shoreline facing the seaside park along King Faisal Highway. All in all, only around 2% of Reef Island, which will be one of the most exclusive residential developments in Bahrain, will consist of public areas. Similarly to the neighbouring Bahrain Financial Harbour and Bahrain Bay, the main goal of the developer of Reef Island is to develop freehold properties for the upper end of the real estate market. Thus, in addition to a high standard of security many leisure facilities have been integrated such as pools and private beaches as well as a Spa Village next to the hotel at the marina. While most parking spaces will be developed as part of individual lots, there will be one separate six-storey car park next to the Icon Tower. Due to the large low-rise residential area the average plot ratio of Reef Island is only about 1.7, which is equivalent to the average plot ratio of the current urban areas of the Seafront District and significantly lower in comparison to the future ratios of its two neighbouring developments.

While the three master-planned projects along the coast will add more than 1 million sq m of developable area to the Seafront District, the undeveloped plots within the old part of the district between King Faisal Highway and Government Avenue offer a developable area of about 364,700 sq m. In addition to these reclaimed and unbuilt areas, there are plans to replace Central Market with a special project, thus adding around 162,000 sq m to the developable area of the Seafront District and bringing it to a total of about 1.57 sq km, which is more than twice the size of the currently built area. In contrast to the Bahrain Financial Harbour, Bahrain Bay and Reef Island developments, which follow approved master plans, there has been no comprehensive master plan for the entire Seafront District.

However, a general zoning plan defines the size of the buildings for each developable lot in addition to several further requirements regarding construction such as the provision of parking spaces. According to the current zoning 82% of the district area can be developed with Investment Buildings Type B-A, which means the permission to develop residential, commercial or administrative buildings. Consequently, the district area, which is mainly occupied by commercial and administrative buildings, has begun to transform into a mixed-use area with an increasing number of residential developments. In addition to the increase in the number of residential buildings in the old area of the district, all three waterfront developments have a high percentage of residential projects. While residential use accounts for around 33% of the developable GFA of the Bahrain Financial Harbour, it accounts for around 60% of the GFA of Bahrain Bay and more than 76% of Reef Island. Consequently, the previously commercial and administrative Seafront District will turn into a 3.5-sq km mixed-use urban centre with around 41% of lots occupied by residential developments while 47% will be commercial and 7% administrative.

In addition to the future expansion of the seaside park in the east of the Seafront District where an area of around 126,000 sq m has been reclaimed, there will be several parks, green areas and public promenades within the three waterfront developments. Around 50% of the development area of the Bahrain Financial Harbour will be green areas including public areas such as promenades at the marina and harbour. Furthermore, there will be two parks at the eastern and western ends of the neighbouring Bahrain Bay project, which together will make up around 14% of the project's development area by covering about 60,000 sq m. Reef Island will offer the least amount of public areas with its five small public green areas of around 2,000 sq m making up only around 1% of the total development area. However, similarly to Bahrain Financial Harbour and Bahrain Bay, it will offer a public promenade along its coast. Thus, the three projects will offer a total of more than 3 kilometres of seaside promenades that will mainly be developed with shops and restaurants running alongside. All in all, public areas both current and future will make up around 16% of the 3.5-sq km district area. Around 50% of this total public area will be occupied by the future 280,000-sq m seaside park that will form a green corridor between the older areas of the district and the new island developments.

Because of the necessity of using cars for transport, one third of the Seafront District will be covered by traffic area, which will include King Faisal Highway and Government Avenue in addition to several smaller access roads and the new outer ring road linking the Bahrain Financial Harbour and Bahrain Bay. While the Bahrain Financial Harbour and Bahrain Bay will be directly connected, there will be no direct link between them and Reef Island. This is illustrative of a general impression that the three new waterfront developments are being added to the Seafront District without consideration of any relationship to each other or to the Seafront District's older urban areas and that they are developing in isolation according to individual master plans. The only attempt to improve the connection between different areas is a plan to link Bab Al Bahrain Square and the Bahrain Financial Harbour by constructing a wide pedestrian bridge connecting all of the

Seafront District's urban areas – the old pre-oil city centre in the souq, the current city centre and the mixed-use master-planned projects along the coast. In order to achieve this plan, the Bahrain Financial Harbour has been developed a few metres above street level on an elevated reclaimed area in order to ease the construction of this bridge. However, there is currently relatively little interest from the private sector in implementing this plan. Similarly, efforts to modernise and restore the old city centre, which was proposed by a study of the Ministry of Municipalities and Agricultural Affairs and the UNDP, have more or less failed because of limited interest from the private sector to invest in old urban areas where the land prices are relatively high. Thus, the current construction of a modern souq development with a canopied pedestrian zone next to Bab Al Bahrain is the only project that is expected to revitalise parts of the old city centre.

Due to a lack of public transport, the growing need for sufficient parking areas has led to the integration of parking spaces on the first floors of buildings or underground. Moreover, separate car parks are often built in the cases of high-rise developments, which have become the favoured typology due to high land prices. While the Bahrain Financial Harbour includes around 14 high-rise buildings with up to 53 floors, there will be about 20 towers with 34 to 59 floors in Bahrain Bay as well as three residential high-rises on Reef Island. All three developments together will add eight commercial towers, three hotels and 26 residential towers to the Seafront District. Apart from around 28 buildings with more than 10 floors in the Diplomatic Quarter which are mainly used for commercial and administrative purposes, two residential high-rise towers with 21 and 23 floors have recently been developed at the western end of the district. Furthermore, the 27-storey building of the National Bank of Bahrain and the 50-storey twin towers of the World Trade Centre in addition to several other high-rise buildings of banks have been developed in recent years. All in all, the Seafront District currently contains a total of about 40 high-rise buildings with more than 10 floors. In the future it is expected that in addition to the 37 high-rise buildings within the three waterfront developments several high-rise projects will follow in the unbuilt areas within the older district area. While the average plot ratio of the Seafront District is currently around just 1.75, the plot ratio in the Bahrain Financial Harbour development will be twice as high at around 3.5, and four times as high in the Bahrain Bay development at 7.3. Due to the development of large low-rise residential projects, Reef Island has a plot ratio of around just 1.7 and thus it will not contribute to the increase of the average plot ratio of the Seafront District. While the average plot ratio of the Seafront District will increase to 3.0 after the completion of the Bahrain Financial Harbour and Bahrain Bay, it is expected to grow even more due to the development of high-rise projects in the older areas of the district. Today, there is a total of around 549,000 sq m of developable area in the Seafront District, which includes 184,000 sq m of special project areas. According to the current zoning plan about 80% of the developable area can be built with Investment Buildings Type B-A, meaning a maximum plot ratio of up to 12.0 and a maximum plot ratio for the whole district of 4.5. However, recent residential high-rise developments generally have a plot ratio of about

Fig. 12: The building heights of current and future projects (2008).

8.0, so the future average plot ratio of the whole district will most likely in reality be between 3.0 and 4.0. Even so, this would still mean a significant increase in the district's built density compared to its current situation.

While the Bahrain Financial Harbour will house around 7,000 future residents, there will be around 25,000 people living in Bahrain Bay and another 5,000 on Reef Island. Furthermore, the expected development of residential projects in the older areas of the Seafront District, where currently only a few hundred apartments have been built, will increase the population of the district to a total of approximately 70,000 future residents, thus creating an urban density of around 200 people living on one hectare. With regard to residential developments that have already been approved and which will be built within the next five years, the urban density of the Seafront District is expected to reach around 106 residents per hectare. Due to high land prices, the Seafront District has become one of Bahrain's prime locations for exclusive real estate developments targeting the upper end of the freehold property market. The consequent high rental prices will turn the district into a residential area for upper-income groups mainly consisting of foreign guest workers. Apart from foreign tenants a certain percentage of apartments are expected to be bought by GCC nationals such as Saudis who often visit Bahrain for business or leisure. In order to attract these very particular future residents the projects have to integrate a high standard of security and accessibility in addition to leisure and shopping facilities. While the low-rise projects on Reef Island will be gated communities, all of the Seafront District's residential towers will be guarded individually. Its location in the

Fig. 13: Future structure of the Seafront District.

centre of the business district and its short distance from the causeway to Saudi Arabia and the causeway to the airport in Muharraq makes the Seafront District attractive as a residential area. Furthermore, each of the three waterfront projects will integrate large retail areas in the form of shopping centres and promenades in addition to exclusive beach clubs and marinas. Apart from its new function as Bahrain's centre of freehold property developments, the Seafront District is currently expanding its role as a financial business district as can be seen by the Bahrain Financial Harbour project and the Bahrain World Trade Centre. Moreover, it will become one of Bahrain's prime tourist centres due to more than 10 large-scale hotel developments and it remains Bahrain's biggest administrative centre with over ten government institutions.

Over the past decades Bahrain's biggest landmarks have been developed in the Seafront District. Today, the two twin-tower developments of the Bahrain Financial Harbour and the Bahrain World Trade Centre with heights from 240 to 260 metres are the country's tallest and most significant landmarks. In the future a growing number of large-scale commercial and residential towers will be developed, for example, the five proposed high-rise projects on Reef Island and in Bahrain Bay, which will have more than 50 floors. By following the design principles of global architecture most of these new high-rise buildings will be developed with hardly any consideration of their local context and thus contribute to the impression of the new urban development areas as being added and detached elements instead of integrated and complementary parts of the overall urban fabric. In addition to the highway and the large seaside park that separate the urban expansion areas of the new developments from the district's older areas, the fact that all three of the new developments are islands contributes to the detached and isolated structure of the urban patchwork of the Seafront District.

With regard to future development, there are currently plans to develop a second row of around six large islands along the coast, which will form an inner bay with the three developments at the waterfront of Manama and a crescent-like reclaimed area at Al Muharraq. Furthermore, there have been proposals to reclaim a large area between the two causeways leading to Al Muharraq in order to expand the business district along the eastern border of the Diplomatic Quarter. The biggest challenges in the future will be the development of links between the different urban areas and the creation of an adequate infrastructural backbone, particularly regarding the expansion of the road network and the development of adequate parking space to meet the demands of the population's high dependency on the car. The rapidly growing number of vehicles has led to heavy traffic congestion within certain district areas such as, for example, the Diplomatic Quarter, where the densely built environment is giving rise to increasing congestion and lack of parking space. Furthermore, King Faisal Highway is threatened by growing traffic jams due to its function as a main access road linking the new urban areas as well as connecting the main island with Al Muharraq. All in all, Manama's Seafront District is a typical example of the current urbanism in Bahrain, which is on the one hand characterised by rapid urban growth due to speculation-driven investments and, on the other hand, an evident lack of comprehensive regulation.

4.2.2 The Seef District

At the time when the Sheikh Khalifa Bin Salman Causeway to Saudi Arabia was developed during the 80s, a new area of more than 5 sq km was reclaimed in the north of Sanabis in order to compensate the Bahraini citizens who had lost their land due to the construction of new highways. Although the area has been subdivided and basic infrastructure was provided, very few lots were built in the following decades. At the end of 20th century two residential areas in the west along the borders of the old fishing village Karbabad and in the south next to Sanabis covered an area of around 500,000 sq m. In addition to low-rise dwellings built by Bahraini families there are two social housing projects built the Ministry of Housing. These residential areas were built according to the guidelines of the first land-use plan, which divided the area into two large residential areas on each side of a central commercial area stretching from north to south on an area of around 196,000 sq m. While the residential areas were restricted to low-rise dwellings, investors were allowed to build medium-rise commercial buildings in the centre of the Seef District. However, most of the district remained undeveloped during the following decades due to a lack of investment from the private and public sectors. One of the district's most significant projects in this early stage of development was the large hotel complex of the Le Royal Meridien on the northern shoreline, the premises of which were later taken over by the

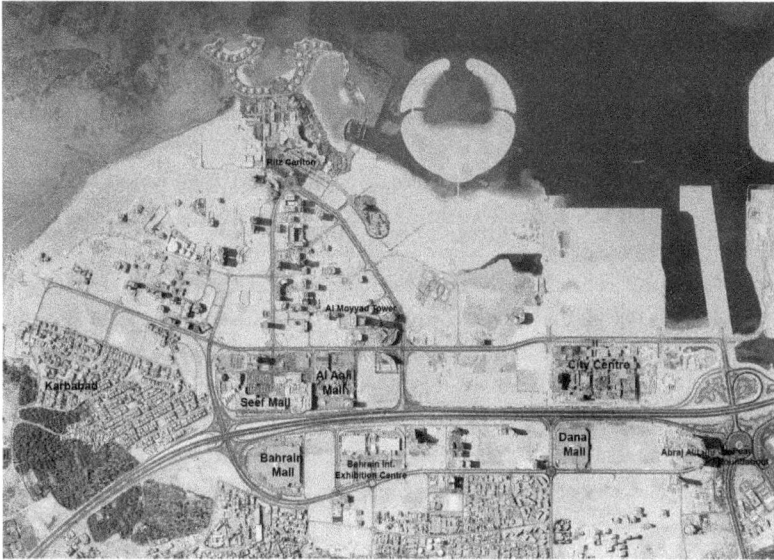

Fig. 14: The Seef District (2008).

Ritz Carlton.

After the Jidd Hafs Governorate was dissolved in 2002 the Seef District became part of the Capital Governorate and thus an urban expansion area of Manama. Since the completion of the King Fahad Causeway to Saudi Arabia, Bahrain's shopping tourism was gradually increasing, which led to the development of several shopping centres, while at the same time the introduction of the freehold property market marked a significant turning point for the development of the Seef District, whose fortunate location and accessibility has led to growing interest from the private sector to develop commercial and residential projects. Thus, in spite the initial zoning plan, which envisioned the district as a suburban area with a medium-rise commercial centre, the Seef District has recently begun transforming into one of Manama's major business districts and fastest growing development areas. Consequently, more and more land was reclaimed and added in the east towards the Seafront District in order to extend the developable area.

The area under study in the Seef District, which includes the northern area of Sanabis, has a total area of around 6.5 sq km stretching four kilometres along the coast and about two kilometres from south to north. The areas along both sides of Sheikh Khalifa Bin Salman Highway are mainly taken up by five shopping malls whose lots cover around 245,000 sq m in addition to a number of residential and commercial high-rises located in the south-east near the Pearl Roundabout. A large area is occupied by the Bahrain International Exhibition Centre and the Bahrain Chamber of Commerce & Industry, which are both government institutions. While in the north there is an area where office towers and exclusive residential medium- to high-rise buildings have concentrated, most of the west and east has remained undeveloped. In the south-west three special project areas, which

171

are as yet unbuilt, separate the commercial north from the low-rise residential south. In the east an area of around 175,000 sq m between the older part of the Seef District and the recently reclaimed areas has been reserved for the development of embassies and other special projects. Apart from a hotel building and the a residential tower under construction, the eastern part of the district has remained unbuilt and according to current plans will be extended by reclaiming the area between the Seef District and the recently developed causeway to Reef Island, which would mean an additional developable area of around 117,000 sq m. In the north the reclamation of a crescent-shaped island with a surface area of about 125,600 sq m has been recently completed and expected to be developed in the near future with a large hotel complex. In contrast to its northern border along the shoreline, the southern border of the Seef District passes into low-rise residential areas that were built in previous decades at the fringes of the village Karbabad and in the north of Sanabis.

With regard to road infrastructure, the six- to eight-lane Sheikh Khalifa Bin Salman Highway divides the Seef District into a large northern part and a small southern part, which includes the northern area of Sanabis. While in the east a fly-over marks the changeover to King Faisal Highway leading to Muharraq, there is another fly-over in the west where Avenue 40 and Avenue 28 join. Both fly-overs were designed to function as the main gateways of the Seef District in addition to an access road in the north and another in the south. While the northern access road leads into King Abdulla the Second Ibn Al Hussein Avenue, which leads to the Ritz Carlton on Bahrain's northern shoreline, the southern access road leads into Avenue 18, which ends at Avenue 28 that runs around the southern development area. In the north Road 2819 starts at the eastern fly-over and runs along the shopping centres towards the western coast. King Mohammed VI Avenue starts at the northern roundabout in front of the Ritz Carlton, runs along the central business district and ends at Road 2819. Today, King Abdulla the Second Ibn Al Hussein Avenue and King Mohammed VI Avenue as well as Road 2819 and Road 3020 in addition to Avenue 40 form the main road grid in the northern part of the Seef District, which will soon be expanded by the construction of a new north-south access road along the western side of the central business area. There are also plans to extend King Abdulla the Second Ibn Al Hussein Avenue toward the western shoreline in order to establish an outer ring road. In the south, Avenue 18 and Avenue 28 are the main access roads connecting the Seef District to the Pearl Roundabout in the east. Other road developments in the expansion areas in the east will complete the main road network, which basically consists of a grid of four- to six-lane main access roads. All these roads are linked to the central highway that connects the Seef District with Manama, the southern parts of the main island and the causeway to Saudi Arabia.

Due to this high degree of accessibility and centrality, several shopping centres have been developed in recent years. The district's first shopping centre complex, called Al Aali Mall, was built on an area of around 52,300 sq m sq m along Road 2819. In 1997 the Seef Mall complex, which includes the 17-storey Seef Tower, was built on a lot next to Al Aali

172

Fig. 15: Commercial developments along the central spine.

Mall. Including an extension that was completed in 2006, the two-storey Seef Mall covers around 60,300 sq m and was Bahrain's largest mall until the recent completion of City Centre in the eastern part of the Seef District. Taking 28 months to construct, this new shopping centre stands on a lot that is around 126,000 sq m and has a built-up area of over 450,000 sq m. It is a three-storey mega mall offering a gross leasable area of 150,000 sq m for around 330 stores in addition to entertainment and leisure facilities, for example, the region's largest indoor water park spreading over 25,000 sq m. It also has a seven-level car park with a capacity for 5,500 vehicles. In order to provide fast access to Manama, an extra fly-over was built to connect City Centre with the highway in the south. Apart from these three shopping centres in the north along the highway, there are two other shopping centres in the south, namely, Bahrain Mall and Dana Mall. Between these two shopping centres the Bahrain International Exhibition Centre occupies a large lot area of around 80,500 sq m. All five shopping centres in addition to the Exhibition Centre occupy a total lot area of more than 493,400 sq m, of which around 55% is occupied by buildings and 45% by traffic areas, mainly in the form of parking sites.

Next to the Exhibition Centre stands the 19-storey high-rise of the Bahrain Chamber of Commerce & Industry as well as eight residential and two commercial towers. Another two residential high-rise projects have recently been completed, one of which is the Abraj Al Lulu development, which has been built on a lot of around 23,000 sq m next to the Pearl Roundabout. This project consists of two 50-storey towers and one 40-storey tower offering 872 dwellings which are sold as freehold properties. The two 50-storey towers, called Gold Pearl and Silver Pearl, each have a GFA of more than 66,000 sq m. The 40-storey Black Pearl tower has a GFA of about 41,200 sq m, so the project's total GFA is more than 107,200 sq m. On top of a four-storey car park, which offers around 1,100 parking spaces, a public space was developed on an area of about 5,200 sq m[18]. Due to the rather large GFA of Abraj Al Lulu, its plot ratio is over 5.5, which is currently one

173

of the highest within the whole Seef District. Apart from the recent development of a 25-storey residential tower, most of the lots near the Abraj Al Lulu project have remained unbuilt. To the south of Abraj Al Lulu a small green area stretches between King Faisal Highway and future development sites. Along the northern border of Sanabis some areas have already been built with low-rise residential buildings. Today, the zoning of a large number of unbuilt lots has changed in order to enable the development of high-rises and freehold property projects.

As well as the shopping centres along the highway the main commercial centre of the Seef District is also in the north, where 21 of the Seef District's currently 32 high-rise buildings with more than nine floors are located. Most of the recent developments have been constructed in the central area between King Abdulla the Second Ibn Al Hussein Avenue and King Mohammed VI Avenue. All in all, there are currently around 103 completed buildings in the northern district area, which includes around 28 villas, 40 multi-storey apartment buildings, of which 11 have been built with more than nine floors, and 33 commercial developments such as the 47-storey Al Moyyad Tower. In addition, there are government institutions such as the Ministry of Industry and Commerce and hotels, for example, the Elite Grand Hotel, in the north of the Seef District.

Due to previous restrictions on the maximum number of floors, the current average height of all 73 multi-storey buildings is between eight and nine floors. In addition to the 47-storey Al Moyyad Tower, which was completed in 2004, the 18-storey residential tower The Palace, the 17-storey Adax Tower and the 17-storey Seef Tower are currently the tallest buildings in the Seef District. While commercial buildings either contain banks or telecommunications companies as well as various other offices, residential developments are mainly leased to foreign guest workers with higher incomes. Today, only around 24% of the developable lot area of around 777,200 sq m in the north of the area under study has been developed. In the east, a new developable lot area of around 768,400 sq m has been recently reclaimed. Thus, the total developable area of the Seef District has almost doubled yet it has remained one of the least dense urban areas in Bahrain. While many unbuilt lots are currently being used as parking sites, recent developments have begun to integrate parking space in the form of integrated car parks on the first floors of buildings or separately built car parks as, for example, in the case of the Al Moyyad Tower and Adax Tower.

Because of different zoning in the past and the slow development of the Seef District during the 80s and 90s, a number of old buildings, mainly villas, interrupt the built environment, which is becoming more and more dominated by the new high-rise projects. According to the current land use plan the zoning of most developable areas in the north of the district has been changed from exclusively suburban residential zoning to mid-rise or high-rise investment buildings that can be developed as both residential and commercial freehold properties. Six lots have been reserved for public service facilities, for example, a water forwarding station that has already been built. Because of the new zoning the district's centre, which offers a developable lot area of about 149,000 sq m, can be developed with Investment Buildings Type B-B, which involves a maximum construction ratio of 750% of the lot area. Further development areas for Investment Buildings Type B-B have been earmarked near Al Aali Mall as well as a 10,000 sq m plot between the low-rise residential area in the south-east and the mixed-use north. While the lots along the main access roads and the borders of the district's centre have been zoned for the development of Investment Buildings Type B-C, involving a restriction to a maximum height of ten floors and a construction ratio of 500%, a large part of the remaining areas in the centre of the Seef District have been earmarked for the development of Investment Buildings Type B-D, which involves a restriction to a maximum of six floors and a construction ratio of 300%. Furthermore, the areas along the western shoreline have been restricted to the development of Investment Buildings Type B-3 and thus to a maximum height of three floors and a construction ratio of 180%. Most of the lots in the eastern part of the Seef District will be developed with Investment Buildings Type B-B. In the south of the Seef District there are three lots stretching along the highway with a total area of more than 74,600 sq m, that can be developed with Investment Buildings Type B-A with a maximum construction ratio of 1,200%. Further inland lots have been zoned for investment building types with lower maximum construction ratios such as types B-B, B-C and B-D in order to prevent major disturbance to neighbouring low-rise residential areas.

In addition to the areas that have been zoned for the six types of investment buildings, there are around 19 special project areas including the lots of the four shopping centres that have already been completed. Because of the new zoning and new investment possibilities certain previously built developments, for example, the Dana Mall complex, are expected to be replaced in the future. Apart from the possible replacement of certain developments, there is currently a maximum developable lot area of more than 2 million sq m. All in all, the study area comprises a total lot area of almost four million square metres. Around 31% of this area is currently zoned for the development of Investment Buildings Type B-B, about 10% for the development of Investment Buildings Type B-D, approximately 8% for the development of Investment Buildings Type B-C and around just 2% for the development of Investment Buildings Type B-A. While the development areas for investment buildings account for around 56% of the Seef District's total developable area, more than 30% is reserved for the development of special projects with no initial restrictions. The remaining 14% of the total developable area is occupied by buildings of public services and low-rise residential areas as well as the two social housing projects of the MoH. In the north, the lot areas of the Ritz Carlton Hotel complex and a new hotel island development will in the future come to around 350,000 sq m, bringing the total area of the area under study to about 6.6 sq km, of which more than 2 sq km will be occupied by the future road network.

Today, a large number of high-rise developments are being constructed, of which around twenty are due to be completed by 2010. Thirteen of these current projects are residential buildings that are similar to the recently completed Abraj Al Lulu and aim for the upper

Fig. 16: The zoning plan of the Seef District (2008).

end of the real estate market. Because of their accessibility and proximity to the central business district of Manama, two major shopping malls and five-star hotels surrounded by private beaches, the Seef District has the highest rental rates of luxury apartments in Bahrain. Due to increasing land prices and construction costs, landlords were forced to raise their rates continuously over the last couple of years. Hence, at the end of 2008, the highest monthly rents ranged, on average, from BD 710 for a one-bedroom apartment to BD 1,200 for a three-bedroom apartment[19]. In the eastern expansion areas two residential high-rise projects, the 34-storey Dream Towers and the 50-storey Era Tower will offer more than 400 apartments. The 30-storey Avare Towers will provide 89 apartments and the 20-storey tower Landmark City View will have around 60 apartments[20]. These four projects are typical examples of the current trend of residential high-rises with up to 50 floors built on a multi-story base used as a car park or as an area for leisure facilities such as a swimming pool or gym. In addition to these residential developments several commercial projects, mainly office towers, are being constructed such as the 27-storey Millennium Tower on King Mohammed VI Avenue in the northern part of the district's centre. Further examples are the 35-storey BFB Tower and the 27-storey Capital Plus Tower, which are both built on top of a multi-storey base used as car park[21]. While most high-rise developments are being built in the north, the construction of a number of projects has begun in the new eastern expansion areas and near the Pearl Roundabout. All residential and commercial projects currently under construction have an average height of more than 16 floors and an average plot ratio of more than 7.0, which is almost the maximum permitted construction ratio of 750% of the lot area.

In spite of recent economic problems in the real estate sector, two large-scale developments are being constructed in the Seef District. The larger of these projects is Water Garden City, the developer of which is Albilad Real Estate. In order to develop the 2.2 sq km project, a large area will be reclaimed on the northern shoreline close to an already reclaimed island reserved for a future hotel complex. It is currently expected that the mixed-use development will house around 40,000 future residents and thus it will be one of Bahrain's most densely populated areas. As well as several residential projects such as the Wharf Housing development Water Garden City will comprise several commercial, retail and leisure projects. In order to accommodate all uses the development is divided into four main districts consisting of the Marina, the Island, the Beach and the Amenities. Each of these districts comprises several projects. The Marina, for instance, will include the 40-storey City Gateway Towers, which will be the future landmark of the whole project, the Conference Hotel and medium-scale commercial and residential developments along the marina. In contrast to the mixed-use Marina the Island will be an exclusive conglomerate of residential projects of different scales from low- to high-rise with heights of up to 48 floors. The Beach district will consist of five projects in the form of the two residential Water Gateway Towers, two-level apartment buildings called Beach Apartments, a group of residential towers in the north called Eco Garden Apartments and two hotels. Last but not least, the integrated amenities of the development will comprise parks, gyms, a primary school and a medical centre[22].

Apart from Water Garden City, the first stage of which is expected to be completed by 2012, construction has begun on the Up Town project on a 167,225-sq m area opposite Bahrain Mall along Sheikh Salman Highway. The project will comprise three high-rise buildings, two with 75 floors and a height of 260 metres and one with 70 floors in addition to a shopping centre, a five-star hotel and further residential towers with lower heights. After the approval of a $190 million loan by the Central Bank of Bahrain, the first phase, which includes the construction of the main utilities and the development of a 26-storey tower, has recently commenced. Construction on the project's 18 plots is expected to be completed by 2014. Because of its overall GFA of around 780,385 sq m and lot area of around 103,573 sq m, the average plot ratio of the project will be about 7.5[23].

Due to the future height of the new projects the average plot ratio in the north of the Seef District will increase from approximately 3.2 to 3.9. With regard to the entire Seef District including northern Sanabis, the average plot ratio will rise from around 2.3 to 2.5. While the lowest plot ratio is currently to be found in the low-rise residential areas in the west and south, where it comes to around just 1.0, the group of ten high-rise buildings in the south of the Seef District between Exhibition Centre and Dana Mall has, at 6.0, a relatively high average plot ratio. Due to previous restrictions the commercial and residential buildings in the north of the district have an average height of six floors and a plot ratio of around 3.2. All in all, a total area of more than 1.5 sq km is currently built, of which 66% is commercial, 24% is residential and around 10% is taken up by public services. In the future, residential developments are expected to cover around 50% of the total developable area of around 3.9 sq km. According to the current zoning plan and the presently available developable area of about 2.5 sq km, a maximum GFA of around 11.5 sq km could be developed in the following decades, which would cause the average plot

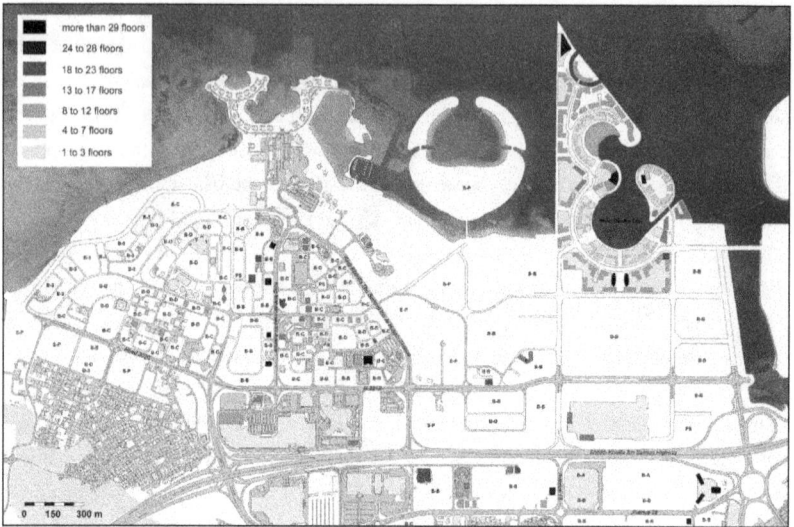

Fig. 17: The building heights of current and future projects.

ratio of the area under study to increase from 2.5 to 3.5. Based on the assumption that at least half of the maximum GFA will be occupied by residential developments and 15% by structure and circulation, the maximum living area of all housing units together in the area under study in the future would be about 4.9 sq km in total. Thus, if the average apartment size of the new developments is taken to be about 200 sq m, the number of housing units would increase from the around 2,560 units of the present time, of which about 1,025 have been developed within the 19 recently completed projects, to more than 27,000 units[24]. Consequently, the urban density would increase from around just 18 residents living on one hectare to more than 190 per hectare, which means a population growth from currently around 12,000 residents to more than 126,000[25]. This development trend is illustrated by thirteen residential high-rise projects currently under construction that will create around 700 housing units and thus living space for at least 2,000 future residents by 2010. Since 1998 the Seef District's GFA has grown by more than 120% and according to this growth rate the Seef District could be completely built up by 2030[26]. However, due to the recent global financial crisis the rapidity and scale of the urban development is expected to be significantly reduced.

Due to the Seef District's development over the last ten years involving the relocation and settlement in the district of several companies, for example, Citibank, Arab Investment Company and Bahrain National Holdings, the Seef District has become one of Bahrain's main commercial hubs. Apart from its function as a centre of financial and telecommunication business, five large-scale shopping centres have turned the Seef District into Bahrain's main location for shopping and thus one of its largest tourist attractions. Because of its proximity to the centre of Manama and its accessibility, the district became the target of large real estate developers wishing to build residential projects to sell as

178

Fig. 18: The future structure of the Seef District.

freehold properties or to lease to foreign professionals working close by. Furthermore, many expatriates working in Saudi Arabia are choosing to live in the Seef District due to Bahrain's more liberal society. Last but not least, it has become a favoured place of GCC nationals visiting Bahrain on weekends, who mainly reside in rental apartments. While its accessibility and proximity to work and to leisure and shopping facilities such as hotels and shopping centres contribute to making the Seef District attractive at the present time, increasing rental prices and traffic congestion could become major threats. Although the current occupancy rate of around 87% is still relatively high and traffic congestion is mainly concentrated around the two main shopping centres during the weekends, certain infrastructural deficits have become more obvious, particularly due to the growing urban density[27]. One very particular reason for this growing need for infrastructure is that the new zoning plan permits the widespread development of high-rise projects, which is in clear contrast to the initial plan to develop low-rise housing areas. Consequently, the previously designed and partly built infrastructure has become more and more insufficient in certain areas. Another serious problem of the district is a growing lack of sufficient parking space, a large amount of which is currently being provided by empty lots. Because underground parking is more expensive to construct, most projects integrate a multi-storey car park on the first several floors. However, this method leads to a decreased quality in the built environment because it reduces the comfort of pedestrians and the additional building height creates higher densities. Furthermore, the dependency on cars is intensified by a lack of integrated facilities such as shops within the residential areas and by a lack of public transport systems. Although plans for a metro have been proposed, there has been no planning and development of wider roads, which will lead to major obstacles for the development of public transport projects in the future.

179

A lack of public areas, which come to around just 15,000 sq m in the form of one playground in the north and one small green area in the south-east close to Sanabis, has become a major threat to the future urban quality of the Seef District. While there are plans to create a pedestrian promenade in the centre of the district, it remains unclear if there will be parks or promenades in seaside areas, which are currently mainly occupied by large-scale hotel developments. All in all, the Seef District is an example of the current urbanism in Bahrain, which is to a large extent driven by the interests of the private sector in the development of real estate. This has led to a rapid increase in land prices and consequently to the granting of permission to develop high-rise projects, which have become more and more profitable. In the case of the Seef District, this has resulted in the emergence of a mixed-use high-rise agglomeration following no comprehensive master plan. This is leading to an increasingly inadequate infrastructural backbone and a lack of public areas and service facilities such as schools and hospitals. Furthermore, the lack of urban design guidelines has led to a rather scattered urban landscape consisting of new high-rise projects that do not respond to their built surroundings and old low- to medium-rise buildings interrupting the growing high-rise conglomerate.

4.2.3 The Juffair District

The Al Fateh District, which is also known as the Juffair District, named after a former fishing village, has become one of the fastest growing urban expansion areas of Manama. Historically, the area has benefited from the construction of the deep water port Mina Salman, which was developed on reclaimed land in the south-east of Juffair in 1962, and the relocation of the US Navy to a former British military base at the end of the 70s. Furthermore, the district's development into a preferred residential area of expatriates was caused by the establishment of the first free economic zone in the south of Manama and convenient accessibility thanks to Al Fateh Highway, which was completed at the end of the 80s. While in the 70s relatively small areas of reclaimed land were added along the northern shoreline of the Juffair village, where for instance a sports complex with an indoor stadium was built, there was a major expansion of the area during the 80s when an area of more than 800,000 sq m was reclaimed in the north as well as around 700,000 sq m in the east between the harbour and an old area called Ra's Al Juffair. One of the first official zoning plans of the Juffair District was prepared in the course of the master plan of Manama from 1988, which included the basic road network in the form of Shabab Avenue. While the plan earmarked most of the reclaimed area for the development of mixed-use projects, it designated three public areas – a playground and two green areas. Service areas of more than 70,000 sq m, including lots for two schools, were established in different locations. Apart from the land use plan, the master plan included the future expansion of the road network by extending Awal Avenue starting at Al Fateh Highway up to Bani Otbah Avenue. Because of the rather slow urban development during the 80s and 90s, Awal Avenue was only partly built and just recently completed after the beginning

Fig. 19: The Juffair District in 1998.   *Fig. 20: The land reclamation in 2008.*

of the new millennium. Due to the expanding military base Awal Avenue could not be extended as far as the harbour in the south and was not connected to Juffair Avenue as previously planned. Hence, instead of becoming part of an outer ring road it turned into a one-way access road linking the northern and eastern expansion areas.

At the same time as the early infrastructure development during the 80s, one of Bahrain's most prestigious modern landmarks, the Al Fateh Mosque, was built opposite Al Qudaybiyah Palace on a reclaimed area of around 150,000 sq m. Beside the mosque, which was completed in 1987, is the Al Fateh Islamic Centre which contains the later established National Library and the Religious Institute. During the 90s several additional areas were reclaimed, adding up to around 303,000 sq m along the northern coast and to about 320,000 sq m in the east, where a small-scale hospital and several low-rise dwellings were developed. It is characteristic of land reclamation in Juffair that rather small areas were gradually reclaimed instead of a large-scale reclamation of one cohesive area. In 1998, all of Juffair's reclaimed areas came to a total of around 2.16 sq km, of which a lot area of around 200,700 sq m was occupied by more than 239 buildings including around 180 low-rise dwellings in the form of villas. While around 118 of these villas were built in the eastern part of the district, around 22 were built in the north and the remaining 40 in the central area. In addition to these villas, which mainly stand alone or in some cases within small compounds, around 46 multi-storey apartment buildings were built offering around 900 housing units. According to the building height permitted at the time, these buildings have up to ten floors but their average height was around six floors. Apart from residential buildings three hotels were built in the north and two schools and several commercial buildings in the south along Shabab Avenue. While 87 buildings, including 33 multi-storey apartment buildings, were built in the centre of the district, which is the district's oldest area, 118 low-rise dwellings and six medium-rise apartment buildings were built in the east of the district, which have mainly accommodated US soldiers, and 34 buildings were built in the newly reclaimed northern expansion area.

Over the next five years further land reclamation extended the district to the north by about 268,000 sq m and to the east by about 179,000 sq m. A large part of the eastern expansion area was reclaimed for the development of the new causeway to Muharraq, the

Fig. 21: Residential projects in the centre of Juffair.

construction of which began at the end of the 90s. Between 1998 and 2003 the developed lot area increased by more than 259,000 sq m to a total of about 460,000 sq m. In addition to around 210 low-rise dwellings in the eastern part of the district, about 104 multi-storey apartment buildings with up to eleven floors were built, of which around 48 were built in the centre, 41 in the north and 15 in the east. Due to the taller average height of the new multi-storey residential buildings of approximately eight floors, the average plot ratio of Juffair as a whole increased from 1.5 to 2.2 in spite of the rapid growth of low-rise dwellings in the east, which occupied almost 43% of the developed area. The new apartment buildings created more than 2,300 housing units and in addition to the newly built 210 low-rise dwellings the total number of residential units in Juffair increased more than three times from around 1,104 in 1998 to more than 4,700 in 2003. The main reason for this rapid growth was the increased presence of the US military in Bahrain since the declaration of the war against terror in 2001, causing an urgent need for accommodation for soldiers. In addition to US soldiers, many expatriates and their families relocated to Juffair because of its accessibility and proximity to Manama's business district and the airport in Muharraq. Consequently, Juffair has developed into a predominantly residential district with relatively few commercial developments with the exception of a few shops and restaurants which are mainly located in the south close to the old Juffair village and the US Navy base. Apart from the large number of residential developments four new hotels have been built, two in the north and two in the central district area, which mainly serve the needs of foreign residents with their bars and night clubs.

Between 2003 and 2008 the area of Juffair was expanded by more land reclamation from 2.6 sq km to more than 3 sq km. Most of the new area was reclaimed in the north between 2005 and 2007 – a large square area of around 92,000 sq m and two areas of 184,000 sq m and 79,000 sq m on the shoreline. This gradual process of reclamation has caused the current patchwork-like appearance and the strangely geometric form of the coastline. Today, Juffair's total lot area, which has been developed since the 70s, is around 1.9 sq km, of which around 39% is built. While the lot area in the north of the district is around 737,525 sq m, of which only about 18% is developed, the lot areas of the centre and east are 473,056 sq m and 690,102 sq m respectively, of which more than 52% are developed.

182

All in all, there are around 850 completed buildings in Juffair, of which 128 are in the north, 182 in the centre and 540 in the east. Of these buildings, 569 are low-rise dwellings with two to three floors, of which there are 44 in the centre, 35 in the north and about 490 in the east. Two-hundred and forty-one are multi-storey apartment buildings, of which 109 are in the centre, 83 in the north and 49 in the east. This comes to a total number of residential buildings of 810 which occupy more than 84% of the total developed lot area. Eleven buildings are hotels, of which eight are in the north and three are in the centre. Two buildings are private schools for Bahraini children, one of which is the Primary Religious Institute. Two are higher education institutions – the Bahrain Institute of Banking and Finance and the Bahrain Society of Engineers. There is one large sports facility, a community centre and a mosque in the oldest part of the district close to the old settlement in the south. Although most of the public facilities have been developed in the centre of the district, the only hospital, the Bahrain Specialist Hospital, is located on the eastern coast. The Civil Service Bureau is in the north close to the Al Fatah mosque and the Ministry of Health recently relocated to Juffair to a newly built 14-storey high-rise building in the centre of the district. Over the last five years several new commercial buildings have been built including two supermarkets – one in the centre toward the northern end of Shabab Avenue and one in the south close to the old settlement and the military base. In addition to several so-called cold stores – small shops providing basic consumer goods – there are several fast food restaurants along the southern end of Shabab Avenue close to the military base.

In total, around 261 buildings were built between 2003 and 2008, which means the development of more than 591 buildings since 1998. Around 83 of the buildings completed during the last five years have been multi-storey apartment buildings with up to 20 floors, the development of which was enabled due to a new zoning plan that permitted investors to build buildings with more than 10 floors. Thus, in comparison to the buildings constructed between 1998 and 2003 the average height of the developments of the last five years has increased from around eight floors to about eleven floors. In 2008 the tallest completed buildings were the twin towers of the Marriot Hotel with 20 floors, which were developed in the northern part of the district. Consequently, the average plot ratio of Juffair has increased from around 2.2 in 2003 to 2.6 in 2008, and if the large area of low-rise dwellings in the east of the Juffair is not included, the average plot ratio has increased from 3.0 to 3.7. While the average construction ratio of the multi-storey buildings constructed between 1998 and 2003 is about 410% of the lot area, the average construction ratio of the multi-storey buildings constructed between 2003 and 2008 is more than 490%. Due to the fact that about 89% of the total GFA in 2008 was for residential use, more than 3,900 housing units were added to the 4,700 dwellings that had already been built as of 2003. Consequently, around 8,600 dwellings units currently occupy a GFA of around 1.63 sq km, and when deductions are made for structure and circulation the average dwelling size is around 140 sq m. Based on the published occupancy rate of apartments in Juffair, which is around 81%, and the assumption that the average living

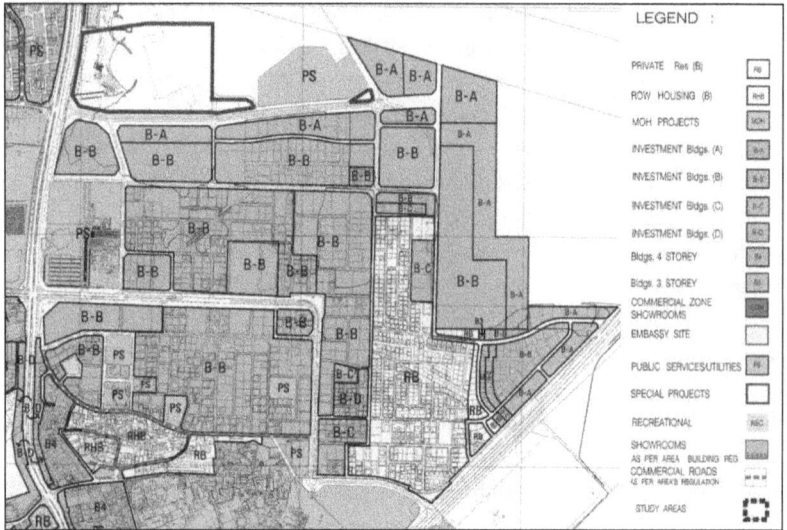

Fig. 22: The zoning plan of Juffair (2008).

space per person is around 50 sq m, about 24,000 people are currently living in Juffair, which means an urban density of more than 8,000 people living on one square kilometre and thus an increase in urban density of more than 45% since 2003.

In 2008 construction began on around 37 residential high-rise developments, which are due to be completed by 2010. While 21 projects are in the north of the district where the most reclamation work has been done in recent years, eight projects are in the centre and eight are in the east. The lots of all 37 projects will cover over 60,000 sq m and in comparison to high-rise developments built between 2003 and 2008, which have an average height of 11 floors, the average height of these recent high-rise developments is around 20 floors. Consequently, their average plot ratio of is 7.4, which is close to the permitted maximum construction ratio of 750% of the lot area. While most areas are restricted to this maximum construction ratio, a construction ratio of 1,200% of the lot area has been permitted in the case of the newly reclaimed areas on the northern coastline. Two projects that have already begun in these new areas are the 30-storey Fontana Towers and the 45-storey Sukoon Tower. Both projects are located along what will be an access road that will be developed in the following years along the northern shoreline. The Fontana Towers comprises four interconnected towers on a four-storey shared base, which will house a car park for about 600 cars and the roof area of which will be used as recreational space for residents[28]. While the Fontana Towers will offer around 400 apartments of various sizes, the Sukoon Tower will provide 592 residential units. The seven-storey base of the 45-storey residential tower will house recreational facilities in addition to a car park offering 758 parking spaces[29]. These projects are representative of a new trend in Juffair of luxury real estate projects sold as freehold properties. While the previous approach was to develop properties that were intended to be leased mainly to

184

companies, the US military or directly to guest workers, the new developments particularly target speculators from GCC countries, who would later trade or lease the apartments. As luxury residential developments all projects comprise certain services and integrated security or leisure facilities such as a 24 hour-guarded lobby, indoor and outdoor pools, health clubs and even cafes. In some cases there will be special facilities such as theatres, spas and mini-supermarkets.

Further examples of this trend in the north of Juffair are the 30-storey Fortune Plaza, the 28-storey Regent Plaza, the 23-storey Terraces of Juffair, the 17-storey DCR Residence and the 18-storey Enjaz Residence. In the centre of Juffair the 18-storey Mal Residence and the 20-storey Juffair Skyview tower are scheduled to be completed in 2010 in addition to several other developments. In the east the construction of a number of residential towers such as the 31-storey Orchid Plaza, the 26-storey Juffair Views and the 34-storey Dream Towers are being carried out near the causeway. The construction site of the 27-storey Crystal Heights tower is located close to the centre of the district. On the eastern shoreline close to the Bahrain Specialist Hospital the 55-storey 360° Tower is being built, which with a height of around 208 metres will be one of the tallest developments in Juffair as well as one of tallest residential towers in all of Bahrain. This new trend of developing high-rises of more than 50 floors is continued in the case of the recently announced project Nomas Towers, which will comprise two 52-storey and two 66-storey towers with heights of up to 239 metres. All in all, the project, which is designed by Atkins, will offer around 930 residential units including several town houses that will be integrated along the northern side of the multi-storey base. In addition to apartments one tower will integrate a five-star hotel, another tower will be occupied by office space and the eight-storey base will house a three-storey shopping mall in addition to a car park for 4,437 cars. The project's lot area of around 35,000 sq m is located in a newly reclaimed area on the northern shoreline and its GFA is expected to be around 400,000 sq m. After completion by 2012 the project is expected to create around 2,500 jobs, mainly within the hospitality, retail and service sector. This expectation is mainly based on plans to establish a one-kilometre long public promenade alongside shops and restaurants in addition to the development of a beach and a marina[30].

All 21 high-rise projects currently under construction will create more than 3,200 apartments. Due to their average height of 20 floors, these 21 buildings will cause the overall plot ratio of the Juffair District to increase from currently 2.6 to over 3.2 by 2010. Furthermore, they will increase Juffair's urban density from around 80 residents per hectare to more than 105 residents per hectare if we assume that the average living space of one person will remain at around 50 sq m. This would mean a population increase from currently about 24,000 people to over 32,000, who would reside in around 11,800 dwellings. Considering that the buildings will be completed by 2010, the total number of housing units in Juffair will increase by more than 37% within only two to three years of development.

According to current plans, more land will be reclaimed in the amount of 937,000 sq m along the northern and eastern shoreline of Juffair expanding the overall district area to about 4 sq km. While today more than 946,000 sq m is already built or under construction, there is a total developable area, including future reclaimed areas, of around 1.9 sq km. About 55% of this area has been zoned for the development of Investment Buildings Type B-B, which have a maximum construction ratio of 750%. Around 21% will be developed with Investment Buildings Type B-A, which have a maximum construction ratio of 1,200%. Due to this permitted high density all Type B-A lots are located outside of the Juffair District's centre along the northern and eastern shoreline where a new multi-lane access road has been planned to connect Al Fateh Highway with Sheikh Khalifa Bin Salman Causeway. A special project area has also been zoned on the north-western shoreline, where more than 224,000 sq m of new land will be reclaimed in front of the Al Fateh Mosque. In addition, in the east of the district there are a number of relatively small lots of around 66,000 sq m in total intended for the development of Investment Buildings Type B-C, about 4,500 sq m for the development of Investment Buildings Type B-D and 9,262 sq m for the development of Investment Buildings Type B-3. In addition to the expansion of Juffair's developable lot area, an area of around 90,300 sq m will be reclaimed and reserved for the development of public services as well as a large seaside park.

The maximum GFA permitted is around 14.7 sq km. Currently about 90% of this GFA is residential developments, and assuming that this does not change in the future this means approximately 10 sq km of new living space will be available in Juffair. Consequently, if we assume that the average apartment size remains about 150 sq m, the number of dwellings will increase from around 12,000 to more than 78,000 units. And according to the current estimate of an average living space of 50 sq m per person there would be a possible population increase to around 230,000 future inhabitants, which is more than 120% of Manama's current total population of around 190,000 people. Thus, the urban density of the Juffair District would grow from currently 80 residents per hectare to around 575 residents per hectare. The average plot ratio would be more than doubled from about 3.2 to around 6.7. However, considering that around 600,000 sq m of GFA was developed in a time frame of two to three years, it would take more than 70 years to develop the maximum GFA, but this hypothetical calculation reflects the very minor restrictions imposed by the current zoning plan on the possible built density of the district in the future.

Because the US Navy Base was one of the main factors that led to the real estate boom in Juffair between 2001 and 2008, landlords are dependent on tenants from the US Navy and its employees. After a minor incident where there was an attempt to attack US soldiers a new regulation was implemented in 2005 stating that any building can only have a maximum of 25% of its apartments occupied by US Navy personnel due to security concerns. This major change led to an increasing demand for new developments and thus a lower occupancy rate, which is currently about 81%. There is also a high rate of tenancy turnover caused by the flexibility of the US Navy's lease contracts, which include the

Legend:
- more than 29 floors
- 24 to 28 floors
- 18 to 23 floors
- 13 to 17 floors
- 8 to 12 floors
- 4 to 7 floors
- 1 to 3 floors

0   150   300 m

Fig. 23: The building heights of current and future projects.

permission to give one-day notice to landlords. Thus, despite higher rents costs have been increasing due to high letting agency fees. Furthermore, there is competition between relatively new buildings offering luxury residential units because of the demand for such apartments from the US Navy Base. Consequently, the average rental prices have increased over the past several years to around BD700 per month for a two-bedroom apartment, to about BD900 for a three-bedroom apartment and to more than BD1,500 for a four-bedroom apartment. There has also been an increase in the development of serviced apartments with daily rental rates ranging from BD37 to BD96[31]. The result of these increasing rental prices has been the relocation of lower-income tenants to more affordable districts. Today, only two apartment buildings, one in the north of Juffair and one in the centre, house labour working in the lower end of the service sector. Hence, most low-income workers live in the centre of Manama and travel by bus to Juffair.

In addition to being the home of the US Navy, Juffair has become an attractive residential area for the staff of companies working in the business centres of the main island or the industrial areas in Sitra and the south of Al Muharraq. This has made Juffair Bahrain's most intensely concentrated area of expatriate residents with middle to high incomes, who enjoy the proximity to Bahrain's business districts and leisure facilities such as hotels and restaurants. In recent years, the biggest challenge of landlords and investors in Juffair has been the relocation of tenants to the Seef District and Sanabis due to their proximity to the business centre in northern Manama. This particularly concerns employees of the financial and telecommunications sectors. Moreover, real estate projects in new and less populated areas such as Amwaj Islands are further increasing the real estate competition in Bahrain. Thus, many of the former tenants of properties in Juffair such as Gulf Air staff

187

Fig. 24: The future structure of Juffair.

have relocated to other districts due to better facilities or more convenient accessibility. Another major threat to the future of real estate in Juffair is that it is still dependent on tenants from the US Navy, which has already started to relocate many of its employees and their dependents and is even considering a complete move to other GCC countries offering lower rental costs.

A further threat to Juffair in terms of how attractive it will be in the future is a growing lack of infrastructure, service facilities and public areas. While the retail strip in the south along Shabab Avenue has been growing continuously in recent years due to new restaurants, coffee shops and retail outlets, the development of public squares, parks and pedestrian zones has been completely neglected. One small playground is currently the only public green area with a size of just around 4,200 sq m located in the oldest part of the district where it was established by the district's very first zoning plan from the 80s. Furthermore, walking in the district is difficult due to partly unpaved walkways and inefficient privatised waste collection, which has led to cluttered sidewalks and thus a growing population of stray dogs and cats. Another reason for the decreasing quality of the built environment is the short distances of sometimes less than five metres between multi-storey buildings causing shaded and badly ventilated areas caused by the fact that the initial physical planning of the district was done for low- to medium-rise buildings. The change in the former zoning plan involving the permission to construct high-rise buildings has led to an insufficiency in infrastructural capacity, particularly regarding the road network, which was designed and built partly on the basis of the old plans. Subsequently, most of the roads in Juffair have just two lanes and are around six metres wide. Due to the allocation of lots on the basis of the old plans, future roads will have to be developed with these outdated dimensions, which is leading to more and more traffic congestion. Furthermore, there is a growing lack of sufficient parking space because of a fewer number of empty lots that can be used as parking sites and the fact that investors are not obliged to integrate more than just one parking space for each apartment. While in the past buildings were often built on pillars and parking areas were mainly located on parts of the ground floors and in areas surrounding the buildings, the recent high-rise projects have begun to integrate multi-storey car parks with up to six floors. This is leading to an even more pedestrian-unfriendly urban environment and a reduced integration of different uses in most areas. Thus, in spite of the development trend of mixing uses by integrating shops on the ground floor in new projects such as the Nomas Towers, large parts of Juffair do not integrate commercial or social facilities. Furthermore, the lack of public transport has added to the traffic inside the district itself, particularly in the areas of the two main supermarkets and along the commercial strip on Shabab Avenue in the south.

Another side effect of the increasing construction of high-rise developments is a growing conflict due to lawsuits between owners of low-rise villas, which were built during the 80s and 90s, and the developers of multi-storey buildings, which interfere in the privacy of the traditional Arab family. This is particularly the case in the south-west of the district close to the old Juffair village, where a small number of Bahraini residents had settled before the recent development commenced. In contrast to other parts of Juffair, large areas of public services such as a large sports facility, a community club, a mosque and educational institutions have been established in this old part of the district. While the planning of the 80s obviously acknowledged the future needs of inhabitants, the new district areas have developed with rather little response to the increasing need for social services and facilities. This particularly concerns a lack of schools for the children of expatriates, who are forced to travel long distances to Isa Town where most of the English-speaking schools in Bahrain are located. All in all, the current speculation-driven development has led to a growing shortage of technical and social infrastructure in addition to an increasing lack of public areas. Although an announcement has been made that a seaside park will be built in the north-west of Juffair, most areas in the centre of the district will be occupied by high-rise buildings with no integration of any open spaces. Thus, as in the case of the commercial area in the south, the majority of the inhabitants would live too far away to be able to walk to parks and leisure areas.

This densification and lack of integration of different uses and service areas will increase traffic, particularly at the entrance and exit points of the district. Until today the four-lane Awal Avenue has remained the main entrance and exit of the district and in spite of the moderate density of less than 80 people currently living on one hectare, traffic congestion has rapidly increased on this avenue, particularly during the rush hours. Although a new main access road has been proposed along the northern shoreline, the current road network will face serious capacity problems if the current speed of construction continues. This rapid growth has also had an impact on the electricity supply, the capacity of which has been exceeded in certain areas. Today, around 16 buildings currently under construction

Fig. 25: High-rise projects in Juffair.   *Fig. 26: Densely built apartment blocks.*

will have no access to the electricity network until a new substation has been built and because this future substation will be located on land that has not yet been reclaimed, it will be over a year until these developments will be supplied[32]. This shortage of supply is due to an evident lack of impact studies and planning in addition to an absence of restrictions and their implementation, for example, the responsibility of landowners to participate in the development of sufficient infrastructure. Furthermore, a lack of urban design has led to a built environment consisting of isolated high-rise projects with almost no response to adjacent buildings. For example, while in the past the colour of buildings was restricted to beige, recent developments do not have to follow guidelines of this kind any more. Consequently, the small number of restrictions on architecture has led to the widespread development of concrete buildings of various forms with large glass façades. This climate-inappropriate architecture has led to a high dependency on air conditioning and thus to an increasing waste of energy. In future however the exponentially increasing need for electricity will be slightly reduced by a district cooling plant, which is currently under development by Tabreed[33].

While the future development of the Juffair District is on the one hand very much dependent on external factors such as the possible relocation of the US Navy base or a decrease in investments due to the international financial crisis, on the other, the lack of urban design and planning has led to major problems with regard to the infrastructural development and the quality of the built environment. Consequently, Juffair's future as one of the most exclusive residential areas of Bahrain remains unclear and will very much depend on future planning and the implementation of clear guidelines that tackle the increasing problems.

4.3 The Consequences of Post-oil Urbanism in Manama

4.3.1 The Development of Additional Districts

In addition to large-scale developments in the Seafront District, Seef District and Juffair District, a large number of real estate projects are being carried out in other districts of Manama such as Al Hoora and Al Mahooz. The two main criteria for attracting investment

190

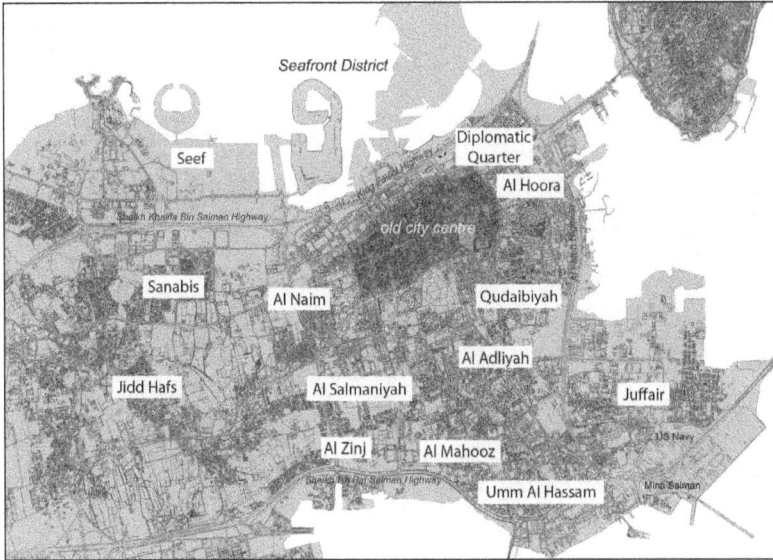

Fig. 27: The main districts of Manama.

in the construction of commercial or residential real estate have been accessibility by car and proximity to the coast. Due to its location close to the business centre of Manama, the district Al Hoora was one of the first areas to attract real estate developers in Bahrain. While the eastern part of the district covering around 620,000 sq m was already reclaimed in the 60s, part of it was developed during the 70s and 80s. The first developments mainly lined up along Exhibition Avenue becoming one of Manama's longest shopping streets where in addition to several restaurants, many hotels with night clubs began to attract tourists, particularly from other GCC countries. Due to its role as one of Bahrain's first tourist destinations, Al Hoora has become attractive for the development of a large number of rental properties, particularly serviced apartments leased to GCC nationals on the weekends. Several freehold property projects were also built in Al Hoora when a new zoning plan permitted the development of Investment Buildings (type B-B). Thus, a total of around 47 residential buildings offering around 1,079 apartments were completed in 2007 in addition to several hotel developments[34]. Similarly to areas such as Juffair, lots in Al Hoora were sold to a large number of different investors in previous decades before the recent construction boom. While the first buildings were constructed with average heights of around six floors, recent developments reach heights of up to 17 floors, leading to an average building height in Al Hoora of around 10 floors. Due to a population density of more than 100 residents per hectare and the large number of tourists on weekends, the district has become more and more congested and currently suffers from an increasing lack of parking space.

In addition to Al Hoora, which has benefited mainly from its prime location in the east of Manama, the Al Mahooz district in the south of Manama has become attractive for real

191

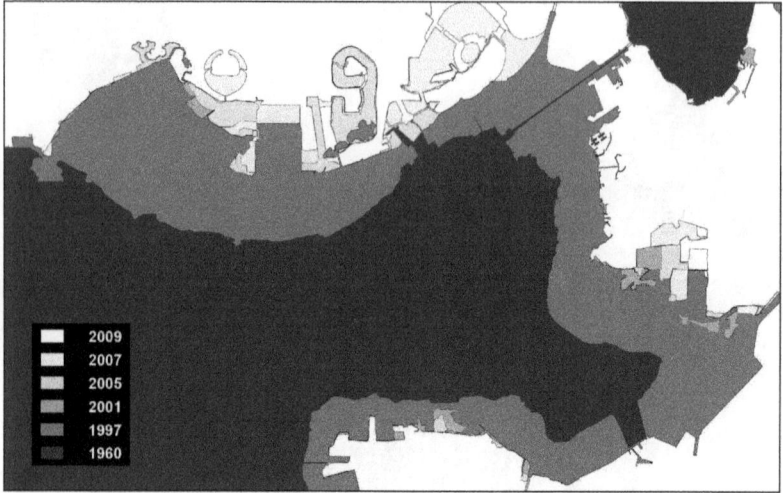

Fig. 28: The land reclamation from 1960 to 2009.

estate developers because of its proximity to the causeway to Saudi Arabia. The recent construction boom in the last four years has led to an increase in the district's overall GFA by more than 22.6%, which has been mainly caused by the development of around 11 luxury residential buildings offering around 600 apartments[35]. However, like Al Hoora Al Mahooz is currently considered to be rather outdated with a large number of buildings aged as much as 15 years and in comparison to newly developed areas such as the Seef District it cannot provide the same level of accessibility by car or integrated leisure facilities. Thus, the increasing trend of focusing on new development areas has caused the neglect of old areas, which have consequently lost their former status as exclusive residential districts. This competition between areas can be observed even within single districts such as Juffair where its centre has become less attractive than the newly reclaimed areas along the waterfront. Hence, the urban structure has become dominated by the interest of speculators to develop new areas instead of modernising or reconstructing old parts of Manama.

4.3.2 The Resulting Urban Structure

The general development trend of focusing on new areas has led to the reclamation of about 1.7 sq km in the Seafront District, about 1.4 sq km in the Seef District and about 800,000 sq m in Juffair since 1998. Thus, the overall urban area of Manama has expanded by more than 15% from around 27 sq km to almost 31 sq km. A further consequence of the recent construction boom has been an exponential increase in land prices, which has again contributed to less investment in old parts of Manama due to low profit expectations. All in all, there has been a clear shift and transformation of the urban development towards more and more densely built areas along the coast with a still high percentage of around

Fig. 29: Manama in 1951 and 2008.

40% of unbuilt land in the old urban areas of Manama[36].

Most of the recent developments have been residential high-rise projects in addition to several commercial towers and shopping centres. While in the north the Seafront District and the Seef District have turned into mixed-use urban areas with large commercial projects and a growing number of residential developments, some areas such as Juffair are mainly residential, with residential projects occupying up to 90% of the GFA. In the case of the Seafront District, the area was almost exclusively occupied by administrative buildings, hotels and banks until recently when the first residential towers were completed. In future, a large number of residential areas will be created as part of the three master-planned developments along the coast, where the GFA is occupied to a large extent by residential projects. While the Reef Island development consists almost completely of residential projects, around 33% of the future GFA of the Bahrain Financial Harbour development will consist of residential units while residential units will account for over 60% of the Bahrain Bay development. In comparison to the Seafront District, which is turning into a mixed-use district due to current residential developments, the Seef District, which was previously designed to be mainly a residential area, has become more and more attractive for the settlement and relocation of companies and shopping centres. Since the completion of the City Centre as Bahrain's fifth and largest shopping centre, around 25% of the developed lot area of the Seef District is currently occupied by shopping centres and 35% by office towers and government institutions.

Thus, the previous land use division of residential and commercial areas has recently begun to change in the north of Manama while it remains unchanged in the case of districts that are almost completely residential such as Juffair. The mix of land uses has been less restricted or controlled due to a general zoning plan that permits investors to develop most areas according to their choice of residential, commercial or administrative projects. The

193

consequence has been a concentration of specific uses or certain kinds of developments in areas where given circumstances attract one kind of utilisation. Examples are the cluster of shopping centres in the Seef District due to the proximity of the causeway to Saudi Arabia or the vast amount of residential projects in Juffair due to the location of the US Navy Base. Thus, in spite of a certain tendency to integrate different land uses in a district on the macro-level, individual parts of districts have remained rather mono-functional with a lack of services and commercial areas within walkable distances. However, in contrast to the publicly zoned urban expansion areas the three master-planned developments have acknowledged the need for integrated shops and restaurants and have started to develop public promenades and even pedestrian zones as in the case of Bahrain Bay but despite the growing integration of commercial and leisure facilities in certain areas, there has been a big lack of development of social facilities and service areas because of increasing land prices. Although a few private schools and small hospitals are expected to settle in the newly developed areas, most residents will still depend on educational and healthcare facilities located in older parts of Manama or satellite towns such as Isa Town.

Apart from a lack of lots for public facilities and social services, very few areas have been reserved for the development of public areas such as parks. While the Seafront District has around 196,000 sq m of green area in the form of the large seaside park in addition to around 200,000 sq m of future green areas, parks and public squares within the Bahrain Financial Harbour, Bahrain Bay and Reef Island, there is only around 14,800 sq m of public areas in the Seef District and just 4,200 sq m in Juffair. Furthermore, in contrast to the integration of promenades within the three master-planned developments along the coast there has been no plan to develop public promenades along the shorelines in Juffair and Seef. This is mainly because of on-going land reclamation and various landowners who prefer to sell seaside lots to private investors. In addition to the lack of development of public areas, infrastructural development has not kept pace with the rapid construction particularly with regard to urban areas where the zoning plan has recently been changed and now permits the development of much higher densities. Hence, the previous infrastructural planning and development, which was based on former zoning plans, has become outdated and insufficient. Furthermore, the highways along the coast, which link all the current development areas, have become more and more congested because of an exponential increase in the use of cars and growing urban densities in areas along the coast. Also, in spite of the development of new access roads around Bahrain Bay and the proposal to construct a multi-lane road along the northern shoreline of Juffair, there has been neither the development of new additional lanes along the main highways nor the development of public transport such as trams or metros. Moreover, the rapid decentralised urban development and lack of restrictions have led to shortages of parking space and electric supply. While in the future several district cooling plants will be built, energy waste has recently increased particularly because of a lack of regulation obliging investors to make their developments more energy efficient. With regard to the lack of parking space, people are getting around the issue by using unbuilt lots or pedestrian

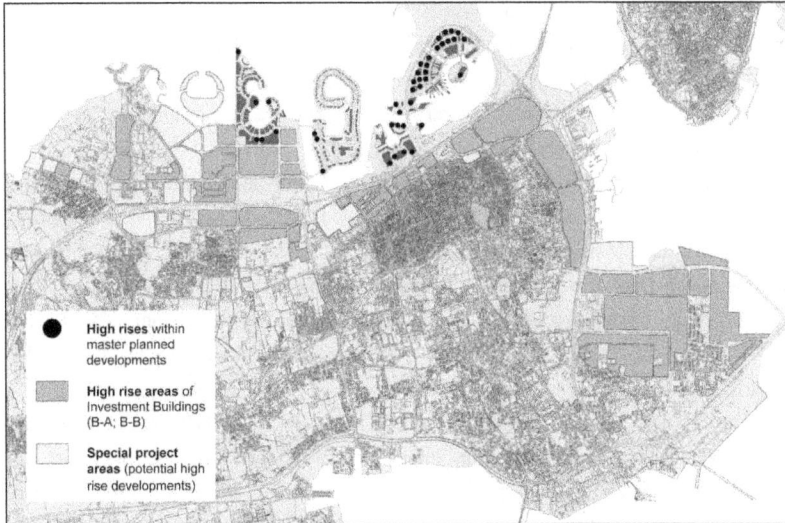

Fig. 30: Development areas for high-rise projects.

sidewalks. In order to solve this problem new developments have begun to integrate multi-storey car parks leading to a growing built density and a rather pedestrian-unfriendly built environment.

Based on the current zoning plan, most of the developable areas along the coast can be built with high-rise projects with construction ratios of 750% and in some cases even 1,200% of the lot area. Consequently, more and more commercial and residential high-rises have been developed. While in the Seafront District including the Diplomatic Quarter around 31 towers have been built with an average height of about 15 floors, there has been the development of around 73 multi-storey buildings with an average height of around nine floors in the Seef District and 241 high-rise buildings with an average height of about 10 floors in Juffair, of which more than 40% have been developed since 2003. At the beginning of the construction boom the average height of multi-storey buildings was about ten floors, which is almost half the average height of the current projects of around twenty floors. Hence, the average plot ratio in the Seafront District is currently increasing from approximately 1.7 to more than 3.0. It is hugely affected by the Bahrain Bay project which has an average plot ratio of 7.25. Furthermore, the plot ratio of the Seef District is growing from around 3.2 to 3.9 and from 2.6 to 3.2 in the Juffair District, where currently over 37 high-rise projects are under construction. Approximately 92 high-rise developments are currently being built in the three main urban expansion areas, of which around 77 are residential towers. Consequently, more than 75% of the currently developed GFA of around 4.9 sq km will be occupied by residential use, leading to an estimated 15,000 new residential units within the next five years and, thus, an urban density increase from currently less than one resident living on one hectare to approximately 102 per hectare in the Seafront District, from around 18 to an expected 21 per hectare in the Seef

195

District and from about 80 to more than 105 per hectare in Juffair. Although this estimated population increase is based on the assumption that the future average living space will remain at about 50 sq m per person and the occupancy rate of currently around 80% will not change profoundly in the next several years, the urban density will increase by around 41% over the next five years.

In comparison to the old districts of Manama that have an average urban density of around 53 residents per hectare, the newly developed areas will become the most densely populated areas in Manama and in all of Bahrain. The Seafront District, the Seef District and the Juffair District together have around 4.76 sq km of area that is still developable and according to the current zoning plan the construction of a maximum GFA of about 30 sq km would be permitted. According to the current development, of the developable GFA in the Seafront District about 60% is occupied by residential projects, about 50% in the Seef District and around 90% in Juffair. Thus, more than 110,000 dwellings with an estimated average size of 150 sq m will be created and the average urban density will increase to around 200 residents per hectare in the Seafront District, about 190 per hectare in the Seef District and around 575 per hectare in Juffair. All in all, a maximum of around 426,000 residents would be accommodated in all three districts, which would cover a total area of around 14 sq km after the land reclamation is completed. This estimate of the possible future population of the three districts is equivalent to more than 210% of Manama's current total population of around 200,000 people. Thus, the current zoning plans permit the development of a rather high urban density with around 300 residents estimated to be living on one hectare.

Most of the recently developed real estate projects consist of luxury apartments with average rental rates of about BD 780 per month for a two-bedroom apartment. Due to these high rents most of the new areas have been populated by expatriates with middle and high incomes. Thus, in addition to the increasing social segregation of income groups, there has been an on-going segregation between the local population and foreign guest workers. In order to attract high-income groups there has been an increasing number of developments offering sea views and integrated leisure facilities such as private beaches, swimming pools and marinas. Furthermore, most luxurious residential developments are designed to be gated communities, for example, the villas on Reef Island or the various compounds in Juffair. In the case of residential high-rise buildings, a 24 hour-operating lobby with guards guarantees the safety and exclusivity of the residences. In addition to the trend of integrating leisure facilities there has been a growing competition to outshine rival developments by designing landmark architecture and having heights beyond 50 floors. Since the completion of the 53-storey Harbour Towers, the 50-storey Bahrain World Trade Centre and the 47-storey Al Moyyad Tower, which are currently Bahrain's tallest buildings, the construction of several new projects with heights of more than 200 metres has begun in all three districts. In contrast to previous landmark projects, which are used as commercial buildings, most of the new developments are exclusive residential towers. While most of the currently developed area in Manama is occupied by multi-storey developments, a few low-rise projects such as the villas on Reef Island and several compounds in the east of Juffair previously mentioned are being constructed. Based on the current zoning and subdivision, most developers are trying to use their lots as profitably as possible, which has led to the densification of most areas.

In contrast to the construction of competing high-rise projects and a more and more densely built urban environment in Seef, Juffair, Al Hoora and the Diplomatic Quarter, clear guidelines have been implemented in the form of separate master plans for the Bahrain Financial Harbour, Reef Island and Bahrain Bay. In addition to specifying the maximum developable GFA, the master plans of these projects, which were designed by individual master developers, include clear restrictions on building height and the use of buildings. Consequently, these master-planned developments have an intentional urban design prepared by mainly worldwide known planners such as SOM or Atkins, which enhances the quality of the built environment. However, despite the detailed planning of these island-developments, there has been a lack of comprehensive planning with regard to any response to adjacent developments and the former urban area. Thus, the waterfront of Manama is composed of three individually designed projects with clearly defined borders and thus they do not contribute to a cohesive and integrated overall urban area. In fact, they are developing into rather exclusive residential areas with integrated commercial developments such as shopping malls or office towers at their entrances close to the coast. Furthermore, all projects include leisure facilities and hotel developments due to an attempt by each developer to participate in the growing tourism industry.

All in all, the recent construction boom has led to the enormous urban growth of Manama, where around 20% of Bahrain's development area as a whole is located. Due to ongoing land reclamation, Manama's urban area has continuously expanded over the past decades. Because of their accessibility and proximity to the shoreline, the coastal areas of Manama became more and more attractive for real estate developers and thus in spite of a large number of unbuilt areas and obsolete buildings in the centre of Manama, the development trend has focused on the coast, leading to a large number of construction sites between the coast and the outer ring road. The density in the old districts however has remained rather consistent due to rising land prices and a lack of interest from investors in developing projects in outdated and overpopulated areas because of lower profit expectations. This competitive aspect of the private sector, which has taken over the main direction of the current urban development, is leading to a more and more densely built frame around Manama and the focus on newly reclaimed areas. Hence, the view of Manama's skyline from the centre to the coast is of increasing building heights from around three floors in the centre to an average of around six to eight floors in the transition zone which was developed between the 60s and 80s and finally to an average of over eleven floors along the coast where a number of landmarks reach heights of up to 50 floors. Furthermore, newly added areas in Juffair and the Seafront District are currently being developed by projects with an average of more than 20 floors. This transformation from the previous development trend during the oil urbanisation wherein mainly low-rise residential areas were built more inland to the current densification along coastal areas due to the construction of high-rises is one of the two main characteristics of the current urbanisation in Bahrain. The second is the development of large-scale master-planned projects in the form of man-made islands shaping a new waterfront.

---

1   Hamouche 2008, p. 207.
2   Congregation hall for Shia rituals.
3   http://archive.gulfnews.com/gnfocus/gibtm_exhibition2007/more_stories/10109162.html, 25.03.2009.
4   Gulf Construction, 2006, p. 69.
5   Bahrain International Insurance Centre
6   http://www.bfharbour.com/html/faq/faq_development.htm, 31.03.2009.
7   Gulf Construction, 2008, p. 109.
8   http://www.bfharbour.com/html/faq/faq_development.htm, 31.03.2009.
9   Gulf Construction, 2009, p. 82.
10 http://www.bahrainbay.com/resources/pdf/BB_Zone1.pdf, 01.04.2009.
11 http://www.bahrainbay.com/resources/pdf/BB_Zone2.pdf, 01.04.2009.
12 http://www.bahrainbay.com/resources/pdf/BB_Zone3.pdf, 01.04.2009.
13 http://www.bahrainbay.com/resources/pdf/BB_Zone4.pdf, 01.04.2009.
14 Gulf Construction, 2009, p. 98.
15 http://www.bahrainbay.com/resources/pdf/BB_Zone5.pdf, 01.04.2009.
16 http://www.bahrainbay.com/resources/pdf/BB_Zone6.pdf, 01.04.2009.
17 http://www.bahrainbay.com/resources/pdf/BB_Zone7.pdf, 01.04.2009.
18 http://www.abrajallulu.com/www/index.php, 22.03.2008.
19 http://web.asteco.com/resources/pdf/Q4/200811_astqrt008_q3_bahrain_qatar.pdf, 16.04.2009.
20 http://www.iriswll.com/bahrain-real-estate projects.php, 16.04.2009.
21 http://www.iriswll.com/bahrain-real-estate projects.php, 16.04.2009.
22 http://www.watergardencity.com, 05.05.2009.
23 http://www.constructionweekonline.com/article-4391-bahrain_construction_looking_to_up-town/, 05.05.2009.
24 http://www.tameer.com/qmr/luxuries-residential.pdf, 17.04.2009.
25 This is based on the assumption that one person has an average living space of around 50 sq m according to official estimates for current developments such as Bahrain Bay.
26 McPolin, 03.12.2008.
27 http://www.tameer.com/qmr/luxuries-residential.pdf, 17.04.2009.
28 http://www.gulfweekly.com/article.asp?Sn=6446&Article=21894, 22.04.2009.
29 http://www.sukoontower.com/SukoonTower-Jan16.pdf, 22.04.2009.
30 http://www.e-architect.co.uk/bahrain/nomas_towers.htm, 27.04.2009.
31 http://www.tameer.com/qmr/luxuries-residential.pdf, 17.04.2009.
32 McPolin, 12.04.2009.
33 http://www.tabreed.org/MediaCenter.aspx?NewsType=News&ID=34, 26.04.2009.
34 http://www.tameer.com/qmr/luxuries-residential.pdf, 17.04.2009.
35 http://www.tameer.com/qmr/luxuries-residential.pdf, 17.04.2009.
36 Nabi, 04.12.2008.

5 The Future Development of Post-oil Urbanism in Bahrain

5.1 Negative Consequences of Recent Developments

5.1.1 Decreasing Urban Qualities

Due to the enormous speed of the current urbanisation, which led to a population growth from around just 391,019 inhabitants in 1998 to 1,046,814 in 2007, the number of vehicles on Bahrain's roads has exponentially increased. While in 1998 the number of vehicles was 198,617, this increased to 293,801 in 2004 and to 359,124 in 2007. Furthermore, it is expected that around 443,576 vehicles will be on Bahrain's streets by the end of 2010[1]. Apart from the growing number of cars, the insufficient development of new roads and a lack of action to reduce car dependency have led to a more and more overcrowded road network. Although Bahrain's road network was continuously expanded from around 3,164 km in 2000 to almost 3,500 km in 2008 including the construction of several flyovers, it has not entirely met the exponentially growing need[2]. Moreover, public transport in Bahrain, which began in 2003 when the Dubai-based private bus company CARS began to provide bus services, has not significantly changed the dependency on cars because the buses are mainly used by low-income groups or students who anyway cannot afford to own a car[3]. Another reason for the high dependency on cars is that current developments in new urban areas tend not to integrate all the services and facilities needed. Thus, most shops are mainly located in large-scale shopping malls and in commercial areas along the highways and main access roads. Furthermore, the lack of integration of schools and other social services within new urban areas forces residents to drive many kilometres to bring their children to school. Consequently, the outdated concept of a car-based and functionally divided urban structure, which is sustained by the current zoning plans, remains one of the main reasons for the increasingly insufficient infrastructure.

Another contributing factor with regard to the lack of infrastructure is the scale and speed of the current urbanisation, which is mainly driven by private investors and their speculative interests. While the buildings are built by the private sector, most infrastructural supply is developed by the ministries, who have not kept up with the exponential demand. Consequently, there have been shortages in roads, water and electricity in densely built areas. This densification of certain districts was made possible due to a change in zoning in recent years which was made under the pressure of increasing speculation. These recently implemented plans have permitted plot ratios of up to 12.0, particularly in the case of the new expansion areas in Manama. Today, an average plot ratio of more than 6.0 is permitted along the shoreline of Manama and therefore a maximum urban density of more than 300 residents living on one hectare. This increasingly dense urban environment, which is in clear contrast to the urban sprawl of the period of early oil urbanisation, has created new development opportunities as well as many challenges. While in the past a large number of areas were developed with suburban settlements, the current land prices are preventing the spread of low-rise typologies in certain areas. Thus, previously suburban areas have turned into the construction sites of multi-storey buildings with an average height of more than 10 floors replacing the old low-rise built environment. Although there have been conflicts between developers of multi-storey buildings and owners of adjacent villas, the new typology has taken over large areas. While the high built density of these areas provides the opportunity to integrate different uses as well as future public transport systems, the lack of restrictions on the private investor has led to a rather uncoordinated development without the integration of sufficient public services and infrastructure. This has become a particular problem in areas where the infrastructure has already been built according to previous zoning plans with an expectation of much lower urban densities.

In addition to the insufficient technical infrastructure such as utilities and roads the lack of parking sites has become a serious problem in many areas due to a lack of building regulations stipulating the provision of an adequate number of parking spaces for developments. This has led to a widespread use of unbuilt lots as parking areas. Although this informal way of parking is still working in many areas, it must be considered as a very temporary solution due to the on-going construction. Hence, in order to meet the growing demand more and more car parks are being constructed on the first floors of high-rises or as separate buildings. These massive car parks with heights of more than six floors are contributing significantly to the growing density of the built environment and the result has been more and more shaded and badly ventilated urban areas.

A lack of development of parks or other public areas such as seaside promenades or public beaches has been a further factor in the decreasing urban quality. While the urban expansion areas in Manama or Muharraq follow a general zoning plan without extensive regulations, most master-planned projects are carried out on the basis of detailed plans, which include guidelines concerning both buildings and public areas. Thus, for instance, there are 22 kilometres of public beachfront in the project Diyar Al Muharraq and around 5 kilometres of seaside promenades in the three developments in the north of Manama as well as many leisure facilities and retail districts, making these developments very exclusive areas. However, these projects seem to be rather detached from the overall urban structure due to their isolated planning and development. In addition to the master-planned developments, which respond very little to adjacent urban areas, a lack of cohesiveness and integration has been displayed by single high-rise developments in certain areas. In the cases of districts such as Seef or Juffair in Manama each developer has focused on the fast construction of his own project without responding to neighbouring buildings. The result has been a rather scattered urban landscape that lacks open spaces. Consequently, most surrounding areas have become rather unsuitable for pedestrians and so residents are forced to leave their districts to find leisure areas by car. Thus, it is not surprising that the biggest and almost exclusive public meeting points are shopping malls, where air-conditioned halls provide visitors with leisure space including cafes, restaurants, cinemas and playgrounds. Although this commercial alternative of a public meeting point is favoured during the hot summer months, it cannot replace a growing need for outdoor public areas close to residences in the form of parks or seaside promenades, the development of which has been neglected due to high land prices. What is significant in this regard is the current amount of coastline that is accessible to the public – it is remarkably small at less than 3% of Bahrain's total coastline due to the high degree of private ownership of coastal land. Furthermore, the maintenance of the few public beaches and seaside parks that do exist has been neglected in contrast to the private beaches and parks of hotels and private residential developments. Thus, while high-income groups enjoy a high standard of gated parks and beaches, a large part of the population has been excluded from high-quality surroundings and leisure facilities.

5.1.2 Endangered Balance of Society and Environment
Due to increasing land prices and a focus on the development of luxury real estate, rental prices have grown rapidly in recent years, which has led to growing social segregation in Bahrain. While in the past social segregation mainly took the form of a clear separation between exclusive suburban areas and the densely populated city centres, the new urbanisation has made social segregation more complex due to the construction of master-planned developments and high-rise clusters. Although there has been an attempt to integrate social housing programmes in projects such as Northern New Town and Diyar Al Muharraq, most master-planned developments will remain exclusively for residents with high incomes. The priority accorded to these exclusive developments has led to major investment in the expansion of the road network such as the two new highways leading to the projects Al Areen and Durrat Al Bahrain. This preferential treatment with regard to infrastructural supply has contributed to an increasingly widening development gap between old districts and new urban areas. In the case of Manama, the old centre, where dwellings are mostly rented by low-income groups, has become more and more neglected while the new urban expansion areas along the shorelines have received more attention and development. Apart from the lack of infrastructure, the liveability of older areas has decreased due to worse ventilation and traffic congestion caused by the expanding high-rise agglomerations along the coasts.

There has also been a growing competition to attract residents with high incomes between new development areas, the major advantages of which are good accessibility and integrated services and leisure facilities. Consequently, the Seef District has recently become one of Bahrain's most attractive locations as well as a number of projects that are part of master-planned developments such as Amwaj Islands in the north-west of Muharraq. While the Seef District has become particularly attractive for foreign employees working in the central business district of Manama or in Saudi Arabia, the district Al Hoora has remained dependent on visitors and residents from GCC countries. The Juffair District in turn relies on a high occupancy rate of residential buildings by the US Navy. The recent relocation of residents from Juffair to the Seef District illustrates the current trend of competing districts that leads to certain areas having decreasing occupancy rates. Although today around 80% of all dwellings within urban expansion areas are rented, it is expected that upon the completion of more developments in the near future the average occupancy rate will be much smaller[4]. Consequently, the rental prices of many old buildings built more than 10 years ago are expected to drop, which in some cases will lead to tenants with lower incomes moving away. In other cases, a large number of older buildings will be replaced if the location of the site is still attractive for real estate, particularly if there is sufficient accessibility.

Due to the increasingly speculation-driven urban development and the consequent increase in land prices, many Bahraini families cannot afford to buy lots and rely on public housing programmes. Today, more than 55,000 families are waiting for a home from the Ministry of Housing, which is having more and more difficulties to meet the growing demand[5]. To solve this problem, public-private partnerships were established for several housing projects, for example, Northern New Town and Diyar Al Muharraq. While the dwellings are partly financed by private investors, the private sector benefits from the fact that the land for the project is reclaimed and financed by the government. Despite this new strategy to supply the local population with housing, the number of families who cannot afford to buy their own homes is still increasing and has become the most serious indicator of a growing social dilemma. On the one hand foreign guest workers are needed due to their higher education or willingness to work in the lower service sectors such as construction while on the other, the growing unemployment rate within the local population and their dependency on government subsidies has begun to threaten the social peace in Bahrain. This is furthermore fueled by Bahrain's long history of social unrest between the Shia population and the ruling Sunni families. Consequently, the current developments are perceived by a large part of the traditional Bahraini population as an aggression against their culture and their own selves. Thus, the current built environment reflects a growing contradiction between local traditions and the invasion of the globalised consumption industry in the form of large-scale shopping malls and high-rise buildings. In addition to the psychological impact of these developments, the living standard of low-income groups has been deteriorating due to rising living costs and a polluted environment, which, for instance, has led to a decreasing fish stock along coastal areas due to continuous land reclamation.

In addition to the environmental damage caused by the on-going land reclamation, the exponential need for energy and water is endangering the ecological balance in Bahrain. Apart from the fact that the huge waste of water has increased energy waste because of the power needed by desalination plants, it has also led to the increasing salination of the sea off coastal areas. This is a particularly serious threat for the marine eco-system due to a naturally high degree of salinity because of complex interactions between hydrography and climate in the Gulf. In addition to the high demand for petrol, the increasing need for energy is fueled by the high number of air-conditioning units, which are needed most times of the year due to climate-inappropriate architecture. All in all, it can be stated that the recent urbanisation has not led to a more effective use of energy although urban densities are increasing and thus new technological opportunities could be implemented such as district cooling or public transport systems. Apart from some district cooling projects and a proposal for a future metro system the most famous project illustrating Bahrain's attempts at a more ecological future is the World Trade Centre with its attached wind turbines. However, only around 11% of the towers' energy consumption, which does not include the waste from air conditioning, will be provided by these three turbines, which therefore should be seen more as a visual statement than a real contribution to less energy waste[6].

Similarly to other places in the world, a lack of investment in an ecological built environment is leading to higher maintenance costs, which in turn increases public expenses and the cost of living for Bahrain's inhabitants. Thus, because the majority of developments are not responding to the climate in order to save on construction costs, monthly electricity bills have been increasing. In some areas the electricity supply has reached its limits, leading to newly completed buildings having no connection to the electric network This leads to investors making a loss because their buildings cannot be sold or rented while the public sector must finance the expansion of the network. In the case of man-made island developments, there has been a growing problem to finance the maintenance of beaches and shorelines due to the movement of sand in and out caused by the current, which is continuously changing due to constant land reclamation. All in all, the increasingly shrinking public budget is hardly able to cope with the increasing costs of infrastructural development and maintenance. Consequently, it is expected that the introduction of taxes will be inevitable in the near future, which again threatens Bahrain's economic future, which is based on its attraction as an economic free zone. Thus, a large number of foreign guest workers will probably leave Bahrain due to decreasing incomes and the relocation of companies to other Gulf countries.

5.2 The New Urban Development Strategy

5.2.1 The Economic Vision 2030

Since its foundation in 2002 the Economic Development Board has been working on a new comprehensive economic strategy for Bahrain, which includes development strategies for the society and environment. The first step towards the implementation of the new strategy was the publication of the Economic Vision 2030, which was released in 2008, in order to introduce the general development goals to the public. The main goal of the Economic Vision is to ensure Bahrain's future economy is globally competitive by allowing it to be driven by the private sector while being overseen and regulated by the government. In this regard, the envisioned future economy of Bahrain is expected to support a broad middle class of Bahrainis who enjoy good living standards thanks to increased productivity. The ultimate aim of the new strategy is to ensure that by 2030 every Bahraini household has at least twice as much disposable income than it does at present. In order to achieve this there will be a gradual introduction of coordinated reforms, particularly concerning Bahrain's transformation from a rentier state based on a large part of the population working in the public sector to a global market place driven by a growing service sector. Today, the private sector is on average creating only around 1,100 jobs per annum with a monthly salary higher than BD 500 for Bahrainis and about 2,700 jobs for non-Bahrainis. This lack of employment of Bahrainis in the private sector indicates an insufficient education system that does not provide young people with the skills and knowledge necessary for the competitive labour market. However, due to recent regional development in the financial, industrial and tourism sectors, there has been a growing opportunity to integrate the local workforce into the attempts to establish Bahrain as one of the Gulf's biggest economic centres. Expected future growth is mainly based on a favourable business environment that provides private companies with zero taxation in addition to Bahrain's geopolitical location between Asia, Europe and Africa. Moreover, Bahrain has few indirect taxes on private enterprises and individuals, a free movement of capital and allows one hundred percent ownership of business assets, including real estate, in most sectors of the economy[7].

One of the biggest current deficits in Bahrain's economy has been identified as a lack of productivity and innovation. While the average worldwide labour productivity has increased by 21 percent over the last 25 years, Bahrain's improvement has been just 17 percent and its innovation output is currently more or less zero. Consequently, the three main guiding principles for economic transformation stipulated by the Economic Vision 2030 are sustainability, competitiveness and fairness. In this regard the term sustainability is mainly understood as a shift from the public sector to the private sector as the main driving force of the economy. In this way, Bahrain is expected to become independent from a public subsidies-based economy and to enhance entrepreneurship and thus innovation and competitiveness, which are the two main factors that should make Bahrain attractive to investment. In addition to an expanding private sector, a high-quality public sector, cutting-edge infrastructure and an appealing living environment will be further important factors in the development of Bahrain's economy towards sustainability. Last but not least, fairness has been included as a guiding principle in terms of the elimination of corruption and the setting-up of a transparent and trustworthy legal framework[8].

In addition to defining Bahrain's main development goals, the Economic Vision contains specifies how the economy, government and society should be transformed. The three keys for robust economic growth are identified as the enhancement of productivity and skills, economic diversification focusing on existing high-potential sectors and the use of future economic opportunities by establishing an innovation-friendly environment. With regard to the transformation of the government, the Vision identifies five main action points. The first is the development of high-quality policies by introducing public-private partnerships on the basis of effective regulations and supervision after privatisation. The second point is the enhancement of productivity and accountability for delivering better quality services via leaner organisations and operations. This will lead to the adjustment of government organisations to increase transparency and the establishment of a performance-management system based on a scheme of rewards for employees. The third point is the setting up of a predictable and transparent regulatory system which will include zero-tolerance policies against corruption and favouritism. The fourth point is becoming free from dependency on oil revenues by reducing inefficient spending, identifying possible new sources of government revenues and deciding how to invest oil revenues to ensure the future prosperity of Bahrainis.

Last but not least, the Vision postulates the development of a world-class infrastructure linking Bahrain with the global economy on the basis of public-private partnerships. With regard to society and its future, the Economic Vision distinguishes five main principles, namely, a high standard of social assistance, a quality healthcare system, an improved education system, a safe and secure environment and a high standard of liveability for all residents. Improving liveability will include measures for protecting the natural environment by conserving green areas, implementing energy-efficiency regulations and investing in ecological technologies. The Economic Vision also aims to integrate more public spaces and enforce laws concerning cultural preservation[9].

5.2.2 The National Planning Development Strategy

In order to achieve the goals of the Economic Vision, the Economic Development Board has engaged several consultants to work on the National Planning Development Strategy (NPDS). The American consultancy Skidmore, Owings and Merill (SOM) has been put in charge of designing the strategy, which includes the master plan for the whole Bahrain. Further consultants who participated in the process of preparing the strategic plan were Atkins, DTZ Pieda Consulting, JMP Consulting, MSCEB and Battle-McCarthy[10]. After the NPDS was completed in 2006, a steering committee was set up to oversee the implementation of the master plan. The design of the NPDS has been based on the initial assumption that Bahrain's population will increase from around 700,000 inhabitants in 2004 to more than 1.2 million by 2030. Due to the ambitious aim to introduce the Gulf's first comprehensive strategy plan designed on the basis of economic, social and environmental development goals, the NPDS has become an extensive document consisting of 10 development themes published in 10 volumes. After years of monitoring the current development 31 projects were selected according to their significance in terms of land area development capacity and reports written on them. These reports were used to create cumulative statistics regarding the impact of these 31 projects on environment and infrastructure in particular. The NPDS was inspired by the development strategies of Singapore and small countries such as Denmark which have been adjusted to the specific needs and requirements of Bahrain[11].

The first development theme is the creation of one plan for the entire kingdom in order to integrate all concerns and replace the previous planning procedure which was focused on certain areas and executed by different parties without regarding interconnected issues. One of the main goals of this 'one plan' approach is to create a balance between conservation and development, between cities and villages, between transportation and community and between modern time and history. Apart from the attempt to create a comprehensive plan targeting all aspects of development, the 'one plan' approach demands the implementation of a planning system which is transparent and adaptable over time. This includes a call for a regulatory system that supports economic development and investment by taking into account social and environmental costs. Consequently, two main initiatives have been defined – the creation of a comprehensive land-use strategy and the development of appropriate planning structures at national, governerate and municipal levels[12].

The second development theme concerns the future establishment of Bahrain as a market economy which is specialised in both regional and global markets. In this context, manufacturing, financial and business services as well as tourism have been identified as the main future economic growth engines. The plan identifies the development of Bahrain's human and physical infrastructure as the most important basis for the achievement of the economic growth goals of quadrupling GDP and doubling employment within the next 25 years[13]. In addition to a programme to improve the existing educational system, there have been proposals as to how to integrate the Bahraini workforce into the private sector. With regard to infrastructural development, an announcement has been made that the International Airport will be extended on reclaimed land and that the harbours in the south of Manama and Muharraq will be expanded. A proposal has also been made to develop two main future industrial corridors in the form of a High-tech Industry Corridor along the north-western part of the main island and a Heavy Industry Corridor along the industrial areas from southern Muharraq to the central and eastern part of the main island. The Heavy Industry Corridor will include a new industrial area on still to be reclaimed land along the future causeway to Qatar. One of the most important centrepieces of the Heavy Industry Corridor will be Sitra Technology City, which will stretch along the eastern part of the main island, where the aluminium plant and the BAPCO refinery are located, along the shoreline and along proposed man-made islands to southern Sitra. Moreover, there are plans to expand Khalifa Bin Salman Port in the south of Muharraq and to establish the Bahrain Investment Wharf, which is currently under construction, as one of Bahrain's most important industrial centres.

The third development theme focuses on strategies to preserve and strengthen Bahrain's environment and natural resources. Thirteen preservation areas have been identified including, amongst others, the Al Areen wildlife reserve, the wadis Buhayr and Hunayniyah, the South-west Coast Wetlands and Arad Bay. The future growth of settlements must be managed to be compatible with Bahrain's natural system as well as preservation measures implemented to protect the environment, for example, the coral reefs, of which 82 are currently at risk. As well as research on the impact of land reclamation, studies have been made on the coastal environment itself in order to define sensitive marine areas, particularly areas of sea grass beds and corals that are the natural habitat of endangered species. In this context, an Integrated Coastal Zone Management plan has been designed, which embraces economic development, social initiatives and the protection of the environment including the creation of five National Marine Preserves. In addition, the plan calls for adopting best practice reclamation techniques and a strict monitoring of projects' coastal impacts. Last but not least, there will be programmes to raise public awareness about protecting the environment and to integrate the tourism industry into the conservation strategy[14].

In order to improve the current situation of transportation in Bahrain a new transport strategy has been designed, which constitutes the fourth development theme of the NPDS. This strategy will concern the development of all different kinds of transportation, including vehicles, water transport and pedestrians. It also aims to ensure Bahrain is competitive in the future as a regional hub for air travel and high-speed rail networks and to create an interconnected and adaptable roadway system and a modern and efficient public transport network that is integrated with other traffic modes[15]. Thus, in 2008 the first plan for a proposed monorail train network was published. It will be about 103 km long and possibly completed by 2030[16]. The first phase of this project will be around 23 km long and will run from Muharraq to Manama, Seef and Isa Town and end at Alba[17]. In order to avoid traffic congestion, a proposal has been made to organise future development around four main transit corridors or so-called 'structural corridors'. All four of these corridors will mainly run through the main island, where they commence at the capital governerate and split up in five directions. While one corridor will lead along the reclaimed islands off the northern shoreline, another corridor will be developed towards the south along Hamad Town and the International Circuit. A third corridor will begin in Sanabis and head south to Isa Town and further along the south-eastern coast. The fourth corridor should connect the Diplomatic Area with southern Manama and Sitra, from where it will continue and join the third corridor heading south. A short fifth corridor will lead from central Manama to central Muharraq.

On the basis of these corridors a strategic road master plan has been designed of motorways, expressways and arterial roads. While most of the roads suggested are to be arterial roads on the northern main islands of Bahrain where two thirds of the current projects are located, there are plans for a so-called 'national loop' to comprise the extension of Sheikh Salman Highway to the south joining it to the recently completed highway along the eastern shoreline. The transport strategy plan also includes a proposal for a future main route between the 26-km long King Fahad Causeway to Saudi Arabia and the recently launched project of the Qatar Bahrain Friendship Bridge Causeway. While the 40-km long causeway to Qatar, which will consist of 22 km of viaducts and bridges, is expected to open by 2015, an announcement has been made to expand King Fahad Causeway by adding new truck lanes[18]. In order to connect these major gateways, the plan presents five possible routes between the Salmabad Interchange, where King Fahad Causeway ends, and the main juncture between Riffa, Isa Town and Sitra, where the causeway to Qatar will be connected. There is also a proposal to develop a high-speed railway running from Saudi Arabia to Qatar along the causeways and across Bahrain, which would consequently make Bahrain a major juncture in such a regional transport network.

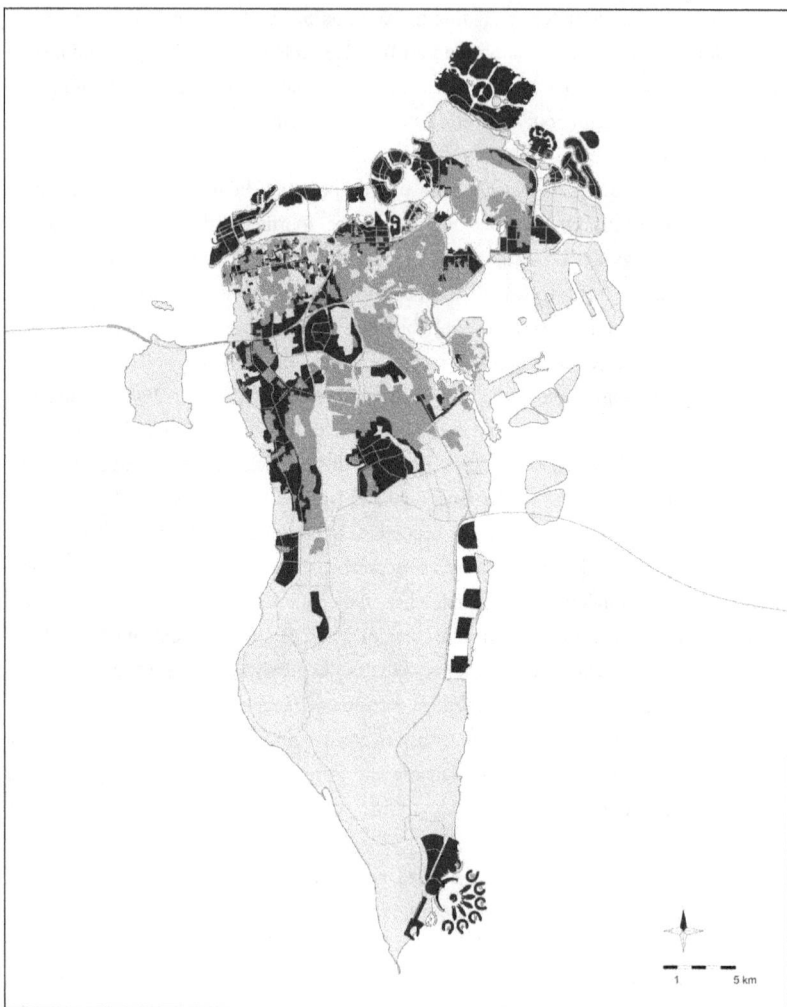

Fig. 1: The vision of Bahrain for the year 2030.

In close relation to the transport strategy plan the fifth development theme concerns the growth direction of settlement patterns. Three main types of settlement patterns are identified, namely, developments in areas within cities and villages, 'new desert communities' and 'outer bank island communities'. According to the new strategy plan, one of the main future priorities will be a focus on unbuilt areas within existing settlements and towns in order to create compact mixed-use neighbourhoods. Furthermore, the development of new desert settlements and man-made islands along the shoreline will shape Bahrain's future urban structure. With regard to new island projects, a new generation of island communities is to be created over the next 100 years in coordination with infrastructure development and marine preservation strategies. Thus, the main goal of the plan is to define the placement of the new islands in order to prevent future problems of infrastructural supply and to conserve marine preservation zones. Consequently, a proposal has been made to reclaim a chain of connected islands along the northern coastline and in the east close to the causeway to Qatar. Today, the West Archipelago covering around 900 hectares and the East Archipelago covering around 2,200 hectares are already being developed in about four large-scale projects. For the future, a proposal has been made to develop the Northern Archipelago stretching over 2,800 hectares in the north of Al Muharraq and the Fasht Al Adhm covering a total area of 3,200 hectares. Thus, the reclaimed area would increase from around 31 sq km to more than 91 sq km. In order to answer the question of which focus has to be chosen regarding the settlement pattern for future urban development, two main scenarios have been proposed. The first scenario suggests a main focus on new island-communities constituting around 50% of the anticipated area to be developed in Bahrain by 2030 with the development of already existing urban areas constituting only around 20%. In contrast, the second scenario proposes that the development of already existing urban areas constitute around 40% of Bahrain's anticipated area and that new island-communities constitute only 30%.

Based on the above, calculations have been made as to the density of future settlements. Not counting new island developments, there is currently a total settlement area of 118 sq km which is occupied by existing urban area, projects under construction and the developable area of existing islands. If the plans for new islands in the north are carried out, Bahrain's total settlement area would increase to 149 sq km and if the islands along the causeway to Qatar are reclaimed, the total settlement area would be 181 sq km. Furthermore, the development of the Northern Archipelago would add 29 sq km to a total area of 210 sq km. While it is expected that around 218,182 dwelling units will be developed for an estimated population of around 1.2 million people within the next 25 years, the total number of dwelling units is expected to increase to around 340,000 for a population of about 1.7 million inhabitants within the next 50 years and to approximately 444,444 units for 2 million people within the next 100 years. Consequently, the average number of people living in one dwelling unit would decrease from around 5.5 to 4.5 and the average density of around 15 dwelling units per hectare would increase to 25 units per hectare. This would be a significant increase in urban density considering that the current density is about 10 dwelling units per hectare within cities and about just 4 dwelling units per hectare within villages.

In addition to these calculations regarding the expected overall urban development, there have been studies about the current and future development of the capital Manama. The area under examination in these studies is around 6,200 hectares, of which around 850 ha are occupied by open space, between 4,000 and 4,300 ha by existing development and about 150 ha by confirmed projects that have not yet been started. Moreover, there are around 700 to 900 hectares of sites inland and sites that are to be reclaimed, and between 200 and 300 hectares of other land available for development. In order to direct future development, a proposal has been made to define eight zones with six main functions.

Fig. 2: The high-rise agglomerations along the coast of Manama.

A large zone will stretch over the central inland area and has been defined as mainly residential (H). A large commercial zone (C) will stretch from Seef to the Diplomatic Quarter. In addition to these two inland zones, six waterfront zones are proposed along the coastline, to be designed designed with four different focuses, namely, commercial (W-C), residential (W-C), cultural (W-Cu) and industrial (W-TH) use. Consequently, the calculation of the total developable area regarding each use comes to 250 to 300 ha of residential areas, 150 to 250 ha of mixed use areas, 150 to 200 ha of open areas and 50 to 100 ha of industrial areas. Furthermore, there will be around 40 to 80 ha for roads and 25 to 50 ha for tourism developments. With regard to future residential development, research has been done to find out what the existing housing densities are and to propose the future housing density. According to the research, conventional housing areas with low-rise residential units usually have a density of 20 to 40 du/ha. With villas, they have a density of 5 to 15 du/ha. There are three different categories of apartments. Areas with high-rises of heights of 20 or more storeys have a density of around 200 to 300 units per hectare, areas with medium-rises of more than eight storeys – between 100 and 150 du/ha and areas with low-rise developments with three to four storeys – 40 to 80 du/ha. Today, at least 5,000 new dwellings will be developed in Bahrain through the current projects but a total of around 50,000 dwellings are expected to be constructed in the future according to further plans, housing more than 200,000 people. As well as defining the future density of each part of Manama, a zoning plan has been made for the overall existing development area. Due to better accessibility and a large amount of unbuilt land, coastal areas such as Seef and Juffair have been favoured for the highest urban densities. More central areas have been earmarked for low-density residential areas in order to prevent infrastructural problems and the conflict of high-rise buildings interfering in the privacy of adjacent low-rise villas.

The fifth volume of the NPDS defines the location of future settlements in Bahrain as well as methods to direct the expected growth in a sustainable manner. Today, the constant reclamation of new islands has led to the transformation and decay of existing waterfront developments and the urban sprawl has led to increasing distances between city centres and the coastlines. Hence, certain initiatives have been proposed to solve this problem. For example, the creation of compact, mixed-use island developments along the current coastline. Furthermore, the existing coastline must be protected so that existing projects keep their waterfront location and that the waterway between islands and the coast can be used as transit area for ships.

The sixth development theme of the NPDS aims to define a precise coastline in order to ensure public access to the coastline and protect waterfront investments. Thus, one of the main goals of this part of the strategy is the improvement of public access to the coast, which is currently rather low considering that less than 5% of the coastline is accessible to the public. Moreover, due to the reclamation of around 30 sq km of land between 1968 and 2000 the waterfront has become further and further away from settlements and cities, which have lost their access to the coastline. This problem will be further exacerbated due to the expected reclamation of more land coming to a total of around 130 sq km by 2030[19]. In addition to the NPDS's suggestion to establish certain points of public access to the waterfront to be equally distributed throughout the entire coastline, it also suggests that a world-class waterfront be built along the coasts of Manama and Muharraq. It would have four distinct areas designed around themes based on the surroundings of each area. While the area called the Seafront Harbour will stretch along the northern shoreline of Manama and Muharraq, the area called the Great Harbour will be located between the eastern coastline of Manama and the western coastline of Muharraq. An area called the Green Harbour will be established at Tubli Bay in the south of Manama along the coast towards Sitra and the Port Harbour will stretch from the harbour area in Manama to the industrial area in southern Muharraq. While the Seafront Harbour would act as Bahrain's front door into the central business district, the Great Harbour will be a landscaped waterfront linking Manama and Muharraq as well as part of a cultural corridor leading from the centres of Manama and Muharraq to the coast. The Green Harbour at Tubli Bay should according to the NPDS integrate the strategic ecological preservation of the natural mangrove habitat[20].

The seventh volume of the NPDS identifies protected archaeological areas that should be conserved as part of Bahrain's cultural heritage and tourism industry and as important centrepieces of future green areas and corridors. In this regard, two main archaeology corridors have been defined. The first corridor leads along the northern and western coast of the main island, where Bahrain's ancient forts are located. The second one starts in downtown Muharraq before crossing the causeway and following Government Avenue towards Budaiya Road and the highway to the south to Hamad Town where areas of burial mounds are located. The eighth development theme of the NPDS concerns how the expanding development areas will impact the military. The ninth development theme of the NPDS consists of a strategy as to how to develop more green areas and public areas in Bahrain. In addition to seaside parks and promenades along the coastlines of Manama and Muharraq, the strategy suggests using a large part of the coastlines of new island-developments for the development of public promenades or beaches and that inland green areas, including agricultural land, should not be made available for use as building land by means of strict zoning regulations. In the case of Manama, a 360-hectare green belt, which is about the size of Central Park in New York, has been proposed in order to connect future seaside parks and the agricultural areas in the west[21]. The last of the ten development themes is called 'Promoting Sustainability' and as well as aiming to increase public awareness about the environment, it includes proposals for ecological designs of future projects such as 'smart islands'. These island developments will be planted with mangroves in order to restore habitats for birds and sea life.

Parallel to the NPDS, several further strategy papers and urban design projects have been developed regarding subordinated planning issues such as the revitalisation of the old centres in Manama and Muharraq. This project has been prepared by the Ministry of Municipalities and Agricultural Affairs and the UNDP. It consists of a research report about the current status of the built environment in the traditional urban centres and proposals of how to restore certain areas. The major goal of this project is to develop a revitalisation strategy that ensures the future urban environment will be suitable for residents, businesses and visitors. Thus, the plans do not intend to turn the old centres into urban museums such as the rebuilt district Bastakiya in Dubai. Hence, the project contains suggestions about the legal basis for alternative programmes and management structures as well as how to raise investments in these areas by establishing public-private partnerships. In addition to the identification of conservation zones which are to be protected, the strategy includes the development of urban design guidelines, rules and codes in order to ensure the quality of the future built environment. While a large number of buildings will be replaced, a certain percentage of the building substance will be restored on the basis of a manual containing suitable practice methods. An important part of this project is the development of a database that will contain economic, social and physical planning data[22]. Although the first phase of the project was completed in 2006, there has been still no sign of implementation due mainly to a difficult investment situation. Due to recent speculation, land prices in central urban areas have rapidly increased and thus it is not considered profitable to invest in the old centres where the restrictions are rather tight. In addition, the old centres are currently the main residence of low-income labour and they lack accessibility due to a narrow road network. Despite these obstacles there have recently been new attempts to investigate the possibilities of restoration.

5.2.3 The Implementation of the NPDS

Although the NPDS has not been published nor officially approved, it has already started to make an impact on urban development because it provides the private sector with a certain reliability and framework for the long-term view of projects. While on the one hand the private sector has been the dominant driving force behind the current development because of a lack of restrictions, it has on the other hand slowly become the victim of a very low level of coordination within urban governance. This lack of coordination has mainly been caused by the absence of a comprehensive plan and a more and more diffuse decision-making process, which has led to a lack of consolidation between the development of more and more buildings and their infrastructural supply, for which the public sector is mainly responsible. Due to inefficient organisation, a shrinking government budget and the current speed of the development, many developers have been forced to wait for the infrastructural supply of their projects, which has meant economic loss. A further problem has been the construction of new developments without consideration of adjacent projects, the investment value of which consequently drop due to the impact of these neighbouring developments, for instance, because of increasing traffic. Thus, one of the main responsibilities of the plan has been ensuring that the private sector's investments are not in danger and that even long-term investments are considerably safe. In this context the NPDS has been a major attempt to restructure the public sector and to implement a new strategy integrating all the interests of the private sector and the future needs of the population. Consequently, a new form of public-private partnership has been established in Bahrain based on cooperation in order to achieve long-term success instead of short-term profits for individual investors.

In order to achieve this ambitious goal the private sector has been allowed to participate in the financing of certain government projects, particularly infrastructural development, and a strong new framework of public regulations has to be implemented. Also, the first steps have been made toward re-organising the public sector wherein ministries will be partly transformed into authorities. These new authorities will be able to raise money from the private sector and thus will not rely solely on the public budget. While the Ministry of Electricity and Water has already been transformed into the Electricity and Water Authority, other ministries such as the Ministry of Works are expected to follow the same transformation within the coming years. Apart from the different way of financing infrastructural projects, these public authorities will be structured like private enterprises. Thus, the outdated structure of a large and bureaucratic public administration would be replaced by dynamic, competitive and profit-oriented authorities, which however would be under strict government observation.

All in all, regulation and liberation have become the two main principles of the current development strategy in addition to innovation by using new technologies such as district cooling, new building standards, congestion management and modern public transport systems. The entire strategy has been built on these basic principles, which are the central themes of four main planning stages, namely, the Economic Vision, the NPDS, Zoning Plans and Action Plans. While the Economic Vision 2030 sets up the targets, the NPDS consists of the major initiatives and guidelines to be implemented in the form of detailed Zoning Plans. In addition to the Zoning Plans, which will function as the basis for the urban development, Action Plans ensure that building permits and master plans of large-scale projects are approved in accordance with the new regulations.

One proposal of a new regulation is the introduction of a development levy, which has been proposed in three different forms[23]. A general levy would require all needed infrastructure to be financed for five years by the developer. A supplementary levy would involve the public sector owning a share of a development in exchange for providing infrastructural supply for the development. A special levy would require that the private sector participate in the building of the needed infrastructure. In this way, infrastructure is expected to match the pace of building either by raising the money from the private sector or slowing down the densification of areas such as Juffair that are facing an increasing shortage of infrastructure. While one positive side-effect of this new regulation would be growing interest from the private sector in investing in infrastructural development, the risk would be that real estate developers would eventually leave Bahrain.

After the introduction of new regulations of this kind, the Action Plan, which consists of a control system, will guarantee their implementation and function as a juncture of infrastructural planning and physical planning. Last but not least, in order to be able to improve and adjust plans, the government has begun to create an integrated database system providing planners with access to reliable data[24].

5.3 Potentials and Challenges of Post-oil Urbanism in Bahrain

5.3.1 Strengths and Opportunities

Bahrain's future development profoundly depends on achieving the main goal of transforming the small kingdom from a rentier state to a competitive global marketplace. Thus, the biggest assets of this small country are its geopolitical location, liberal policies, already established industries, the current development of an efficient regulatory system and a modern infrastructure with growing capacity. Geopolitically, Bahrain is located in the centre of the Gulf and therefore it has been one of the most important harbours and transportation hubs in the region from very early on. Based on the introduction of free zone policies Bahrain has developed to a potential global trading centre and an attractive place for companies to relocate, particularly firms from the financial and telecommunications sectors. Furthermore, due to more liberal policies Bahrain has become an important regional tourism centre, which has rapidly grown since the 90s after the completion of the King Fahad Causeway to Saudi Arabia. While Bahrain's tourism industry benefits from the proximity to Saudi Arabia, where the population is rather restricted in their choices of lifestyle, the Formula One Circuit has brought international attention to Bahrain as a travel destination. Thus, it is currently expected that the number of tourists will increase from currently around 4.7 million annual visitors, which is relatively high considering the size of countries such as India with similar numbers, to more than 16 million by 2030[25]. In addition to the growing financial and tourism sectors, Bahrain's industry has expanded in aluminium production, logistics and oil-related industries. A further asset of Bahrain's economic development has always been its modern infrastructure of an international airport and a deep water port, which are expected to remain important factors for future economic success due to planned extensions.

This potential for future economic growth has led to increasing interest from investors in developing real estate in Bahrain, which since the introduction of new laws in 2003 can be sold as freehold properties to foreigners. In the following years Bahrain underwent a construction boom, particularly as a result of master-planned developments such as man-made islands. However, several reasons have prevented Bahrain from matching the number and scale of developments in Dubai, where currently more than 21 projects are scheduled to be completed by 2010 and cover almost 90 sq km. By contrast, only around 10 projects, which occupy around 29 sq km, are currently under construction in Bahrain and most developments will not be completed before 2015. One of the biggest factors in Bahrain's slower development is the fact that there is a scarcity of cohesive developable land suitable for large-scale projects. Thus, more than 80% of the current projects are built on reclaimed land along the northern coast, which attracts investors due to the proximity to the urban centres. Another important factor in the focus on land reclamation is the fact that reclaimed land belongs to the government, who acts as a shareholder in projects by providing the land. Moreover, in contrast to Dubai, a large amount of unbuilt land on Bahrain's islands is currently not owned by the government and thus developers are forced to negotiate with many different land owners in order to get a suitable piece of developable land. In addition, in cases where areas are still undeveloped developers would have to convince the government, who is usually less interested in developing land belonging to different owners, to provide the main infrastructural supply. The second main reason for a slower urban development in Bahrain is the regulations of the local banking system, which in contrast to Dubai restricts banks in order to prevent excessive speculation. Although after the international financial crisis in 2008 applications for building permits fell by 50%, there have already been signs of recovery and a slight increase in applications by April 2009[26]. This continuing trust in Bahrain's future growth is mainly due to the combination of a still relatively small scale of development and efforts to implement an efficient regulatory system within a comprehensive strategic plan. Today, the National Strategic Development Plan can, despite its delayed implementation due to several adjustments, be considered to be key in Bahrain's future development in a way that allows economic prosperity without harming the social and environmental balance. This new approach of managed growth based on one comprehensive plan instead of exponential growth based on de-centralised urban governance is of urgent importance because of increasing development problems caused by the recent urbanisation. The lack of central planning and building regulations has led to growing traffic congestion and a large need for parking space as well as a lack of sufficient public areas and social infrastructure. In addition to the decreasing quality of the built environment, the natural environment, particularly coastal regions, has been endangered due to ongoing land reclamation, desalination and pollution. In order solve such problems the NSDP intends to integrate all aspects of urban development within one plan. While this plan has not yet been implemented, it has already started to generate a certain level of interest and trust in Bahrain's future.

While the construction boom has led to more and more densely built districts along the shoreline of Manama and exclusive residential developments, the increasing land prices have caused a growing need for social housing programmes for the Bahraini population, who are becoming less and less integrated within the current development. Thus, one of the central tenets of the Economic Vision 2030 is to develop Bahrain's future on the basis of promoting the local workforce and gradually reducing dependency on foreign guest workers, who currently run most of the private sector. Consequently, a labour regulation has been introduced which imposes a tax of 1% on private businesses in order to raise money for education programmes for the local population. In comparison to other GCC countries the Bahraini workforce can be seen as one of the country's biggest assets for future development due to the fact that Bahrainis have been less subsidised and are thus generally more willing to work in the service sector. With regard to the fact that more than 50% of the current population is still Bahraini and that the future immigration of foreign guest workers will be limited, there is a reasonable chance that the local population could take over the role as the driving force of the more and more privatised economy.

5.3.2 Weaknesses and Threats

The future success of the current development strategy of Bahrain is endangered by several weaknesses, particularly in relation to a lack of consolidation of the current urban development. The lack of comprehensive planning has led to serious problems in the current built environment, particularly regarding the development of sufficient infrastructure and public areas. In turn, the decreasing quality and liveability of the built environment due to traffic congestion and highly dense residential areas has begun to pose a major threat to the future environment, society and economy of Bahrain. While the ongoing land reclamation is endangering the environment along the coasts, growing social segregation is leading to the decreasing liveability of residential areas where middle- and lower-income groups are settled. Furthermore, Bahrain's economy suffers from the decreasing quality of the built environment, particularly traffic congestion, and its future development has become uncertain because of the possible relocation of certain economic sectors and a reduction in investments from the private sector.

In addition to the decreasing liveability of the built environment, there are several other obstacles endangering Bahrain's future. For instance, due to the lack of education of the Bahraini workforce, private companies still prefer to employ foreign guest workers. More and more restrictions forcing these companies to employ Bahrainis might lead them to relocate to other Gulf countries. The implementation of a development levy system could pose a further obstacle by forcing private investors to participate in the funding of the country's general infrastructure. While on the one hand each developer is a victim of the increasing lack of infrastructure, an obligation on the other hand to financially participate in the development of the overall urban quality would mean increasing costs and thus less opportunity for short-term profits. Hence, the major threat to future development is if investment leaves Bahrain for neighbouring countries offering less restriction. In this context the successful privatisation of infrastructural development profoundly depends on the private sector having confidence that investment in the construction of infrastructure will be profitable despite the chance of a decreasing number of real estate projects in Bahrain. Furthermore, this privatisation would inevitably mean favouritism of developers' areas of choice in terms of infrastructural supply, and thus a growing risk of more and more segregated urban areas, in turn endangering the social peace in Bahrain, which has already proven to be fragile. Therefore, the integration of all development concerns will be one of the most important challenges regarding Bahrain's future.

The successful implementation of the NSDP depends to a large extent on the development of Bahrain's neighbours. The increasing competition between GCC countries has led to enormous pressure on each country to open up to private investments and speculation-driven mechanisms. While Bahrain is currently the only Gulf country to have worked on a comprehensive strategy to integrate all development concerns, many other Gulf countries are just at the beginning of evaluating possibilities that could solve their growing infrastructural problems. Although the approach of creating 'One Plan' and implementing it by reforming public-private partnerships is a reasonable choice in theory, it remains doubtful that one country on its own can succeed in the current situation of accelerating regional competition. On the one hand, Bahrain has the unique chance to become the new role model of post-oil urbanism in the Gulf, choosing smart growth based on the actual productivity of its inhabitants instead of exponential growth based on speculation. On the other hand, it is very much endangered by becoming less attractive to investment than its regional competitors and thus it would be forced to continue its development trend of recent years. The result would be a segregated urban landscape of high-rise agglomerations and man-made islands along coasts and highways in addition to increasingly dense inland settlements with no access to the coast or sufficient infrastructure. Consequently, the damage to the social, environmental and economic balance would endanger the whole future of Bahrain as a potential global market place.

All in all, it can be summarised that Bahrain's future depends on the fact that a major shift to a more sustainable urban development would only be possible if all GCC countries recognise the hazardous direction of the current urbanisation and the danger of working on short-term and rapid urban growth instead of a long-term and consolidated urbanisation. In spite all its economic damage the international financial crisis can be considered as a chance to rethink the development direction in the Gulf and in the case of Bahrain it can be seen as an opportunity for the small kingdom to take over an important role as an alternative development model of post-oil urbanism.

1 http://www.gulf-daily-news.com/NewsDetails.aspx?storyid=2113 02, 30.04.2009.
2 http://indexmundi.com/g/g.aspx?v=115&c=ba&l=en, 30.04.2009.
3 http://www.transportation.gov.bh/en/modules.php?name=Content& pa=showpage&pid=119, 30.04.2009.
4 http://www.tameer.com/qmr/luxuries-residential.pdf, 17.04.2009.
5 http://www.bahrainrights.org/en/node/2431, 08.05.2009
6 http://www.worldarchitecturenews.com/index.php?fuseaction=wanappln. projectview&upload_id=2133, 01.05.2009.
7 http://www.bahrainedb.com/uploadedFiles/BahrainEDB/Media_Center/Economic%20Vision%202030%20(English).pdf, 03.05.2009.
8 http://www.bahrainedb.com/uploadedFiles/BahrainEDB/Media_Center/Economic%20Vision%202030%20(English).pdf, 03.05.2009.
9 http://www.bahrainedb.com/uploadedFiles/BahrainEDB/Media_Center/Economic%20Vision%202030%20(English).pdf, 03.05.2009.
10 Al Kalali 2005, p.1.
11 Al Kalali 2005, p.10.
12 Al Kalali 2005, p.11.
13 Al Kalali 2005, p.11.
14 Al Kalali 2005, p.12.
15 Al Kalali 2005, p.14.
16 Gulf construction January 2009, p. 57.
17 http://www.gulf-daily-news.com/NewsDetails.aspx?storyid=230865, 05.05.2009.
18 http://www.constructiondigital.com/MarketSector/Civil-Construction-and-Engineering/Construction-of-Qatar-Bahrain-Friendship-Bridge-to-start_38445.aspx, 28.12.2009.
19 Al Kalali 2005, p.22.
20 Al Kalali 2005, p.23.
21 Storch 2009, p. 94.
22 EDB 2006, p. 6.
23 McPolin, 03.12.2008.
24 McPolin, 03.12.2008.
25 McPolin, 03.12.2008.
26 Al Ghazal, 07.05.2009.

III
Present and Future of Post-oil Urbanism

6. More Centres of Post-oil Urbanism

6.1 Centres with Rapid Urban Growth

6.1.1 Abu Dhabi and the Northern Emirates

Abu Dhabi City:
Like many other coastal settlements along the Gulf, the relocation of Bedouins from inland oases led to the foundation of Abu Dhabi in the south of the Persian Gulf. In 1761 part of the Bani Yas tribe settled on a small island close to the coast, where sufficient water resources enabled the establishment of a small fishing village consisting of a cluster of around 20 barasti huts. Due to the increasing threat of Wahabi forces at the inland oases, the Bani Yas tribe decided to move entirely to the new settlement at the coast, which led to the construction of about 400 houses at the end of the 18th century. In the following century the settlement did not expand significantly and thus the population remained rather low. After a short increase due to the pearl trade at the beginning of the 20th century, it again declined to around 1,500 inhabitants during the 50s. About one decade later it rapidly grew to around 4,000 people when oil production began. However, despite the rapid modernisation the settlement remained underdeveloped for a long time and many inhabitants moved to Dubai due to the conservative policies of the ruler Sheikh Shakhbout, who opposed modern urbanism. However, in 1966 he was removed by his brother Sheikh Zayed, whose rulership marked a new beginning for urban development in Abu Dhabi[1].

In 1968 the Halcrow plan from 1962 was modified in order to develop the first modern road network and the first basic land-use plan. In addition to the dredging of a canal around the island, the barasti huts were torn down and people were re-housed in newly constructed dwellings. After the founding of the United Arab Emirates in 1971, the attempt to settle all nomads led to the development of large low-rise residential areas at the outskirts. Furthermore, the immigration of guest workers led to the construction of high-rise clusters in the former settlement area in the north of the island. During the 70s the centre of the traditional settlement was completely replaced by a group of four small markets, also known as Central Market, and buildings with eight to ten floors which were mainly occupied by apartments for guest workers. Furthermore, a modern cultural centre was built close to the restored Hosn Palace, which has remained the last relict of the pre-oil settlement[2].

Because of higher land prices, which were caused by the increasing oil price in the middle of the 70s, the building code for the maximum height was changed to 13 floors for the

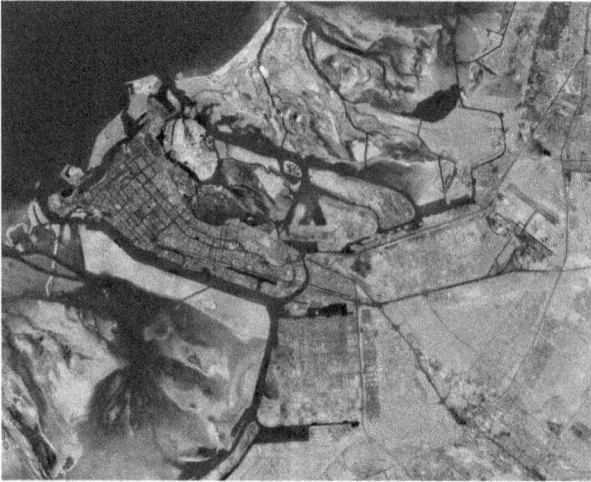

Fig. 1: Satellite image of Abu Dhabi City.

centre of Abu Dhabi City. In 1977 development slowed down due to over-development, which led to around 15,000 empty apartments. Parallel to the first phase of modern urbanisation, land was reclaimed along the shorelines from the 70s on and expanded during the 80s and 90s leading to an increase in the island's land area from 60 sq km to 94 sq km. In 1988 the Abu Dhabi Town Planning Department in cooperation with Atkins and the UNDP prepared the first Abu Dhabi Master Plan, which comprised five main phases up to 2010. Due to the limited area on the island and the speed of urban growth, the plan recommended several areas of possible extensions including the smaller neighbouring islands, particularly Saadiyat and Hadriyat. The plan also suggested growth towards the mainland along two axes in the form of the highways leading to Dubai and Al Ain[3]. While during the 90s the urban development spread towards the mainland, where the new International Airport was established in 1982, the islands Saadiyat and Hadriyat remained undeveloped in the 20th century. A large part of the urban development was carried out by the Khalifa Committee, which was installed in order to distribute state-owned land to Emirati citizens. The committee was also responsible for building a large part of the city, constructing around 200 apartment blocks a year through the 80s and 90s. Consequently, the built environment in the city centre is dominated by apartment blocks with up to 20 floors, built within a strict grid pattern and often in clusters around a central square, which was turned into a parking area. Away from these densely built areas, land is primarily occupied by low-rise dwellings of the local population and government buildings. The most important industrial areas are located in Mussafah in the south-west and at Port Zayed in the north-east. With regard to the main road grid, the Corniche, Airport Road, Sheikh Zayed, Khalifa and Hamdan have become the principal thoroughfares, where specialised businesses are located. While Hamdan is the main shopping street, Khalifa is lined with banks and Sheikh Zayed, also known as Electra, is the favoured location of

Fig. 2: The contemporary centre of Abu Dhabi.

electronic stores[4].

The year 2004 marks the beginning of another development phase in the emirate's history caused by the death of Sheikh Zayed and the succession of his son Sheikh Khalifa. Subsequently, many members of the cabinet were replaced by young, Western-educated ministers, most of them belonging to the ruling family, in order to establish Abu Dhabi as a regional and global centre in the Gulf. In contrast to its neighbour Dubai, Abu Dhabi has for a long time resisted the introduction of freehold property laws because of Sheikh Zayed's fear that increasing speculation would lead to negative consequences. However, his son did not share these fears to the same extent and due to his vision of a competitive marketplace instead of a welfare state, his government introduced new property laws permitting foreigners to own land for 99 years. Consequently, new real-estate developers were founded including Al Maabar, which is a joint venture of Abu Dhabi's four major property developers Al Qudra Real Estate, Surouh Real Estate, Aldar Properties and Reem Investments. Apart from local projects, Al Maabar has begun several developments abroad in Marocco, Jordan and Qatar. Due to the rising construction boom, two new landmarks in the form of the Emirates Palace, one of the most exclusive hotels worldwide, and the 37-storey ADIA building were developed along the Corniche. This recent development trend of large-scale projects is a sign of a new approach from the once conservative emirate, which in previous years left the global business to Dubai and concentrated on its function as an administrative centre of the UAE.

As a consequence of the above the Urban Planning Council, with the support of international consultants, modified the old master plan and published 'Plan Abu Dhabi 2030 – Urban Structure Framework Plan' in September 2007. The plan is based on the assumption that the city will grow from its current approximately 890,000 inhabitants to over three million people by 2030. One of the main themes of the plan is that Abu Dhabi will be a 'contemporary expression of an Arab city' built on measured growth instead of purely speculation-driven development. In order to reach the ambitious goal of a sustainable Gulf city with a strong identity, the plan specifies land uses, building heights and transport plans as well as the expansion of the business district and the creation of new business centres[5]. The plan comprises four Urban Structure Framework Plans regarding the future environment, land use, transportation, public open space and business centres. It focuses on the design of certain districts such as the Central Business District and suggests the construction of certain types of residential areas which are strictly divided into 'Emirati Communities' (residential developments for Emiratis) and 'Urban Communities' (residential developments for guest workers). Last but not least, the plan puts forward various policies regarding land use, transportation, building typologies and urban design as well as environmental, social and economic development policies.

In spite of the government's ambitions, the recent urban development has led to several problems such as an acute housing shortage, growing traffic congestion, increasing cost of living and a lack of appropriate housing conditions, particularly for low-income groups[6]. The current urbanisation can be characterised into two main development types – the development of master-planned projects along the borders of the former urban area and the development of projects, mainly in form of single high-rises, on plots within the former urban area. Thus, for instance, 85 old buildings were demolished by the government in 2006 in order to construct modern skyscrapers. Furthermore, the development of the Central Market project, which was designed by Foster & Partners, was placed on the area of the old souq, which was built in the 70s. The project, which will cover a total built-up area of around 490,000 sq m, will consist of a three-storey shopping mall and three high-rise buildings. While the design and layout of the mall will be influenced by an old traditional souq, the three towers, which consist of a 52-storey hotel, a 58-storey office tower and an 88-storey residential tower, have been designed as modern high-rises made of glass and steel[7].

The largest development on the main island is currently Capital Centre, which is being developed by the Abu Dhabi National Exhibition Company. It will spread over 148,000 sq m on an area next to the Abu Dhabi National Exhibition Centre (Adnec), which will be gradually expanded. The development will consist of a 35-storey landmark tower, a mixed-use 'Micro-City', which will comprise around 23 towers, and a marina with waterfront leisure zone. All in all, it is expected that the development will offer around 950 apartments by 2011[8].

With regard to other master-planned developments, a joint venture between three real-estate developers Sorouh, Reem Investments and Tamouh will develop the mixed-use development Reem Island on a 633-hectare natural island located about 500 metres off the north-eastern coast. After completion, the project will offer accommodation for around 200,000 residents in its Residential District as well as several commercial projects to be located in the Commercial District and Central Business District. Each of the three master developers is working on its own project in addition to the main infrastructure, which will be completed by 2010. Tamouh, for instance, is in charge of the Pearl of the

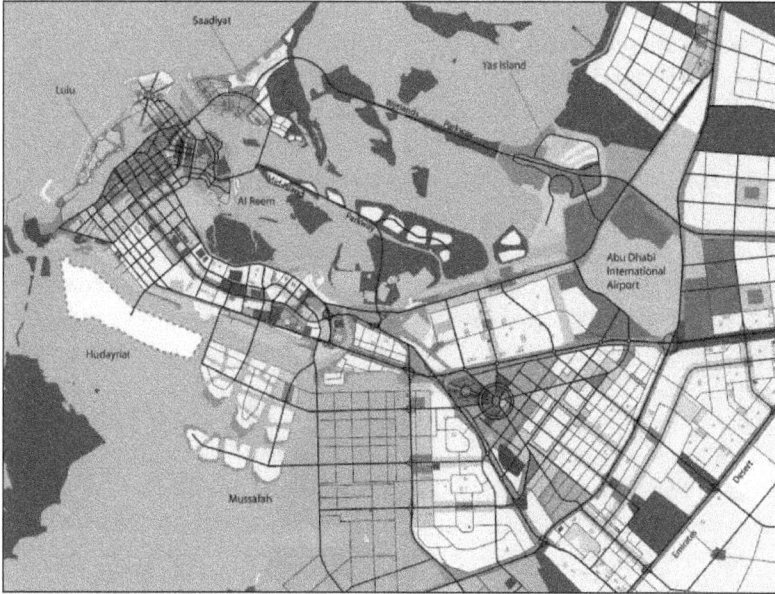

Fig. 3: The Urban Structure Framework Plan (2030).

Emirates development, which consists of the Marina Square and Addax Port projects. While Marina Square will be 70% residential and 30% commercial developments in the form of 14 high-rise buildings surrounding a central marina, Addax Port will consist of one commercial and four residential towers[9]. The developer Sorouh is responsible for the Shams Abu Dhabi development, which will cover around 25% of Reem Island. Around 90% of all the projects of this development will be residential-use offering around 22,000 residential units within about 100 high-rise buildings. The entrance of the development will be the Gate District consisting of eight high-rises and one of Reem Island's biggest landmarks – the 83-storey Sky Tower[10]. The third master developer, Al Reem Investment, is in charge of the Najmat Abu Dhabi development, which will occupy around 1.49 sq km (about 20% of Reem Island). In the future, the project is expected to house around 80,000 residents, mainly accommodated in high-rises which will be surrounded by three marinas and open spaces such as parks covering around 55% of the overall development area. The largest landmarks of the development will be the 80-storey twin towers of the Bay Centre Marina project[11].

Another project that will be developed on one of the natural islands surrounding the main island is Saadiyat Island, a project of the Tourism Development Investment Company (TDIC). With a size of around 27 sq km, it is currently the largest single mixed-use development in the Arabian Gulf and is expected to be home to more than 150,000 people after its completion by 2018. The development will consist of six districts – Al Marina (4.4 sq km), Cultural District (2.7 sq km), Saadiyat Park (6 sq km), Saadiyat Beach (4.33 sq km), South Beach (2.68 sq km) and The Wetlands (5.23 sq km). While the

223

Cultural District and Al Marina will be the urban centres in the north-west of the island closest to the main island, the remaining four districts will mainly be occupied by low-rise residential projects, hotels and large open areas in the form of parks and lagoons. The future centrepieces of the overall development will be five museums in the Cultural District designed by the architect icons Tadao Ando, Zaha Hadid, Norman Foster, Jean Nouvel and Frank Gehry[12]. All five projects will be built along the eastern shoreline facing the main island about 700 metres away. Due to a multi-lane highway, the Saadiyat Island development will be connected to both the central business district on the main island and Reem Island. The highway will also link Saadiyat Island with Yas Island development and the International Airport on the mainland.

Only around 2 sq km smaller than Saadiyat Island, Yas Island is the second largest mixed-use development in Abu Dhabi. Its master developer ALDAR has chosen this strategically located natural island close to the coast and the International Airport. The island was recently expanded by the reclamation of a 1.7-sq km area in order to develop residential projects, theme parks, shopping malls, hotels, marinas and the Formula 1 Circuit – the centrepiece of the development[13]. It is currently expected that the development will house around 110,000 residents and that more than 300,000 people will visit the island at peak times after completion by 2014[14].

The Masdar development is located on a 6.4-sq km area to the west of the International Airport. It is an ambitious project that has been marketed worldwide as the first 'zero carbon and zero waste city' by its master developer Abu Dhabi Future Energy Company, a subsidiary of Mubadala Development Company, which was engaged by the government of Abu Dhabi to develop this project. Designed by Foster & Partners, the project will cater to around 50,000 future residents and about 40,000 daily commuters, who will work in around 1,500 expected businesses, mainly in the form of commercial and manufacturing facilities. It will also be the location of a university, the Masdar Institute of Science and Technology (MIST). Due to the plan to ban automobiles within the city, public mass transit and personal rapid transit systems will replace the usual way of transport[15]. In addition, various renewable power sources will be used such as a solar power plant already under construction in addition to future projects such as a wind farm, a geothermal power plant and a hydrogen power plant. A solar-powered desalination plant will be used to provide the city's water needs and it is estimated that around 80% of the water will be recycled. In addition to the extensive use of grey water, the city will attempt to reduce waste as much as possible by introducing several recycling techniques. Last but not least, low- to mid-rise buildings will be built with state-of-the-art technologies, for example, photovoltaic modules on rooftops to reduce the energy waste arranged in clusters, inspired by traditional oasis settlements[16].

Although the planning is more or less complete, it is not yet clear if investors are willing to follow all of the project's restrictions and if there must be no future adjustments and compromises in order to satisfy the needs of the private sector. It is also rather unclear if there will be the expected number of people interested in residing in the rather restricted environment of Masdar, particularly regarding the use of cars. Moreover, the introduction of certain technologies such as photovoltaics has turned out to be problematic due to the very specific situation of the desert environment, where thin layers of sand, a high degree of humidity and high peaks of heat reduce or harm the functionality of these technologies. Apart from these general obstacles to creating an ecological city, the development has led to many controversies, for instance, the fact that the project is expected to be financed by the Clean Development Mechanism, a system which was introduced by the Kyoto Protocol allowing industrialised countries to invest in projects that reduce emissions in developing countries instead of more expensive emission reductions in their own countries. In the case of Masdar, the plan is to certify and sell so-called Certified Emissions Reductions, which are based on the difference between the normal average energy waste of a town in the UAE and the waste of the Masdar development. Due to the fact that the UAE has the second largest amount of average emissions per capita, the project would ironically contribute to a high degree of energy waste elsewhere[17].

Next to the Masdar project the master developer ALDAR has begun the construction of the Al Raha Beach development on an area of around 5 sq km stretching along 11 kilometres of coastline. The development, which is one of the first areas where non-UAE nationals can invest in leasehold properties, will offer residential units for around 120,000 future residents by 2019[18].

In addition to the many developments of the private sector, the Abu Dhabi Municipality is in charge of developing Mohammed Bin Zayed City, which will cover around 5 sq km and consists of 349 residential towers offering apartments for around 85,000 people[19]. An integral part of the emirate's tourism ambitions is the redevelopment of Abu Dhabi International Airport, which will be expanded in order to increase its annual passenger traffic from around 9 million currently to more than 40 million travellers[20]. Furthermore, Mubadala is the investor and developer of the $2 billion Khalifa Port and Industrial Zone, which will be located midway between Abu Dhabi and Dubai. In addition to the port, which will have a cargo capacity of 33 million tonnes, the new industrial zone covering around 100 sq km will cater to various industries, for example, aluminium production, with the world's largest single-site aluminium smelter complex[21].

All these recent developments have begun to transform the former urban structure of Abu Dhabi City, which is expanding on the neighbouring islands Saadiyat and Reem and on the mainland, mainly close to the International Airport and towards Dubai. Due to the redevelopment of the harbour area, the Central Business District will be expanded in the north, where many proposed bridges will link to the business districts in Saadiyat and Reem. In addition to the CBD extension, the new master plan proposes the future development of a second downtown called Capital District, which will occupy an area of around 49 sq km on the mainland. The mixed-use development will be able to house about 370,000 future inhabitants and it will comprise high-density communities and lower-density Emirati neighbourhoods[22]. There have also been suggestions to develop the Grand Mosque District and Lulu Island District. Except for these plans for the future, the developments that are already under construction cover an area of around 74 sq km offering residential projects for around 715,000 future residents by 2020. While certain developments will have high urban densities, for example, the Raha Beach development with about 240 people living on one hectare, other projects such as Saadiyat will have average densities of just 55 residents per hectare.

Due to the large number and variety of projects, Abu Dhabi has recently become one of Dubai's biggest competitors regarding almost every economic sector from tourism and trade to financial services. It seems that the long-term symbiotic relationship of these two emirates wherein Abu Dhabi has been the conservative administrative centre outshone by Dubai as the up-coming global city, has recently come to an end. The new ambition of Abu Dhabi's rulers to transform the emirate into a global business centre has inevitably led to economic strategies that are similar to those of Dubai, for example, the construction of a large-scale airport or harbour in order to establish tourism and trade as major driving forces in the post-oil era. Despite all the up-coming rivalries, Abu Dhabi, one of the oil-richest places in the world, has bailed out Dubai with about $10 billion to $15 billion in 2009 due to Dubai's struggle with insufficient liquidity caused by the international economic crisis. This may mark another turning point in the relationship of both emirates, in which Dubai is losing its economic autonomy and Abu Dhabi is gaining a new degree of influence over its neighbour[23]. Consequently, Abu Dhabi's biggest asset for its long-term economic success is the fact that it is still one of the top ten oil producers in the world. Although certain projects such as Masdar illustrate a new attempt to establish a more ecological urban development, the overall development pattern and the structure of most projects do not differ from those in other places such as Dubai. While Dubai expects a population of 3 million people by 2015, the government of Abu Dhabi is planning for around just 1.3 million inhabitants by 2013[24]. However, like Dubai, where currently less than 12% of the population are Emiratis, Abu Dhabi expects that the share of its local population of currently around 21% will drop. Although Plan 2030 has many suggestions for smart growth development, because of the current development trends and construction growing annually at a rate of 25%, it is doubtful if the current master plan will be able to direct development and enforce consolidation of the built environment[25]. All in all, there are few signs that the urban development of Abu Dhabi will be different from the development of its neighbour Dubai. The pressure to catch up in combination with an inevitable liberalised investment climate with few restrictions will lead to a similar path of development causing similar problems of consolidation.

The Capitals of the Northern Emirates:

Apart from Abu Dhabi City and Dubai City, which are the centres of the two largest emirates of the UAE, the remaining five emirates in the north have gradually developed into urban centres, benefiting from Dubai's enormous growth in recent years. The development of the emirate of Sharjah in particular, stretching along the borders of Dubai, has been enormously effected by the growth of its neighbour, where increasing rental prices have forced thousands of guest workers with middle and lower incomes to move. Consequently, large residential high-rise clusters have been developed along the border of Sharjah and Dubai. Today, around 890,700 people live in Sharjah, which, at around 2,600 sq km, is the third largest emirate of the UAE. The districts along the border with Dubai and the urban centre at the Corniche are the most densely populated areas.

In contrast to the rulers of Dubai and Abu Dhabi, who are related members of the Ban Yas tribe, the rulers of Sharjah are descendants of the Al Qasimi tribe, which moved from the eastern coast of the Arabian Peninsula to the Arabian Gulf in the 17th and 18th centuries. Another major difference from its neighbours is a conservative attitude that has persisted with regard to restrictive rules such as the prohibition of alcohol. However, the emirate has recently joined the trend of liberalising the real-estate market, leading to the launch of several projects. In addition to many small-scale developments, mainly residential towers, the Nujoom Islands project, which is currently under construction, is the largest mixed-use development in Sharjah spreading over 5.6 sq km along the coast close to the border with Umm Al Quwain. Nujoom Islands will create around 30 km of new developed coastline. Around 60% of the development will be occupied by green areas. The remaining 40% will consist of various residential and commercial projects. In addition to around 40 residential and commercial high-rise buildings, there will be around 145 apartment buildings and about 1,400 villas. All in all, the developer Al Hanoo Holding Company expects that the project will house around 40,000 future residents[26]. The Saudi-based real-estate company Al Hannoo is also developing Emirates Industrial City, which covers around 8 sq km and is intended to attract logistics companies[27]. Another development is Sharjah Investment Park on an area of around 3 sq km close to Emirates Road. The mixed-use development consists of an Industrial Park for light and medium industries and a Logistics Park for warehouses as well as labour and staff accommodation in the form of medium-rise apartment buildings. In charge of the project is master developer Snasco, who is also responsible for the Al Basateen project, a mainly residential development, in the outskirts of Sharjah[28].

At around just 259 sq km, Ajman is the smallest emirate in the UAE and surrounded by the emirate of Sharjah to its north, east and south. Today, around 360,000 people live in the small emirate, which has lately benefited from increasing immigration by guest workers working in Dubai, particularly low-income groups, due to increasing rents in the neighbouring emirates Dubai and Sharjah[29]. Due to the growing interest in cheap residences, Ajman was the third emirate after Dubai and Ras Al Khaimah that offered the possibility of developing freehold properties[30]. Consequently, there are plans to build 15 mainly residential projects along the Emirates Road, such as Emirates City, Marmooka City and Al Humaid City. The projects mainly consist of residential towers with 20 to 50 floors surrounded by green areas and small commercial centres in the form of malls and supermarkets. Emirates City, for instance, will consist of around 72 of these high-rise buildings and a central park with lakes. Marmooka City will contain around 200 residential and office towers and Al Humaid City will contain about 69 high-rise projects[31].

In addition to high-rise developments, several low- to medium-rise projects such as Al Ameera Village and Ajman Uptown are also being constructed. While Emirates Road has become the new centre of real-estate developments in Ajman due to the accessibility to the neighbouring emirates, an announcement has been made to develop the project Ajman Marina at the emirate's waterfront as well as several single high-rise buildings in the centre of Ajman. Furthermore, the development of the first International Aiport began in 2008 in the Al Manama area in order for Ajman to join the regional aviation boom[32].

Umm Al Quwain, the UAE's second smallest emirate, occupies an area of around 777 sq km along the northern border of the emirate Sharjah. With only around 66,000 inhabitants, who mainly live in a small city at the coast, it is currently the least populated emirate. Nevertheless, four large-scale developments have been begun after new property laws were introduced which allow non-GCC nationals to own property without, however, owning the land they are built on[33]. The developer Emaar is responsible for the project Umm Al Quwain Marina, which will cover around 6 sq km. The project, which will be partly developed on reclaimed land, will consist of around 6,000 villas, 2,000 townhouses and around 1,200 resort and hotel rooms along a 23-kilometre long waterfront comprising beachfronts and marinas[34]. Another development along the coast is the White Bay project, which is being developed by the Emirates Sunland Group on an area of around 1.77 sq km. The project will offer around 8,000 leasehold residential units in the form of villas, townhouses and multi-storey apartment buildings with up to 15 floors[35]. The largest development in Umm Al Quwain is currently Tameer's Al Salam City, which will cover around 20.5 sq km. Al Salam City will consist of a number of residential districts, a downtown area, parks and entertainment centres. The downtown area of the project will contain a shopping mall, a 50-storey hotel tower, 20 residential and commercial towers with 20 to 25 floors each and 1,000 residential low-rise dwellings and 200 residential buildings with five to ten floors. All in all, the project is expected to accommodate half a million residents after the completion of all three development phases by 2020[36]. In addition to these large-scale residential developments with integrated commercial projects, the development of the Emirates Modern Industrial Area is another project of Tameer. The free-zone development, which occupies around 4.65 sq km on Emirates Road, is a self-contained project integrating warehouses and light industrial facilities in addition to commercial developments and labour accommodation[37].

Further north, the emirate of Ras Al Khaimah covers an area of around 1,700 sq km and like its southern neighbours it has recently become an attractive location for several developments. Today, around 263,000 people live in the fourth largest emirate, whose rulers decided early on to liberalise the property market. One of the first developments was Al Hamra Village, which is located partly on reclaimed land in the south of the small town Al Jazirah Al Hamra. The project consists of 1,350 residential units, a marina, a shopping mall and three hotels[38]. Another beachfront development is The Cove, comprising 134 villas and a hotel on an area of more than 200,000 sq m[39]. While these two developments have already been completed, three other waterfront projects are scheduled to be completed by 2011. The largest of these is Mina Al Arab, which is being developed by RAK Properties on an area of around 2.8 sq km spreading over 13 kilometres along the coast in the form of two islands and a beachfront strip on the mainland. After completion the project is expected to offer around 3,500 apartments and 388 villas as well as medical and recreational facilities[40]. The second largest waterfront development is Al Marjan Island developed by Rakeen and located close to Al Hamra Village on a reclaimed area of around 2.7 sq km. The coral-shaped islands are occupied by around 50 large villa sites, 10 hotel sites and a theme park. The third waterfront project is Saraya Ras Al Khaimah, which is being developed by Saudi investors on three man-made islands covering an area of around 1.4 sq km. The three islands in addition to Saraya Village on the mainland will offer around 4,800 residential units and hotels and a marina[41].

In addition to these five projects, Ras Al Khaimah Eco City is an ecological city project designed by Rem Koolhas, which will be developed in five phases on an area of around 37 sq km. The first phase, which comprises a 1.2-sq km integrated city that will service, supply and supplement the capital city of the emirate, is expected to be completed by 2012. Similar to Masdar in Abu Dhabi, the project will be developed on a square area and will integrate modern technologies such as photovoltaics. The design of the project is inspired by the density of traditional oasis settlements and there are plans to use local construction materials. The biggest contrast to Masdar will be the development of a large number of high-rise buildings, which will create a high density and short distances between different uses and services[42]. The project is based on a conceptual design from 2006 by OMA known as 'City In The Desert', which was designed for any location in the UAE on an area of around 43 sq km for around 150,000 future inhabitants[43].

In the north-east of the UAE in the Gulf of Oman, the emirate Fujairah at around 1,150 sq km is the fifth largest emirate in the UAE and has a population of around 130,000 people. Fujairah's efforts in the freehold market have been limited to a small number of projects which include the 43-storey and 170-metre tall Al Jabar Tower offering 270 apartments and developed by Al Jabel Contracting[44] as well as the Al Dana project and Fujairah Paradise offering around 1,000 villas on an area of around 650,000 sq m along the coast[45]. All in all, despite the different scales of projects, each of the five northern emirates has been influenced by the recent developments in Dubai. Due to the exponential increase of land prices in Dubai, many investors have found the northern emirates to be an investment alternative as well as because of the connection of all the emirates by a modern highway network, particularly Emirates Road. Consequently, thousands of guest workers have relocated to the northern emirates because of cheaper rents, particularly Sharjah and Ajman. This exodus of Dubai's working class has led to a growing number of dormitory settlements and thus inevitable traffic congestion into and out of Dubai during rush hours. Despite the increasing coalescence of the various emirates' urban areas, there has been little effort at strategic cooperation. Thus, the Dubai Metro, for instance, will not be extended to Sharjah and the emirates further north in order to ease commuter traffic. This is mainly caused by Dubai's fear of losing more residents if they move to the northern emirates, leading to a decreasing real-estate market in Dubai. In this regard, it remains doubtful that the announced railway system interconnecting all urban centres in the UAE will be constructed in the near future. Hence, the dependency on highways is expected to increase, which will lead to more and more traffic congestion. Apart from the development of residential projects for Dubai's workforce, several small-scale industrial projects are also underway such as Sharjah Investment Park and the Emirates Modern Industrial Area. In addition, many leisure and hotel projects have been integrated within developments along the coast. Last but not least, ports and airports have been planned, built or expanded. Even the smallest emirate Ajman has recently begun its own airport development, which will lead to the creation of a conglomerate of around four international airports within a distance of less than 170 km.

Until recently the built environment of the northern emirates was influenced by the oil urbanisation of previous decades. Large areas were occupied by suburban low-rise residential areas and downtown areas mainly consisted of the housing areas of low-income labour. New developments are changing the former urban landscape with their high-rise conglomerates along the highways and exclusive low-rise residential developments with integrated landscaped leisure areas, particularly along the coast and on reclaimed land. All in all, around 14 large-scale master-planned developments are under construction on a total area of around 49 sq km offering approximately 190,000 residential units by 2020. Through this development trend, the whole coast from Abu Dhabi to Ras Al Khaimah could develop into one urban agglomeration consisting of a large number of individually designed projects, leading to a patchwork-like urban structure. The biggest challenge of the future will be how to resolve old rivalries between the emirates in order to establish a cooperative urban development instead of increasing competition and thus a growing lack of integration and consolidation.

6.1.3 Qatar's Capital Doha

Parallel to the recent developments in the UAE and Bahrain, Doha, the capital of Qatar, has also become a centre of the current real-estate boom. During the pre-oil era Doha was a small oasis settlement known as Al Bidaa on the eastern coast of the Qatari peninsula. Like other places in the Gulf, its pre-oil economy was based on agriculture, fishing and trade. In the 19th century, the long-term conflict between the tribe Al Khalifa, which had left Qatar in order to settle in Bahrain, and the tribe Al Thani, which remained on the peninsula, ended with the intervention of Great Britain and Qatar's new political situation as a British protectorate[46]. At the beginning of the 20th century the pearl trade led to Qatar's first economic upswing and a population increase to around 12,000 inhabitants living in eight distinct settlements along the bay covering an area of around 1.25 sq km. Due to the collapse of the pearl trade in the 30s, Qatar's population dropped from around 27,000 to 10,000, caused mainly by the move of many Qataris to eastern Saudi Arabia looking for jobs[47]. After the first discovery of oil during the 40s and the beginning of oil

Fig. 4: The main projects in Doha.

production after World War Two, Qatar's economy quickly recovered and the oil boom during the following decades led to the first phase of modern urbanisation, transforming Doha from a group of small settlements into an exponentially growing oil city. Within only two decades the urban area increased tenfold to around 12 sq km and in 1970 the population was about 85,000 with foreign guest workers constituting around 67%[48]. Thus, the period between 1949 and 1969 witnessed a population increase of about 600% and an increase in foreigners of more than 1,000%. In addition to many guest workers, many Qataris from rural settlements moved to Doha during the 50s and 60s. Consequently, the city was structured into four main districts which were characterised by their different inhabitants. While the Qataris settled in the northern and eastern districts, foreign labour was accommodated in the western and southern districts.[49]

Although Doha's basic infrastructure was developed in the first phase of its urbanisation in the form of the first road network including Al Corniche Road along Doha's natural bay, its urban development was not based on a comprehensive plan because there was no urban planning institution at that time. It was not until 1970 that a modern government was installed, which led to the first public planning authority – the Public Works Department. Its first major challenge was to solve the housing crisis, which was increasing after Qatar gained independence and Doha became the official capital in 1972[50]. In order to accommodate the population growth, the government commissioned its first foreign planning consultant, the UK-based Llewelyn-Davies, to design a master plan for Doha extending to 1990. One of the recommendations of this early plan was the redesign of the centre by replacing the old buildings with modern multi-storey buildings and shopping

231

districts. Consequently, the government began to acquire land from Qatari citizens, who owned approximately 90% of all lots at that time. In the following decades the Qatari families moved to newly developed suburbs in the north, west and south while more and more guest workers rented apartments in the centre. This development was accelerated by the government, who decided to pay higher than market prices for the houses of Qatari families, which also led to a redistribution of the oil wealth to the local population. While a large number of buildings in the centre were replaced, the few remaining ones were used as accommodation for guest workers with the lowest incomes. During the time of rebuilding the centre and the development of more and more suburban settlements, the first modern business district was established along the six-lane Grand Hamad Street leading from the International Airport through the centre to the harbour area at the coast. However, only a few banks, hotels and shopping centres settled along this highway due to the reclamation of a large area in the north-west of the city which soon became Doha's most attractive development area[51].

In 1974 the reclamation of over 2,000 hectares began along the bay, shaping the natural bay into a circular corniche, which became a new symbol for modern Doha. Hence, the most iconic government and commercial buildings were built along the Corniche and the basic structure of the city was formed by a system of ring roads intercepted by several radial roads[52]. A large part of the construction activities during the 80s took place in the new district along the Corniche, where the university, a diplomatic quarter, residential blocks for expatriates, a football stadium and the Sheraton Hotel as a new landmark were developed. In this time the capital city area grew to around 20 sq km and the population to approximately 200,000, of which nearly 70% were foreign guest workers. In the following years the population continued to grow, reaching around 400,000 people in 1995[53]. While the 70s and 80s were mainly dominated by basic infrastructural projects and urban sprawl due to ever-increasing suburban settlements, the end of the 90s marked a turning point in the urban history of Doha when the government announced the transformation of the capital into a global destination for leisure, business and sporting events. Thus, the Qatar Tourism Authority worked out a plan regarding the development of hotels, lifestyle resorts, cultural projects, sports facilities and a more capable infrastructure. While in 2004 around 500,000 tourists visited Qatar, the government expected this number to more than double, reaching around 1.5 million visitors by 2010. Similarly to the recent strategy of Abu Dhabi, Qatar's rulers decided to invest in the construction of museums and other cultural institutions in order to make Doha more attractive as a travel destination. One of the most prestigious projects, the Museum of Islamic Arts, was designed by I.M. Pei and developed on a reclaimed area close to the harbour area[54]. In addition to the development of museums the Souq Waqif was reconstructed in the old city centre in order to create an authentic and traditional-looking market consisting of shops, cafes and restaurants. In order to extend the market and its pedestrian zone, the demolition of old buildings, which were mainly constructed during the 70s, has continued[55].

While the old buildings in the former city centre have been replaced by low-rise buildings

Fig. 5: The high-rise conglomerate in West Bay.

with traditional-looking architecture, the opposite side of the Corniche has developed into Doha's main CBD – a growing high-rise agglomeration known as West Bay. The centre of this modern business district of Doha consists of a the large-scale City Centre mall, which is surrounded by about 60 mainly commercial towers with up to 50 floors. Consequently, the waterfront at the Corniche has become more and more dominated by the intended antagonism of modern high-rise buildings in the north and traditional-looking buildings in the south such as the Museum of Islamic Arts and the Souq Waqif. While development in recent years was mainly focused on the coast, the government has initiated the construction of Sports City around 11 kilometres inland. The centrepieces of this 130-hectare development are the Khalifa International Stadium and the 318-metre tall Aspire tower, which is currently Doha's tallest landmark. The main reason for the new sports facilities is the government's aim to establish Doha as a host of global sports events such as the Asian Games in 2006[56]. In addition to the goal of attracting tourists with sports events and cultural projects, Education City was created in order to establish Doha as a centre of academic education and knowledge-based industries. Education City is located on a 14-sq km area about 5 kilometres to the north of Sports City. The initiative of Education City goes back to 1995 when the Qatar Foundation was established with a mandate to develop the large-scale project, which contains facilities for the education of children and adults up to postgraduate studies as well as institutions for applied research such as the Qatar Science & Technology Park and the Qatar National Research Fund. All in all, there are currently six universities, most of which are partners of US universities such as the Virginia Commonwealth University[57].

In addition to the above cultural, sports and educational projects, investment has been put into several real-estate developments. Since 2004 foreign ownership of properties has been permitted on a lease basis for 99 years in 18 areas across Doha and on a freehold basis in the three freehold zones Pearl Qatar, West Bay Lagoon and Lusail. In 2006 the construction of the Pearl Qatar was begun by United Development Company, which is currently Qatar's largest private sector shareholding company. After two years, the reclamation of the 4-sq km island was completed in the north of Doha, which created around 32 kilometres of new shoreline. The project's shape was inspired by pearls, which were previously found at the location of its site. In contrast to many other developments,

Fig. 6: The Pearl project in Doha.

the properties on the Pearl Qatar are sold to foreigners on a freehold basis in order to attract investors. All in all, the development comprises five main precincts in addition to several low-rise residential developments. The two largest precincts – the Porto Arabia and Viva Bahriya precincts – are located along two circular bays, contain many residential high-rise buildings and townhouses as well as large retail areas, leisure facilities and marinas. In between these two precincts the Medinat Centrale is the island's centre for amenities and facilities, including a large retail district and parks. While the Abraj Quartier has the tallest high-rises in the north-west of the island, the Qanat Qartier, which is next to it, was inspired by the architecture of Venice and consists of medium-rise buildings along several canals. Further east, the Costa Malaz contains several hotel and spa developments as well as several villas along the smallest of the three bays of Pearl Qatar[58]. The remaining areas are occupied by seven low-rise residential developments, which include the exclusive Isola Dana project on nine separate islands in the east. After its staged completion starting in 2009 and ending by 2011, it is expected that around 41,000 residents will move to the development. More than half of the future population will live in around 66 high-rise residential towers offering approximately 12,000 residential units[59].

To the west of the Pearl Qatar, West Bay Lagoon sits on an area of more than 3 sq km. Several hundred buildings are being developed along a man-made lagoon extending over 2 kilometres towards inland. In addition to many residential projects, which offer housing units for around 35,000 future residents, there will be hotels, a commercial centre and several recreational facilities. Today, the almost completed 34-storey Lagoon Plaza towers, which contain 748 apartments, are the biggest landmarks of the development and due to their unusual shape they are known locally as the 'dancing towers'[60]. To the north of West Bay Lagoon, the Lusail development – a new waterfront city and currently Qatar's largest project – covers more than 21 sq km. The large-scale mixed-use development will contain several residential areas, commercial districts and shopping malls as well as two marinas and two golf courses. The project will also integrate schools, mosques and medical centres. It is expected that more than 200,000 people will move to the development, which will offer properties to foreigners on a 99-year lease basis. Lusail's developer Diar Real Estate Investment Company, which is fully owned by the Qatari government, is planning to build the project in several stages by 2017. In addition to the Marina District – a mixed-

234

use area on the seafront – the projects Fox Hills and Energy City are being constructed in the centre of Lusail[61]. While the mixed-use development Fox Hills will cover around 1.6 sq km with mainly medium-rise buildings with four to seven floors, Energy City will be the first hydrocarbon industry business centre in the region, which will act as an oil and gas hub[62]. The first landmark project of Lusail will be Lusail Towers – a group of four towers designed by Foster & Partners[63]. Together with the Pearl Qatar and the West Bay Lagoon, the Lusail development will be an urban agglomeration of around 28 sq km.

Close to Sports City and Education City, the developer Nasser Bin Khaled and Sons Group is constructing Al Wa'ab City on an area of around 1.2 sq km. The mixed-use development integrates a variety of residential projects offering around 2,000 housing units and several commercial projects including one big shopping mall and several health facilities. Several parks and green areas will also be integrated within the development area. After completion by 2011, it is currently expected that around 8,000 people will move to the development, which will have one of lowest urban densities in Doha. In addition to several medium-rise residential buildings, which will provide around 1,450 apartments, over 450 villas will be constructed[64]. While there are currently around four large-scale real-estate developments in Doha, there are also a number of relatively small projects, for example, West Bay Complex or Barwa Al Doha, which is being built by the developer Barwa. This project will cover around 55,000 sq m along Muntazah Road close to the old city centre. While the project contains several apartment buildings and around three residential towers offering together around 6,000 housing units, a large area of around 127,000 sq m will be occupied by retail space[65]. Barwa, which is one of Qatar's largest developers, is constructing several other developments such as the Barwa Financial District, which includes ten high-rise buildings with up to 45 floors. Another current development is Barwa Ain Khalid – Commercial Avenue, which is a low-rise mixed-use development along 8 kilometres of a new road leading from Doha's centre to the industrial area in the west offering around 600 retail spaces and about 850 residential units. Barwa is also constructing a mixed-use development called Barwa City, which will cover around 2.7 sq km in Musameer on the outskirts of Doha. The project will contain around 128 medium-rise residential buildings offering more than 7,000 apartments for an estimated population of around 20,000 people by 2010. In contrast to many other developments which target the upper real-estate market, the project is intended to be an economic alternative for middle-income groups[66].

Barwa's largest development covers an area of around 5.4 sq km in Al Khor, a coastal town about 50 kilometres north of Doha. After completion by 2015, the mixed-use development, known as Urjuan, will contain several residential projects in the form of villas, townhouses and apartment buildings offering around 24,114 housing units for an estimated population of around 60,000 people. It will also have two hotel complexes, a shopping mall, four schools, a mosque, an international golf course and around 250,000 sq m of office space[67]. In addition to Barwa's project in Al Khor, a development called Mesaieed Industrial City (MIC) is being constructed by the Qatar Petroleum Company around 40 kilometres south of Doha in Umm Said. The project will be a new centre for various industrial, commercial and residential projects close to the old industrial area where the Qatar Petroleum Company established a tanker terminal in 1949[68]. In the past decades several oil-related industries have settled close to the port while only a few workers' accommodations were built in the north. Today, the whole area will be turned into a mixed-use self-contained city, the current master plan of which includes several projects including a Medical Centre, Gabbro Import Facilities, the Centralized Office Complex and Al Afjah Heritage Village[69].

Parallel to these real-estate developments, there has been increasing investment in infrastructure projects such as New Doha International Airport and the expansion and relocation of ports. The new airport will be developed on reclaimed land in the east of Doha right next to the area of the current airport, which will be turned into an urban expansion area. After completion by 2011 the airport will be able to handle around 24 million passengers, 1.4 million tonnes of cargo and over 360,000 aircraft movements annually[70]. Apart from the construction of a new airport, the port will be relocated to a new area close to Mesaieed Industrial City in order to develop a modern port that will meet Qatar's requirements beyond 2030[71]. In addition to the new port development in the south of Doha, LNG Port in Ras Laffan is being expanded, where the world's largest natural gas terminals are located[72]. Last but not least, in cooperation with the Kingdom of Bahrain, a 40-kilometre long causeway from the small island to Qatar is being constructed, which is expected to have a large economic impact on both countries. However, due to the international financial crisis the project has been delayed and its construction did not commence on the scheduled date in January 2009[73].

One of the biggest consequences of the recent construction boom in Qatar has been an increasing population because of immigrating guest workers, particularly in Doha, where the population has more than doubled from around just 339,847 inhabitants in 2004 to more than 998,651 in 2008[74]. In order to cope with the rapid urban growth, the preparation of a new comprehensive master plan has been launched. As well as the international consultancy Pacific Consultants International (PCI) from Japan, who was been commissioned to design the plan, the Urban Planning & Development Authority (UPDA) has been engaged as a public authority to direct and oversee the plan's preparation. The plan, which is known as the 'Master Plan for the State of Qatar', is expected to guide the physical development of Qatar to the year 2025. Its main intentions are to ensure that future development is timed to infrastructure and road capacity and that adequate land is allocated for residential, commercial and industrial projects and public services. The new plan also seeks to evaluate the recent developments in order to minimise negative impacts in the future. Its major goal is to implement a physical development strategy that promotes a vision for Qatar as an academic, sports and tourism centre in the region and for Doha to be a capital city well renowned for its international influence in politics and corporate business. The master plan will be updated every five years in order to guide physical development in accordance with changing social and economic needs. The plan also aims to improve the integration and coordination of development projects to replace the current practice of decentralised and isolated decision-making processes by various agencies. In order to meet all the major goals there are five different levels of planning which are all related and complementary to one another – the National Plan, the Regional Structure Plans, the Local Area Plans, the Detailed Area Plans and the Urban Design Plans[75].

Although the scale and number of Doha's projects is not comparable to those of Dubai, both Doha and Dubai have developed a strong affinity to each other over the past few years. The very particular relationship of Doha and Dubai was established through marriages between their ruling families and by the unified currency, the Qatar-Dubai riyal, which was in use until the early 70s. While strong ties remain, an expression of which is the 80-storey Dubai Towers Doha, there has been growing competition between the two cities. In many cases Doha has followed Dubai, as for instance in the liberation of the real-estate market for the development of large-scale freehold property projects such as the Pearl Qatar, which was announced two years after the Palm Jumeirah in Dubai. Qatar also copied Dubai in the establishment of free economic zones, for example, the Financial Centre of Doha was established two years after the Dubai International Financial Centre. However, in other cases Doha has been a forerunner in development. For example, the state-owned news channel Al Jazeera was established in 1996 long before Dubai's launch of its equivalent Al Arabiya and the development of Education City commenced long before the Knowledge Village project in Dubai. Education City is also larger than Knowledge Village, which compelled Dubai's rulers to initiate Dubai Academic City on an area of almost 12 sq km. Another economic sector the two cities are competing for is the aviation industry, with new large-scale airports being developed in both Doha and Dubai and the expansion of the airlines Qatar Airways and Emirates[76]. All in all, Qatar and its capital Doha have proven to be a serious competitor in the region and thanks to the remaining wealth of natural gas and oil Qatar's rulers have a fortunate starting position to establish Doha as a regional and global economic centre. Because of the ambitious development goals the built environment has begun to transform the face of the former oil city Doha more and more. In addition to the growing high-rise cluster in West Bay, an agglomeration of three master-planned projects in the north and the redevelopment of the old city centre to the south of the Corniche as well as high-profile developments such as Sports City and Education City are shaping contemporary Doha.

6.2 Centres with Inhibited Development

6.2.1 Kuwait City
The city-state of Kuwait is located in the north of the Arabian Gulf on an area of around 17,818 sq km. In 1672 Kuwait City was established when inland tribes from Saudi Arabia moved to the coast. These included the Al Sabah clan, who has ruled Kuwait since the

middle of the 18th century. At this time the population grew to around 10,000 and in 1793 the British East India Company established a base in Kuwait, which led to a long-term relationship between Kuwait and Great Britain. Thus, in 1899 the Al Sabah family concluded a contract with Great Britain in order to receive protection from the Ottoman Empire. In addition to its function as an important harbour, Kuwait became a centre of pearl diving, particularly during the 20s when more than 10,000 sailors and divers worked on around 800 ships[77]. The socio-economic conditions started to change in the 30s when the first municipality was established in 1930 and the first oil was found in 1938. Consequently, oil production, which began in 1946, transformed the tribal society that had mainly lived from fishing and trading into a typical oil-city society with a growing number of guest workers and a large dependency on public subsidies[78].

Along with oil wealth, modernisation began and in 1951 the first master plan, designed by the British firm Minoprio, Spencely and McFarlane, was introduced. The main goals of this plan were to provide a modern road system, to create a zoning plan and to modernise the town centre. In the following decades this plan was implemented by the Ministry of Public Works under the supervision of the Kuwait Development Board, which was established in 1950 and headed by the Amir of Kuwait. In order to provide land for the construction of new roads and buildings, the city walls and old residential areas were demolished. New residential areas were developed in the south and due to the widespread integration of schools, shops, mosques and other services the new neighbourhoods were rather self-contained regarding services. Most workplaces however remained in the centre, which turned these neighbourhoods into dormitory settlements[79]. Each of these new neighbourhoods accommodated around 12,000 residents and freeways on all four sides provided sufficient accessibility[80].

During the 50s the old city centre became the new CBD and most traditional buildings were replaced with modern cement buildings. Despite the large-scale demolition of old urban areas, the old oval form of the city centre is still recognisable for a road was constructed along the line of the old city wall. Since the 18th century the wall has had to be removed and reconstructed twice because of the gradual urban expansion. Thus, the first wall, which was built in 1760, was replaced by a new wall in 1811 because of the growth of the urban area from just 11 hectares to around 72.4 hectares. Finally, the third wall was built in 1921, when the urban area had grown to about 750 hectares, which is approximately the size of the current centre[81].

In addition to the development of modern administrative buildings and commercial projects in the city centre, a new harbour was built about 3 kilometres from the city centre to the west and along with it a large industrial area. After Kuwait's independence in 1961, the new International Airport was completed in the south of Kuwait City and gradually expanded in the following decades. In this time, new residential areas mainly grew towards the south, particularly along the coast and the new highway leading to Al Ahmadi, which became the centre of the oil industry. At the end of the 60s Colin Buchanan & Partners was engaged to design a second master plan in order to improve the

Fig. 7: The City of Silk.

road system, which led to a focus on the construction of a road network along the eastern shoreline and thus the city expanded from north to south. After this master plan was at first reviewed by Shankland Cox in 1977, who suggested the development of two satellite towns, Colin Buchanan & Partners adjusted the plan because of the rapid urban growth in 1983. Later on, in 1990 the third master plan was developed by the Kuwaiti firm SSH and W.S. Atkins, which was again reviewed in 2003 by the Kuwait Engineering Group in cooperation with Colin Buchanan & Partners[82]. The designs of these first master plans, which were mainly focused on accommodating the needs of cars, have contributed to a massive urban sprawl. Furthermore, the vast immigration of guest workers who have been employed in the lower service sector has caused increasing segregation.

While in 1961 Kuwait had a population of about 321,621, of which 49.7% were foreigners the population increased to around 2,152,775 in 1997, of which 65.4% were foreigners. Today, around 2.7 million people live in Kuwait, including around 1.3 million non-Kuwaitis, whose number has decreased due to the recent conflict in Iraq. During the years of the oil boom a certain settlement pattern evolved in which the local population moved to the outskirts, particularly to Ahmadi and Jahra, and foreign guest workers were mainly accommodated in the centre and in Hawali. While guest workers resided in low- and medium-rise apartment buildings, the Kuwaitis preferred one- or two-storey villas, which currently occupy most of the settlement area of Kuwait City. Consequently, the urban area has become dominated by a rather monotonous repetition of low-rise residential areas and only a few public buildings, for example, the parliament building, stick out due to their architecture, designed by world renowned architects. Until today, the biggest landmark of Kuwait City is the Kuwait Towers on Gulf Road – an arrangement of three towers, of which two are used as water containers and the third is a lighting pole. In addition to its water reservoir, the tallest tower with a height of around 180 metres contains a restaurant and a rotating observatory[83].

Mainly outer circumstances such as the third Gulf War which began in 2003 have

239

prevented Kuwait City from developing in the same way as, for example, the UAE, Qatar or Bahrain. Due to their proximity to the conflict in Iraq, Kuwait's rulers have preferred to invest a large part of their country's wealth in neighbouring states or foreign countries than in developments within Kuwait itself. However, the recent success of the real-estate market in the region has led to the announcement of several large-scale projects in Kuwait. While the largest project currently under construction is the 412-metre tall Al Hamra Tower, which will be Kuwait's tallest building after completion by 2010, there are plans to develop an entire new city on the north side of the Kuwait bay area. The project, which is known as Madinat Al-Hareer or City of Silk, will cover an area of around 250 sq km where around 700,000 future residents are expected to settle within the next 25 years. The centrepiece of the development will be the 1,001-metre tall Burj Mubarek Al Kabir. Other developments in Madinat Al-Hareer are Finance City and Silk Road Free Zone, which are both designed to attract international trade, finance and commerce. There has also been the announcement of Leisure City, Culture City and Ecological City, which are mainly focused on developing Kuwait's tourism industry. Apart from the integration of projects geared toward the tourism, trade and financial sectors, there have been plans to develop the health and education industry through Health City and Education City[84]. In addition to Education City in the City of Silk, New University City will be developed in Shidadiyah within the next 10 years. The 5.2-sq km project will provide a modern campus with state of the art facilities for around 40,000 future students[85]. Another project of the near future is the Heritage Village development, which includes the construction of a traditional village in the old city centre in addition to the preservation of a few historic buildings[86]. Lastly, there have been plans to turn Falaika Island, which lies 20 kilometres east of Kuwait City, into a major tourist destination by developing new resorts[87].

Despite all these projects, which were announced in 2003, Kuwait has not yet become a major attraction for real-estate developments. Apart from some high-rise projects such as the 240-metre tall Arraya Office Tower, very few projects have been completed or even begun. The main reason for the lack of interest is the on-going conflict in Iraq, which has created a rather unstable investment climate. Furthermore, the recent outbreak of the global financial crisis, which has led to a general economic downturn in the Gulf, has made many plans for large-scale developments rather unrealistic. Consequently, Kuwait cannot be seen as a major competitor to the UAE, Qatar or Bahrain, which can be considered the current centres of post-oil urbanism. However, the still existing oil wealth of Kuwait is an important basis for its future economic development, and while its location close to the biggest conflict in the region is currently hindering its urban development, it would have a development advantage if political stability and peace were to be established in Iraq as it is one of the most wealthy oil nations.

6.2.2 Oman's Capital Muscat

The Sultanate Oman, whose territory covers an area of around 309,500 sq km, is the second biggest country of the GCC. Furthermore, it is the only GCC country with a long history of political power due to its centuries-old role as one of the biggest trading points to India and East Africa. Apart from the 200-year colonial reign of the Portuguese until the end of the 17th century, the sultanate was more or less able to sustain its political independence. In the first half of the 19th century Oman's political power reached its peak when it took a large part of East Africa including Zanzibar under control. Since 1967 Oman has belonged to the oil producing countries of the Gulf, which has led to a high economic dependency on oil. Thus, in the 80s and 90s almost 70% of all state income came from the export of oil and gas. In comparison to its neighbours, Oman has one of the lowest oil productions and until the present day one third of the population still works in agriculture. Hence, in addition to the growing capital Muscat, many rural settlements have expanded in Oman's agricultural region in the north[88]. Consequently, many regional centres outside of the capital such as Nizwa have been developed with modern infrastructure. While the capital Muscat with a current metro population of around 1 million people has become the political centre of Oman, Salaleh in the south has, with almost 200,000 inhabitants, become the second largest city and the biggest industrial centre. Heavy industries have recently started to expand due to the introduction of a new free economic zone in Salaleh, which has developed into the biggest transshipment hub for container shipping in the south of the Arabian Peninsula[89].

Similarly to other oil cities, Muscat has witnessed urban sprawl in its low-rise residential settlements on the outskirts. However, due to the very specific topological condition of the Omani coastal environment, there has been no development of one cohesive settlement area. Canyons and hills divide the settlement areas, which therefore have developed into rather self-contained districts. The largest settlement area is along the even coastal area towards the west where Seeb International Airport is located. In addition to the large-scale expansion of the airport, which is scheduled to be completed by 2011, a large real-estate project called the Wave is underway close to the airport along the coast. The Wave project is one of the first freehold property projects in Oman after the introduction of freehold ownership laws for 'tourism designated areas' at the beginning of 2006. The large-scale development is a joint venture between Oman's Waterfront Investments (representing the Government of the Sultanate of Oman), the Majid Al-Futtaim Group and the National Investment Funds Company (representing the Omani Pension Funds)[90]. Designed by Atkins, the project will stretch for about 7 km along the coast covering an area of around 2.5 sq km, which will be partly reclaimed. A large canal will also be developed along the southern border of the project and a large golf course will stretch along its eastern part. In addition to around 4,000 residential units, mainly villas, there will be four luxury hotels and several office buildings as well as retail and restaurant outlets. The commercial district will be mainly concentrated on the reclaimed land at the marina in the east where around 40,000 sq m of office space will be developed[91]. All in all, the whole development is divided into five different sectors, which are expected to be completed within five phases by 2012[92].

The biggest real estate development in Oman is currently a completely new satellite town of Muscat known as Blue City Oman. Like the Wave project, the initiator of this development is the government of Oman itself, who intends to diversify Oman's economy, particularly by attracting tourism. Blue City Oman, also known as Madinat A'Zarqa, is located on a natural peninsula around 70 kilometres to the east of Muscat. After its completion by 2020 the development is expected to house around 200,000 inhabitants on an area of around 32 sq km stretching along 16 kilometres of shoreline. The first phase of the development has been designed by Foster & Partners and includes 5,500 dwellings for a future population of about 27,000 residents. In addition to residential units, the 2.2-sq km first phase will also integrate hotels, schools and several leisure facilities such as an 18-hole golf course, a marina and a park. The master developer Al Sawadi Investment & Tourism Company (ASIT), who has the endorsement of the government of Oman to develop this project, founded the sub-developer Blue City Company 1 (BCC1), which is in charge of the development of phase one. Similarly to the Wave project, high-rise buildings will not be permitted and the design of both settlements and single buildings should respond to the local climate and culture[93].

Although these two large-scale developments are a sign of Oman's ambition to be part of the current race to become a centre of post-oil economies in the Gulf, the strict building guidelines and restrictions on investment, which are limited to special development areas, have prevented a construction boom like that in places such as the UAE. Oman however has benefited from the growth of the neighbouring country due to the rising tourism industry, which has led to more and more tourists visiting Oman because of its nature and history. Thus, Oman has become a tourism alternative to Dubai, providing visitors with the possibility of discovering a more traditional side of the Arabian Peninsula. One of the biggest reasons that local traditions have been maintained in Oman is that in contrast to most GCC countries the locals have remained at about 70% of the 2.8 million population and are thus the absolute majority in Oman. The Sultan's strategy to turn the local inhabitants into the driving force of the future economy by providing them with education and, more importantly, time to adjust to the new economic circumstances has contributed to Oman's relatively slow development. The emphasis on a climatically and culturally appropriate built environment is another important characteristic of Oman's current urbanisation. Thus, despite its very small oil resources, which have forced Oman to change its economic strategy that has long been dependent on oil, it has not undergone rapid privatisation and decentralisation for the sake of fast economic growth. While the restrictive policies of the current sultan, Qaboos bin Said Al Said, are most probably the main reason for this development, another important cause is the fact that investment pressure from the private sector has been rather low due to Oman's shrinking oil resources.

6.2.3 The Urban Centres in Saudi Arabia

Saudi Arabia, which is the largest country in the Arabian Peninsula, covers a territory of more than 2 million sq km stretching from the Red Sea to the Persian Gulf. Before Saudi Arabia became independent in 1932, most of its western part, including Mecca and Medina, was under the reign of the Ottoman Empire. The turning point towards the foundation of an independent state at the beginning of the 20th century was marked by the successful unification of most tribes under the leadership of the tribe Al Saud. This tribal federation was made possible by the spread of a new Islamic sect known as Wahabism, marriages between tribes and in several cases by force. In contrast to all other GCC countries, Saudi Arabia has remained a conservative Islamic country with many restrictions on lifestyle because of the establishment of Wahabism as the state religion.

The oasis town Riyadh, which in 1902 was recaptured by the tribe Al Saud, is the capital of the independent kingdom and its main administration centre. Apart from the traditional centres Riyadh and Jeddah, a new growing urban centre developed in the Eastern Province of Saudi Arabia when oil production began during the first half of the 20th century[94]. The oil settlements led to the foundation of the three new towns Dhahran, Dammam and Al Khobar, which together form Saudi's third biggest urban agglomeration with around 1 million inhabitants. Due to its expanding administration, which attracted the relocation of many families, the capital Riyadh became with 4.7 million inhabitants the largest city in Saudi Arabia and the entire Arabian Peninsula. At the Red Sea, Jeddah has remained, with 3.4 million inhabitants, Saudi's largest metropolis due to its harbour and proximity to the pilgrimage town Mecca. While today the population of Saudi Arabia has reached more than 28 million people, of which around one third lives in the three major urban centres, less than 4 million people lived in Saudi Arabia in the middle of the 20th century. This rapid population growth during the oil boom was caused by a rapidly decreasing mortality rate because of better living conditions and the widespread practice of polygamy. In addition, a large number of guest workers immigrated to Saudi Arabia in order to work in the private sector, particularly the lower service sector. Today, between five and six million people are foreign guest workers, mainly from South Asian countries such as Pakistan[95].

Today, Saudi Arabia is the world's largest oil producing country and thus its economy is still mainly dependent on oil, which has led to a rather subsidised local population mainly working in the public sector. The rapid population growth however is becoming a major threat to the social balance due to increasing unemployment causing growing poverty. Due to the recognition that the decreasing oil resources cannot be the economic basis of the future, the government has started to initiate several programmes to diversify the economy of Saudi Arabia. However, due to religious restrictions, it was not possible to integrate certain economic sectors on the same scale as the more liberal GCC countries did. This particularly concerns the tourism and entertainment industry, which is tightly connected to the current real estate boom in the Gulf. Despite this development obstacle, several new projects have recently been launched in order to relocate knowledge-based

Fig. 8: King Abdullah Economic City.

economies to Saudi Arabia. In this regard, there have been new strategic development plans for cities such as Riyadh where a growing part of the population faces economic problems.

It is currently expected that the population of Riyadh will grow to 10 million inhabitants over the next 12 years. Therefore, the Ar Riyadh Development Authority (ADA) has issued a new general master plan for 2021. Due to the previous oil urbansiation, Riyadh has undergone an exponential urban sprawl leading to an urban area of more than 1,000 sq km. In order to direct and limit urban growth, the current master plan includes three future satellite towns in the north, east and south. In contrast to previous master plans, which designed one main centre for the entire city, new sub-centres along the outer ring road and three main urban centres in the downtown area have been planned in order to decentralise the urban structure in future. Furthermore, the construction of the 22-km long Prince Abdullah Corridor, which is already underway, leading from the CBD in the west to new urban expansion areas in the east, will transform the previous urban structure which has been dominated by the Central Spine leading from north to south[96]. In addition to this new commercial spine along the Prince Abdullah Corridor, construction has begun on an airport city at King Khaled International Airport, which includes several projects related to the cargo industry as well as other industries such as information technologies. Since the completion of the 267-metre tall Al Faisaliah Tower and the 302-metre tall Kingdom Centre in the financial district of the Central Spine known as Olaya District, the development of several new large-scale high-rise projects has begun. While plans for the Al Jeraisy Towers, also known as Jewel of Saudia, have recently been proposed, the construction of the 430-metre tall Al Rajihi Tower has already started. One of the most important infrastructure projects is the proposed development of the Al Riyadh Monorail, which will connect the centres along the main spine.

Although there are no freehold property laws in Saudi Arabia, the recent developments have attracted investment from the private sector, which is expecting rapid growth in the local markets. The biggest example in this regard is King Abdullah Economic City (KAEC) on the coast of the Red Sea around 100 km to the north of Jeddah, which has become the largest single development of the private sector in the entire Gulf. The master developer Emaar, The Economic City, which is a subsidiary of the Dubai-based real-estate company Emaar, has been put in charge of the development. The Saudi Arabian General Investment Authority (SAGIA) is the main public development partner. It has chosen the KAEC to be part of the ambitious 10x10 programme, which aims to place Saudi Arabia in the world's ten most competitive places by 2010[97]. After the completion of all phases by 2020, the KAEC will cover around 173 sq km and provide residential units for a projected population of about 2 million people. The first phase of the KAEC, which involves the three districts Sea Port, Industrial Zone and City Districts, is already expected to be completed by 2011. The most ambitious goal of the development is the creation of 1 million jobs within post-oil economic sectors, which will be mainly located in the Industrial Zone[98]. KAEC will also have a Central Business District, which will offer more than 3.8 million sq m of office space including the Financial Island project. In addition, an Educational Zone will be occupied by several universities and so-called Research & Development Parks in order to establish the basis for a knowledge-based economy. Last but not least, the Sea Resort district, which is the first tourist development aimed at attracting local and international tourists to Saudi Arabia, will integrate several hotels and leisure facilities such as a golf course[99].

The KAEC project highlights Saudi Arabia's efforts to diversify its economy and provide more jobs for one of the world's youngest nations with a median age of just 22 years. Due to growing social problems based on an increasing unemployment rate and widespread restrictions due to Islamic laws, the current post-oil urbanism in Saudi Arabia has been different in scale from developments in other GCC countries. Nevertheless, examples such as the KAEC point to the possibility of a new future in which certain centres will have the chance to become competitive places, and because of the remaining oil wealth these new centres would have the opportunity to even outshine the currently booming centres in the Gulf. However, this would require a major transformation of the Saudi Arabian society and economy that is currently still rather unrealistic. Decades of oil urbanisation have led to a very difficult starting position for socio-economic transformation, which will need time in order to become more sustainable and competitive. However, with regard to architecture and urban design, the current urban developments hardly differ from developments in other GCC countries. High-rise agglomerations and exclusive residential districts with integrated leisure facilities such as golf courses and marinas are the two main characteristics of the current urbanisation in Saudi Arabia.

1   Elsheshtawy 2008, p. 266.

2   Elsheshtawy 2008, p. 269.

3   Elsheshtawy 2008, p. 270.

4   Elsheshtawy 2008, p. 273.

5   Urban Planning Council 2007, p. 38

6   Elsheshtawy 2008, p. 277.

7   http://www.centralmarket.ae/main.html, 09.05.2009

8   Gulf Construction August 2007, p. 80.

9   http://realestate.theemiratesnetwork.com/developments/abu_dhabi/al_reem_island.php,
09.05.2009

10 http://realestate.theemiratesnetwork.com/developments/abu_dhabi/shams_abu_dhabi.php,
09.05.2009

11 http://realestate.theemiratesnetwork.com/developments/abu_dhabi/najmat_abu_dhabi.php,
09.05.2009

12 Gulf Construction August 2007, p. 72.

13 http://yasisland.ae/Default_en_gb.html?1, 11.05.2009.

14 http://www.halcrow.com/html/our_projects/projects/yas_island.htm, 11.05.2009.

15 http://www.masdar.ae/en/Menu/index.aspx?MenuID=48&CatID=27&mnu=Cat,
11.05.2009.

16 http://www.masdar.ae/en/Menu/DescriptionPage.aspx?MenuID=48&Catid=27&cat_
id=27&SubcatID=25&mnu=SubCat, 11.05.2009.

17 http://www.masdar.ae/en/mediaCenter/newsDesc.aspx?News_ID=65&fst=mc&nws=1,
11.05.2009.

18 http://www.alrahabeach.com/main.php, 11.05.2009.

19 http://www.constructionweekonline.com/projects-276-mohammed_bin_zayed_city_devel-
opment/, 11.05.2009.

20 http://www.abudhabiairport.ae/theairport/index.asp, 11.05.2009.

21 Gulf Construction August 2007, p. 53.

22 http://www.ameinfo.com/193154.html, 11.05.2009.

23 http://www.economist.com/finance/displaystory.cfm?story_id=13186145, 11.05.2009.

24 Elsheshtawy 2008, p. 276.

25 Elsheshtawy 2008, p. 275.

26 http://realestate.theemiratesnetwork.com/developments/sharjah/nujoom_islands.php,
11.05.2009.

27 http://www.ei-city.net/, 11.05.2009.

28 http://www.sic.ae/SIC_Brocure.pdf, 11.05.2009.

29 http://www.oxfordbusinessgroup.com/country.asp?country=63, 11.05.2009.

30 http://realestate.theemiratesnetwork.com/articles/freehold_property.php, 11.05.2009.

31 http://realestate.theemiratesnetwork.com/developments/ajman/, 11.05.2009.

32 http://archive.gulfnews.com/uae/ajman/more_stories/32581.html, 11.05.2009.

33 http://realestate.theemiratesnetwork.com/articles/freehold_property.php, 11.05.2009.

34 http://www.emaar.com/index.aspx?page=umm-al-quwain-marina, 11.05.2009.

35 http://realestate.theemiratesnetwork.com/developments/umm_al_quwain/white_bay.php,
11.05.2009.

36 http://realestate.theemiratesnetwork.com/developments/umm_al_quwain/al_salam_city.php,
11.05.2009.

37 http://realestate.theemiratesnetwork.com/developments/umm_al_quwain/emirates_mod-

ern_industrial_area.php, 11.05.2009.

38 http://realestate.theemiratesnetwork.com/developments/ras_al_khaimah/al_hamra_village. php, 11.05.2009.

39 http://realestate.theemiratesnetwork.com/developments/ras_al_khaimah/the_cove.php, 11.05.2009.

40 http://realestate.theemiratesnetwork.com/developments/ras_al_khaimah/mina_al_arab.php, 11.05.2009.

41 http://www.sarayarak.com/#7, 11.05.2009.

42 http://www.inhabitat.com/2008/03/10/ras-al-khaimah%E2%80%99s-gateway-city-to-rival-masdar/, 11.05.2009.

43 http://www.oma.eu/index.php?option=com_projects&view=project&id=443&Itemid=10, 11.05.2009.

44 http://realestate.theemiratesnetwork.com/articles/freehold_property.php, 11.05.2009.

45 http://www.fujairahparadise.com/inde.htm, 11.05.2009.

46 http://www.nationsencyclopedia.com/Asia-and-Oceania/Qatar-HISTORY.html

47 Adham 2008, p. 225

48 Adham 2008, p. 226

49 Adham 2008, p. 227

50 Adham 2008, p. 227

51 Adham 2008, p. 229

52 Adham 2008, p. 233

53 Adham 2008, p. 234

54 Adham 2008, p. 239

55 Adham 2008, p. 240

56 http://www.weavingideas.com/en/applications-products/architecture/projects/singleview/ seiten/0/mode/1/hsref/aspire-tower.html

57 http://education.theemiratesnetwork.com/zones/qatar_education_city.php

58 http://www.thepearlqatar.com/SubTemplate1.aspx?ID=165&MID=115

59 http://web.asteco.com/resources/pdf/Q4/200811_astqrt008_q3_bahrain_qatar.pdf

60 Gulf Construction March 2008, p. 88

61 http://www.fdimagazine.com/news/fullstory.php/aid/2585/Qatar:_The_cost_of_growth.html

62 http://realestate.theemiratesnetwork.com/developments/qatar/lusail_city.php

63 http://www.fosterandpartners.com/Projects/1461/Default.aspx

64 Gulf Construction March 2009, p. 62

65 http://www.barwa.com.qa/english/LocalProjects.aspx

66 http://www.barwa.com.qa/english/LocalProjects.aspx

67 http://www.barwa.com.qa/english/LocalProjects.aspx

68 http://www.qp.com.qa/qp.nsf/web/bc_mic

69 http://www.mic.com.qa/mic/web.nsf/web/mic_masterp#

70 Gulf Construction March 2009, p. 54

71 Gulf Construction March 2009, p. 49

72 Gulf Construction March 2009, p. 75

73 http://www.arabianbusiness.com/542962-qatar-bahrain-causeway-unlikely-to-start-in-january

74 http://dohatraveller.info/

75 http://www.up.org.qa/upeng/modules.php?name=News&file=article&sid=7

76 Adham 2008, p. 247

77 http://www.kuwaitpast.com/era.html

78 Heck and Wöbcke 2005, p. 86

79   Mahgoub 2008, p. 156
80   Mahgoub 2008, p. 157
81   Mahgoub 2008, p. 156
82   Mahgoub 2008, p. 161
83   Mahgoub 2008, p. 167
84   Gulf Construction December 2007, p. 78
85   http://liskw.wordpress.com/2008/07/21/kuwait-university-furute-plans/, 14.06.2009
86   Mahgoub 2008, p. 179
87   http://www.arabnews.com/?page=6&section=0&article=56605&d=26&m=12&y=2004, 14.06.2009
88   Heck and Wöbcke 2005, p. 171
89   Scholz 1999, p. 170.
90   http://thewavemuscat.com/Development/Developers.aspx, 23.08.2009
91   http://realestate.theemiratesnetwork.com/developments/oman/the_wave.php, 23.08.2009.
92   http://thewavemuscat.com/Overview/Masterplan.aspx, 23.08.2009.
93   http://www.bluecity.me/HOME/tabid/36/Default.aspx, 23.08.2009
94   Heck and Wöbcke 2005, p. 45
95   https://www.cia.gov/library/publications/the-world-factbook/geos/sa.html#People, 24.08.2009.
96   ADA Report 2004, p. 44.
97   http://www.kingabdullahcity.com/en/AboutKAEC/Visionaries.html, 24.08.2009.
98   http://news.bbc.co.uk/2/hi/middle_east/7446923.stm, 24.08.2009.
99   http://www.kingabdullahcity.com/en/CityInProgress/CityPhases.html, 24.08.2009.

7 The Contemporary Character of Post-oil Urbanism

7.1 General Characteristics

7.1.1 The New Form of Urban Governance

Urban governance is changing in the Gulf due to the new role played by the private sector in urban planning, which was previously the exclusive responsibility of the public administration. Since Dubai has proven that the introduction of an open real-estate market in combination with a generally tax-free environment can cause rapid growth, many cities in the Gulf are currently adopting this development strategy. Consequently, there has been widespread liberalisation of property markets permitting regional and foreign investors to invest in local real estate, which has become the object of speculation. The official goal of this speculation-driven development is economic diversification in order to secure wealth in the upcoming post-oil era. In addition to the trade and finance sectors, tourism and knowledge-based industries are targeted by the current diversification strategy. The successful implementation of this strategy is becoming more and more dependent on successful town marketing in order to spread awareness about economic potential. The consequence is growing investment pressure on the market that leads to more and more large-scale projects that are planned and built by new and specially founded real-estate companies, who are the decisive force behind the current urban development.

These real-estate companies, which are officially private enterprises, are often partly or completely owned by the state rulers. This very particular circumstance is made possible by the fact that most unbuilt land is owned by the ruling families, who initiate joint ventures with the private sector to develop projects. In addition to these joint ventures between political rulers and private partners such as investment banks, there are cases, for instance Nakheel in Dubai, in which real-estate companies are completely owned by a ruling family. Despite their ownership by political rulers, the structure of these companies remains one of a profit-oriented private corporation. Due to the scale, speed and quantity of new developments these companies are often authorised to design and implement their master plans with hardly any general guidelines or restrictions by the administration. This lack of restriction has led to outdated zoning plans and the decentralisation of urban governance with developers responsible for large urban areas. Subsequently, the public administration is losing more and more control over the management of urban growth.

Despite the official legal system that requires any project to be approved by the appropriate public authority, the decision-making process has been simplified by allocating more and more responsibilities to developers in order to prevent a slow-down in the rapidly growing real-estate market. Now, it is master developers that have the power to approve the plans of the sub-developers for their projects. Thus, the job of public authorities has in several cases been reduced to the formality of simply issuing the official building permit. In some rare cases such as the JAFZA in Dubai the authority that issues building permits has completely shifted to a private enterprise. Thus, developers have to a large degree taken over much of the responsibility for town planning in recent years, particularly in the case of large-scale developments. In addition, the private sector has an increasing influence on zoning plans, which are often adjusted by the public administration in order to accommodate the interests of investors.

While the private sector, namely, developers, has become the actual driving force of urban development, the public sector has lost its leading function as a central authority. The administrative organisation and its large number of employees, who are often not sufficiently qualified, have given rise to a rather static public sector that is unable to cope with the rapid developments. Therefore, attempts are being made, for instance in Dubai and Bahrain, to turn municipalities and ministries into authorities that are structured and managed like private enterprises. The desired goal of this restructured and semi-privatised public sector is to improve the supply of services, particularly infrastructure, through public-private partnerships. In addition to the on-going restructuring of old public institutions, many new institutions have been set up, mainly development agencies and committees. This phenomenon is because old institutions can only be restructured slowly, which is mainly due to the difficulty of releasing employees and the urgent need for establishing an effective central institution capable of working out the general development strategy. These new authorities are not only responsible for the design of new comprehensive master plans for the entire urban area, they are also in charge of formulating the general development goals regarding the society, economy and environment.

Due to these new authorities, where decision-making responsibilities lie between new and old public institutions has often become unclear and is an additional factor weakening the public sector. All in all, it can be observed that the public administration is hardly able to deal with the rapid growth and dynamics of the urban development. This is on the one hand caused by the above-mentioned lack of structure and capability of public institutions and on the other hand by the nature of the decision-making process wherein the rulers themselves allow the projects of the private sector despite their contrariness to existing plans or guidelines. Thus, it is the rulers who remain the ultimate decision-making force despite the on-going privatisation and decentralisation.

The very specific role of the rulers as political leaders on the one hand and as chairmen of profit-oriented enterprises on the other has led to a new and very specific form of urban governance in the Gulf. The old governance structure, which was established during the first period of oil urbanisation, has become outdated due to its outdated approach of supplying the nation's inhabitants with all the needed infrastructure and services without considering growth or profit. The introduction of laws permitting foreign investments in local real-estate has led to rapidly increasing prices of unbuilt land. Consequently, the rulers, who own this land, have a new function as the issuing authorities of large-scale real-estate developments and thus they have increasingly left their old role as tribal patriarchs allocating land and services to their people. Saeed al-Muntafiq, head of the Dubai Development and Investment Authority, once said, "People refer to our crown prince as the chief executive officer of Dubai. It's because, genuinely, he runs government as a private business for the sake of the private sector, not for the sake of the state"[1].

7.1.2 The Transforming Built Environment

Due to the changes in urban governance, the built environment is currently undergoing a profound transformation. While in the past most cities were dominated by large monotone suburban neighbourhoods, the new master-planned developments of the private sector have imported various new elements such as the widespread integration of leisure areas, for instance Golf courses, and the construction of large residential high-rise agglomerations. In order to create an identifiable image for a new development, each master plan is based on a particular theme. While a large amount of these themed developments mainly consist of residential projects, there is a widespread tendency to integrate tourist resorts such as hotels or theme parks. Parallel to this trend there has been a growing interest in investing in economic free zones or industrial parks, which are usually developed as mixed-use projects in which the respective economic sector is the basis of the specific theme (for example, Dubai Silicon Oasis) and the residential component is a subordinated part of the development. More recently, developers have launched projects intended to integrate all the aspects of a self-contained city. In contrast to large-scale residential developments, which usually integrate services and shopping facilities, these new 'cities' have their own downtown with a CBD.

The preferred location of recent developments is along the coasts, where accessibility is usually provided by the main road network and large areas can be reclaimed. In addition to the high market value of seaside real estate and the possibility of developing large-scale projects in proximity to the existing main infrastructure and business centres, land reclamation has become an attractive marketing tool due to the possibility of shaping islands in iconic forms. For instance, it was Nakheel's first Palm project in Dubai that brought widespread international attention to the real estate market in the Gulf. Because of the success of waterfront projects, man-made lakes and canals are popular with developers.

While most projects have been placed close to the sea for the previously mentioned advantages, a number of projects are placed in the deserted inland, mainly along major highways. In contrast to the waterfront developments, land prices away from the coast are usually lower, which has led to the widespread construction of low-rise villas and thus to much lower urban densities. Generally, it can be observed that the rapidly growing land prices have caused a new kind of urban sprawl in the Gulf. The focus of investors on building on cheap land has led to the on-going development of new areas. The result is a more and more scattered urban landscape with large unbuilt areas between single developments. In addition, the decentralised and isolated nature of the process of designing master plans for developments has led to a rather patchwork-like urban structure. In order to prevent counter this lack of integrated urban areas, there has been an initiative to launch large-scale mixed-use developments that are intended to function

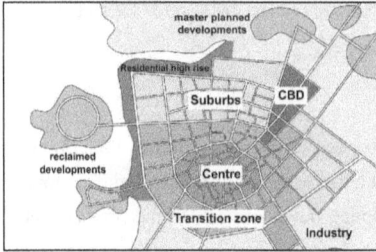

Fig. 1: Typical structure of current Gulf cities.   Fig. 2: The coastal typologies.

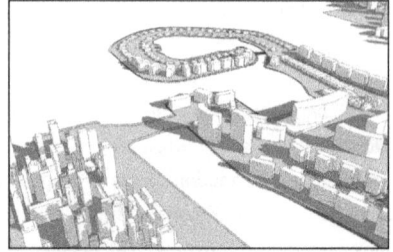

as self-contained cities within the city. In some cases such as Waterfront City in Dubai, these 'cities' are placed at a large distance from former urban centres in order to become independent entities.

While the recent urbanisation is dominated by large-scale master-planned projects that are placed either at the sea or in the desert, many urban areas are being built according to plans of the public sector. The construction of buildings usually follows zoning plans that were designed before the recent real-estate boom. Thus, these plans are often adjusted in order to accommodate the interests of investors, particularly concerning the maximum building heights. The result of the new regulations is a growing number of high-rise conglomerates mainly consisting of residential buildings in which apartments are leased or sold as freehold properties. In contrast to the more recent master-planned projects in unbuilt areas, areas already zoned have a relatively long history and thus a number of lots are already built. Today, the age of buildings can generally be distinguished by their differing heights ranging from one floor to over fifty. Before freehold property laws were introduced, most buildings were either low-rise villas, often built within compounds, or low- to medium-rise apartment buildings that were leased to companies for their employees. The recent construction of a large number of high-rises that mainly offer exclusive apartments for the property market has caused a rapid increase in urban densities. These dense agglomerations, which are usually not based on any comprehensive master plan, are built by a large number of different investors and not by single large-scale developers.

Despite the growing attempt to develop new centres and mixed-use projects, the widespread division between residential areas in the outskirts and business districts along the main urban axes remains. The two main reasons for the continuing concentration of offices in the main CBDs and their extensions are the infrastructure supply and services as well as the notable address. Consequently, the CBDs are expanding along the main growth corridors, which are defined by multi-lane roads. Thus, these CBDs can be best described as 'central business spines' because of their linear development. Similarly to offices, the main shopping districts, namely, large-scale shopping malls, are located at certain junctions of these spines providing high accessibility by car. In addition, there has been a recent tendency to integrate small shopping malls and retail districts within recently developed projects. A very particular phenomenon is the widespread development of seaside promenades within waterfront projects integrating shopping and leisure facilities. While these areas are often part of gated communities and thus rather exclusive, the city centres remain the main shopping and business centres for lower income groups. In addition to the concentration of commercial buildings, the public administration has in most cases remained in the centres as have the main services such as hospitals and schools. Thus, most new development areas are still dependent on the social services provided in the old urban centres.

While there has been no major change in the general land-use distribution of Gulf cities, there has been a visible transformation in recent developments of typologies. This particularly concerns the new tendency to construct residential high-rise buildings in large agglomerations which, in contrast to the past, are even close to low-rise suburban areas. In addition to the much taller height of the current developments, another significant difference to older apartment buildings is the targeted income groups. Most of these high-rise buildings are exclusive residences with integrated leisure facilities. While residential high-rise projects are mainly developed close to the coast due to higher land prices and sea views, most low-rise residential developments are either developed inland or in the form of the most exclusive waterfront properties on reclaimed peninsulas and islands with private beaches. In addition to the widespread low-rise and high-rise developments, there are more and more examples of medium-rise residential projects in order to serve mid-income markets. These large-scale projects are often developed towards the inland and close to industrial areas.

Apart from certain exclusive residential projects that integrate leisure areas such as promenades and small parks, there has been a general lack of open space and public areas in most urban areas, particularly within city centres and their fringes. While the private sector has recognised the important marketing factor of green areas and thus integrates them in order to enhance the value of their residential projects, the public sector has been unable to develop more open space within urban areas despite the growing urban density. Consequently, the lack of integrated public leisure areas intensifies the already existing tradition of using shopping malls as the main public meeting points.

In addition to the lack of public areas there has been almost no integration of public transport. In most cities public transport is limited to private bus companies mainly serving lower income groups. Although Dubai has started a metro-project, it can be generally observed that investment in public mass transport systems is still rather low and generally limited to the preparation of plans. Thus, the road network has had to gradually expanded in order cope with the exponentially growing traffic. While the main arterial roads have usually been planned before land is allocated to developers, there are often difficulties with developing sufficient service roads within urban districts that were initially designed for much lower densities. Moreover, the lack of available data and coordination between developers and the public sector is causing unexpected traffic problems in certain areas. All in all, it can be stated that dependency on cars will continue into the coming years due to the absence of mass transport systems and the general lack of land-use integration.

Despite the large amount of unbuilt lots within city centres, the current urban development is mainly concentrated at the urban fringes because of the absence of large cohesive areas, general accessibility concerns and high land prices in the centres in addition to the widespread demand for waterfront properties. Thus, similarly to the development during oil urbanisation, the current urbanisation is mainly dominated by urban sprawl to new unbuilt areas. Thus, the old centres remain mixed-use areas while their fringes are a mixture of low-rise suburbs and medium-rise developments. The recent increase in the plot ratio caused by the construction of high-rise agglomerations mainly takes place along the coasts and along certain main development axes. Urban densities however are usually still highest within the oldest part of city centres where lower income groups are accommodated. This situation would change if the large-scale high-rise agglomerations become fully occupied by residents, thus shifting urban densities from the old centres to the new expansion areas along the coasts.

While coastal areas are transforming into dense agglomerations, reclaimed island developments are usually built with low-rise buildings in order to reduce the inner traffic and the general costs for infrastructure. Inland developments can be distinguished between exclusive low-density neighbourhoods and agglomerations of medium-rise apartment buildings for lower income groups. In the case of large-scale projects of many square kilometres, different building typologies are often mixed but usually in accordance with the general trend of high-rises close to the coast and main arterial roads.

In addition to the expected shift of urban densities, the current transformation process is leading to the import of a variety of urban and architectural designs. Instead of using an orthogonal grid, most new developments are based on an ornamental layout in order to be easily identified from differing neighbouring projects. Reclaimed peninsulas and islands make particular use of iconic shapes in order to attract attention but even less exclusive developments, which often consist of just one building type, use ornamental urban design layouts and thus contribute to the diverse patchwork appearance of the urban landscape. Similarly to urban design, there is a widespread attempt to build iconic landmarks in order to be recognised among the large number of projects. In this regard, height is the most important identifying feature as illustrated by Dubai's 828-metre tall Burj Khalifa – the world's tallest building.

Architectural design is rather diverse and apart from the widespread integration of oriental ornaments it is generally rather detached from the locality and its climatic conditions. In the case of large-scale developments, identical architectural designs are usually multiplied many hundreds of times, leading to a monotonous urban landscape. Subsequently, the current built environment is mainly characterised by the rapid speed of development and the scale of projects rather than a very particular design approach. This phenomenon is mainly the result of a missing link between architecture and the local environment and culture. In fact, the architecture itself is expressive of the current speculation-driven development in which the common denominator lies between the most economical design and the marketing theme. Thus, one of the main characteristics of the current built

254

environment in the Gulf is the generation of a superficial identity by developing projects in iconic forms such as man-made islands or construction superlatives.

7.2 Post-oil Urbanism in Context

7.2.1 The Historic Context

Despite the socio-economic transformation taking place, the political structure of one ruling family that dates back to pre-oil times has remained unchanged in all GCC countries. The main difference from the time of the oil urbanisation is the new role of the rulers as general managers of their cities enforcing new economic development strategies in order to enhance economic growth and diversification. By contrast, in the past their role was mainly reduced to that of governors of rentier states allocating the oil wealth to the population. In addition to the expansion of traditional economic sectors such as trade the new economic strategy focuses on the financial sector, tourism and knowledge-based industries in order to transform the Gulf cities into global service centres. The basis for the realisation of this new development approach has mainly been the introduction of tax-free zones and the liberalisation of local markets, particularly real estate. In combination with the already existing infrastructural backbone and free-zone policies, the open real estate market has led to enormous urban growth.

The oil boom has led to a very particular socio-economic structure in the Gulf, in which the local population has been mainly employed in the public sector and guest workers have been immigrating to work in the private sector. Therefore, many hundreds of thousands of foreign guest workers have come to Gulf cities in recent years as new economic sectors in the context of the real estate boom have been booming. In addition to a large number of South-Asian guest workers, who have mainly been employed in the lower service sector such as construction, large numbers of educated guest workers have also immigrated to work in the advanced service sector. Subsequently, in some extreme cases such as Dubai the 90% of the population is foreign and the society has rapidly become a more and more diversified mix of many different nationalities wherein the South-Asian workforce from the Indian subcontinent is the largest group. A direct consequence of this is that a large part of the local population is not very involved in the current development and thus continues to be dependent on public subsidies as in previous times. Thus, it can be stated that a high percentage of the local population is not participating in the current process of economic diversification. This is mainly caused by a lack of education and a long history of dependence on public subsidies.

In contrast to the oil era, urban governance has become increasingly decentralised through the allocation of more and more decision-making responsibilities to the private sector. Moreover, the public sector itself has been partly restructured in order to reform the outdated administrative structure. Hence, urbanisation has become more and more dominated by investors and their preferences instead of the former practical approach of serving the local population with a functioning modern town. The result has been a transforming urban structure expressed by large-scale master-planned projects spreading in deserts and on reclaimed land and growing high-rise agglomerations along the coasts. The previously monotonous structure of large suburban settlements, mixed-use and dense urban centres and a modern CBD along the major development axes has become increasingly replaced by a diversified urban landscape of different shapes and designs competing for attention. Construction superlatives and the mass production of real estate form the patchwork-like new structure of Gulf cities.

While in the past the image of the oil city was equivalent to the image of a villa behind walls, it is now an image of sparkling high-rises and pictorially shaped islands in the sea. In comparison to the oil urbanisation, the goal of which was modernisation and the construction of basic infrastructure by investing the oil wealth, post-oil urbanism is not limited to the building of a functioning city. It is rather the idea of building cities for the purpose of marketing and trading real estate. While previously the city mainly had to satisfy the needs of the local population, the current urbanisation is almost detached from this development goal. Today, real estate has become a commodity, the appearance of which is adjusted to the demands of the speculation-driven market. Consequently, the unbuilt desert has become one of the most valuable commodities and the previously common allocation of lots to the local population has been reduced. While in the past urban sprawl was mainly caused by ever increasing low-density suburban areas, the current urban sprawl is caused by real-estate developers searching for large cohesive areas of land at low prices. Thus, instead of large areas of low-dense suburban neighbourhoods with a high percentage of unbuilt lots, the current urban landscape is dominated by a patchwork-like structure with large unbuilt areas between single developments.

7.2.2 The Regional Context

Due to the special conditions in the Gulf related to a still existing wealth of fossil resources and the relatively stable political climate, the current development is rather unlikely to spread in the same form and scale to other countries in the Middle East. However, there are several projects that are comparable to the developments in the Gulf in countries such as Egypt, Jordan and Turkey. Some of these projects such as Sheikh Zayed City in Egypt are being developed by real-estate companies from the Gulf. One of the most prominent examples of these 'globalised' companies is the developer Emaar, which in addition to its base in Dubai has several new subsidiaries in various locations such as Egypt, Marocco, Tunisia, Syria, Turkey, Pakistan and Jordan[2]. Developers from non-GCC countries in the region often imitate the projects of these more globalised Gulf real-estate companies. For instance, the Lebanese real-estate developer Mohammed Saleh has proposed reclaiming a 3.3-sq km island in the form of a cedar tree off the coast of Beirut[3]. This phenomenon, which is often called 'Dubaification' or 'Dubaiisation', is a direct consequence of the recent economic success of real-estate speculation in the Gulf. Subsequently, other countries in the Middle East see the Gulf cities as models and reference points for new lifestyles and points of view in the region.

Despite the economic problems due to the world financial crisis, Dubai and its competitors have established an image as modern and 'progressive' Arab cities. In this regard, they are an expression of the current inner Arab conflict between adopting Western images and lifestyles on the one hand and maintaining Islamic traditions on the other. Thus, Gulf cities have become both model and provocation at the same time. While there is widespread admiration and pride in Arab countries about the superlatives of their development, there is much scepticism and fear about losing their own identity. This conflict is furthermore fuelled by their geographical location between the two most conservative countries in the region – Iran and Saudi Arabia. Today, regional emotions towards the current developments in the Gulf can be described as an oscillation between euphoria, incomprehension and disgust.

7.2.3 The Global Context

The very particular socio-economic situation in the Gulf, which is mainly based on a high dependency on a foreign workforce, is rather unique in the world. It is most comparable with the cities along the border of Mexico and the United States, which have become known as 'border towns'. In both cases, modern infrastructure, which is mainly planned and built by the developed north, meets cheap labour coming from the south to work in the lower service and manufacturing sectors. While in the case of North-American border towns Mexican labour works mainly in industrial production, the South-Asian labour in the Gulf is mostly engaged in the booming construction sector and other areas of the lower service sector. In both places transnational companies benefit from reduced labour laws and low salaries. Hence, present post-oil urbanism has to be seen in the context of the current globalisation in which the Gulf has begun to define its own very particular role by copying and adopting existing strategies.

In addition to the oil wealth the very fortunate geopolitical location between the markets of Asia, Africa and Europe has enabled the current economic development of the Gulf countries and attracted global attention. This has led in the Gulf to the development of large logistic centres, several new harbours and a large number of international airports as well as the establishment of new airlines. Similarly to other parts of the world, airports have become an important economic basis due to global trade. Another factor that has attracted regional and global investment to the Gulf is the existence of authoritarian political structures in combination with very small subsidised local populations leading to a relatively stable political situation. Moreover, like places such as Switzerland or the Bahamas the Gulf has benefited from the introduction of offshore banks and a liberal financial market, which has led to its dubious fame as centre of money laundering. In addition, the setting up of free economic zones, used worldwide to attract developments, has led to a growing global interest from various industries in settling in the Gulf. One of

257

the most successful sectors in recent years has been tourism due to the widespread lack of leisure facilities in the region. Due to liberal policies there has been a similar effect as in the case of Las Vegas in the United States, which has become economically a highly successful centre of 'forbidden joys'.

While the current economic development strategies of the Gulf cannot be seen in themselves as unique in the world – in fact, they are mainly following global tendencies – it is rather the scale, speed and concentration of all these strategies in one place that make the Gulf a very specific global phenomenon. In addition, one significant difference to most other places is the governance structure in the Gulf in which the central decision-making force has remained in the hands of the ruling families despite privatisation and liberalisation. In many cases, rulers have entered into joint ventures with the private sector to develop unbuilt land, which is their legal property. This direct liaison between monarchs and private investors has reached a new scale in the Gulf that is not comparable with any other place in the world. Despite their growing dependency on the private sector, each ruler is able theoretically to prevent any development at any time. Furthermore, most large-scale developments are even based on the direct initiatives of rulers themselves, who like to express their visions and ideas. This construction of personal visions might remind one of the monuments of other authoritarian rulers elsewhere but they are different in that their general goal is to be commercially successful. Thus, a personal poem by Dubai's Amir in the form of small islands for instance is intended to be sold as freehold properties. Although the Gulf rulers remain the main decision-making force, the current form of urban governance in the Gulf has led to a weakened public sector that has lost its role as the main planning authority due to the increasingly decentralised nature of urban development. This is again similar to other regions in the world where investors have become the driving force of urban development. This phenomenon has been described as the 'post-modern urban condition' by the author Michael J. Dear.

The typical urban structure of Gulf cities is mainly based on master plans of the 60s and 70s that were designed by Western architects and thus in accordance with the general planning ideas of that time. The functional division of land uses and the focus on the car as the most important means of transport has led to the import of a type of city in the Gulf that is rather detached from the region itself, particularly with regard to its climate and culture. Thus, the basic structure of these cities is more comparable with examples in the West such as desert towns in North America. In contrast to other development countries the Gulf has no long-term history of urbanisation due to the difficult environmental conditions. The traditional oasis towns have been relatively small and due to their urban structure, which is inadequate for modern development, they were widely replaced during the first phase of oil urbanisation. This in turn led to the absence of authentic traditional elements in Gulf towns. Subsequently, except for the large-scale individual expressions of rulers, most of the current built environment in the Gulf can be seen as replaceable with that of other places, especially places where large investor groups are developing towns from scratch. This is particularly the case in Asia, for example, China, where similar forms and scales of themed developments can be observed.

7.3 Liveability and Sustainability of Resulting Urban Structures

7.3.1 The Main Factors Reducing Liveability

While the new developments have improved the quality of certain areas, particularly exclusive residential projects with integrated leisure areas, more and more factors are reducing liveability in most of the urban area. This is mainly caused by the speculation-driven aspect of the current urbanisation in which pressure from the private sector has led to the decentralisation of urban governance in order to cope with the extent and speed of developments. Thus, the increasing influence of the private sector on the decision-making process has led to a weakened public sector and a general lack of comprehensive planning and outdated zoning plans. Consequently, there has been a growing lack of integration regarding the development of infrastructure, the distribution of land uses and the introduction of new building regulations. In particular, the construction of the infrastructural network has not been able to keep up with the speed and scale of real-estate developments. This can be observed in the case of both large-scale developments on unbuilt land and already developed areas that are becoming more dense due to the preference for high-rises. Despite the ongoing privatisation, the public sector (either municipalities or ministries) is usually still in charge of developing the main infrastructural supply such as roads, electricity, water and sewage. In the case of master-planned projects of the private sector, the developers are usually responsible for planning and constructing the infrastructure within their projects while the public sector has to develop the main network linking the single developments to the remaining urban structure. In the case of already developed urban areas the public sector is mainly responsible for the supply of infrastructure and services. However, due to the scale, speed and quantity of projects, there is a lack of infrastructure because the private sector does not participate in investment in the main network. The difficulty of planning and developing a sufficient infrastructural network is further compounded by the absence of accurate and cohesive data and the unforeseen launch of projects in certain areas. Consequently, in addition to the lack of supply for new developments, the existing networks, particularly roads and electricity, have become overloaded, causing frequent breakdowns.

The absence of data and the ever-growing and changing projects of the private sector have also hindered the development of a comprehensive zoning plan. In the case of master-planned projects the private sector is usually in charge of developing the general plan and public authorities are usually only responsible for the official approval. This approval however is normally not based on existing plans or guidelines and often follows irregular decision-making processes. Furthermore, the method in which impact studies are conducted, particularly regarding traffic and the environment, is in many cases based on outdated data and does not account for neighbouring developments. Thus, both

259

infrastructure and land uses are often planned within an isolated planning process leading to an unconsolidated urban development.

In the case of urban areas that have already been planned and developed in the past, zoning plans have been often adjusted to meet the demands of investors to develop high-rises. Due to this permission to construct high-rises in areas that were once designed to be suburban neighbourhoods and thus did not integrate large public areas, certain districts have transformed into dense urban clusters with a growing lack of open spaces. Similarly, there has been hardly any integration of sufficient social services such as educational or medical facilities. Another profound deficit in the current built environment in many places is the absence of sufficient public transport systems that were not integrated in initial plans due to inaccurate expectations based on past experiences of urbanisation. Furthermore, the high costs, the general lack of acceptance by upper income groups and the still very low average urban densities due to urban sprawl are the main reasons for the current absence of public transport.

In addition to the lack of comprehensive planning regarding infrastructure and land use, the general lack of building regulations, particularly regarding building heights, has led to a rather low urban quality in certain areas with buildings spaced close to one another and regulations regarding the development of parking spaces insufficient. Hence, in recent years the tendency has been to construct multi-storey car parks integrated within multi-storey buildings, which has led to higher plot ratios and an unsuitable environment for pedestrians.

Thus, today, the largest factor reducing liveability in the Gulf can be identified as the continuing dependence on the car, for which there are many reasons including the lack of land use integration, the absence of public transport, the habits of the population and the former urban structure. This dependence in combination with the lack of integrated land uses has caused exponentially increasing traffic, which is expressed in large traffic congestion particularly during rush hours. In addition to the waste of time caused by traffic jams, the noise, pollution and the increasing problem of finding parking lots have diminished the liveability of Gulf cities.

Another factor reducing liveability is the unstoppable rise of rental prices and general living costs, which are forcing many people of low and middle income to either share apartments with others or to move to more affordable areas, which are often located at far distances from business centres and less supplied with services. This development trend has led to growing social segregation and the risk of certain areas transforming into ghettos with high urban densities and growing social problems. In addition, there is a widespread lack of public open spaces and service areas in order to increase a project's developable area.

The next factor that reduces liveability is the quality of buildings and their surroundings, which concerns both exclusive master-planned projects and low-budget developments. Due to the lack of building regulations and the general tendency of investors to maximise their profits, most buildings are constructed with cheap materials and catalogue architecture, which neither responds to the climate nor to the culture of the future inhabitants. Furthermore, the lack of anticipated planning and absence of restrictions has led to the development of dense high-rise clusters along the coasts, endangering the general ventilation of central urban areas. In the case of reclaimed islands and man-made canals or lakes, stagnant water creates an unpleasant living environment for residents due to the growth of algae.

The design of most projects in large scales and iconic forms often also contributes to a diminishing liveability of the built environment. Recent urbanism has favoured the idea of creating a certain image for the aeroplane or satellite rather than to please the human perspective on the ground. Thus, many reclaimed projects such as the Palm Jumeirah have rather monotonous surroundings that are hardly associated with the image from above that promises an interesting seaside location. Moreover, in the case of mass produced buildings from individual developers, large areas have turned into repetitive and mono-functional places with hardly any identity. The privatisation of most urban areas has furthermore caused an urban structure with hardly any centres and identification points in cities. Subsequently, shopping malls have remained the most important public areas but this rather replaceable commercial alternative can hardly replace real city centres and thus do not generate the needed identity.

All in all, the current urbanisation has led to five main factors that diminish liveability, namely, increasing traffic, rising living costs, a lack of integrated services, a low quality of the surroundings and a lack of identity. All these factors are rooted in the generally unrestricted speculation-driven development, which has transformed the built environment on the basis of short-term interests of investors instead of a public concern to create liveable cities.

7.3.2 The Main Factors Endangering Sustainability

Today, the decreasing liveability of the Gulf cities has begun to affect their economic sustainability due to less investment and companies relocating. In particular, traffic congestion and high living costs are putting off companies more and more since they are mainly dependent on foreign workers. While most educated guest workers who have moved to the Gulf did so for the economic possibilities and higher living standard, rising costs and traffic jams are impinging on these initial motivations. If a large number of these immigrants, who are the driving force of the private sector, move away, this would mean great economic loss. Another negative consequence of this dependence on guest workers is the fact that most guest workers stay temporarily and do not reinvest their entire salary within the local economy. This also regards low-cost labour from South Asia who generally work in the Gulf in order to support their families in their home countries. Due to the increasing costs caused by rising prices of land and building materials, real-estate companies are often forced to delay the payment of salaries, which again leads to strikes and further economic loss.

Thus, on the one hand, the speculation-driven development has caused enormous growth expressed by the construction of large-scale projects, but on the other hand, the missing implementation of regulations to guide this growth has led to a lack of consolidation of urban areas. Consequently, the delayed integration of the necessary technical and social infrastructure is in many cases contributing to overdrawn state budgets. Furthermore, many projects ignore rising maintenance costs in the future. This problem concerns both cheaply built apartment blocks, villas or high-rises, which depend on excessive air conditioning, and man-made island developments, which involve a lot of maintenance costs due to erosion and the deposit of sand along their shorelines. In addition, energy costs are rising because of growing waste and decreasing resources, posing a serious problem for the future. These growing maintenance costs are clear signs of the short-term ambitions of investors. Due to growing regional debts the deregulated banking system has caused the Gulf economy to become increasingly dependent on international financial markets. Thus, the international financial crisis that began in 2008 has had a serious impact on the region, whose economy is more and more reliant on speculators. Subsequently, any break within the circle of speculation should be seen as a major economic threat.

In addition to the economic factors endangering sustainability, there are several threats to sustainability with regard to the development of the Gulf societies. Since foreign guest workers usually stay for a short period of time and so the population is frequently changing, this leads to the serious dilemma of establishing a cohesive society. This kind of ever-changing 'airport society' consisting of different cultures is further encouraged by the absence of a legal right for foreign workers to stay for a longer period of time and of political rights to be active members of society. Therefore, most guest workers do not feel they are an integrated part of the local society and rarely develop identification with the host country. While the educated workforce from Europe, USA and Australia benefit from higher salaries and many amenities, the low-cost labour has rather low living standards and hardly any possibility to improve their situation. The new role of Gulf cities as border towns between the Third World and First World is causing deep social segregation and is leading to the risk of growing social unrest, crime and violence.

A further threat to social stability is the increasingly difficult economic perspectives of the local population, who is still used to receiving public subsidies. The situation during previous decades when the GCC was basically a group of oil-funded rentier states has resulted in a population that is often neither sufficiently educated to compete with foreigners nor prepared to work in lower service sectors. In general, it can be observed that the competitive private sector has been avoided by locals, who prefer to work in the public administration. Thus, the growing influence of deregulated markets and widespread privatisation has led to a completely new situation for the local population, who finds itself more and more excluded from current developments. Consequently, unemployment has become a major problem for local inhabitants, particularly for the younger generation. This is a dangerous situation, which is expected to worsen due to the high birth rate of the local population. All in all, recent developments have created more and more threats to social stability, which includes the difficult issue of alienation and a growing lack of identity as well as social segregation and the dangers of social unrest.

In addition to the factors endangering future sustainability due to the current socio-economic conditions, the environment has become increasingly polluted and exploited, which again threatens liveability and sustainability in the Gulf. The high energy consumption of the region has given Gulf cities the biggest ecological footprint in the world as a result of three factors, namely, growing traffic, air conditioning and water production by desalination. This high waste of fossil energy resources, which are still being provided by local regional production, has led to increasing air pollution due to an exponentially growing number of cars, fossil-fuelled power plants and energy-wasting industries such as aluminium production. In addition to the air pollution caused by traffic and industry, the sea, particularly in the case of coastal regions, has suffered from salt water intrusion, untreated sewage and the impact of land reclamation. While desalination plants need more and more energy to meet the growing demand for water, they also produce a large amount of salt, leading to salt water intrusion in coastal areas. This increasing salinisation in turn leads to more and more energy consumption in order to continue producing water. Thus, water production should be seen as a serious threat to the ecological balance in the Gulf.

Another environmental risk is the high degree of contamination of coastal waters by untreated waste from industries and households, which is mainly caused by a lack of restrictions and control systems governing services, which are becoming more and more privatised, for example, waste collection. In Dubai, for instance, a large number of sewage lorries have emptied their tanks into storm drains due to long waiting hours at the city's only sewage treatment plant, resulting in polluted beaches[4]. Last but not least, the widespread land reclamation contributes to an increasingly unbalanced eco-system by destroying a large part of the sea flora and fauna. Due to the dredging procedure, sand deposits on the seabed and thus disturbs the growth of plants and the reproduction of sea fauna. In addition to turbidity, the water quality might change as a result of an alteration in its chemical composition caused by heavy metals and organic contaminants within the moved sediment. Thus, toxic algae would be able to grow, threatening the balance of the eco-system and human health[5]. Consequently, the current urban development is leading to serious environmental problems, which in the long term will be a major threat to the economies in the Gulf, particularly in relation to the expanding tourism industry. The enormous growth of recent years has caused an increasing imbalance between the fragile desert environment and human settlements, which are increasingly wasting natural resources.

It was once oil production that made modern urbanisation first possible for Gulf cities during the second half of the 20th century, and it is again oil that should be considered the main reason for the rapid growth of Gulf cities in recent years. Large state budgets thanks to the export of oil in combination with local oil production for growing domestic use have led to fortunate conditions for rapid urban development. However, instead of introducing restrictions and modern technologies to develop ecological future-oriented urban settlements, the private sector has been allowed to carry out large-scale developments targeting short-term speculative profits. The result has been a patchwork-like urban structure with an obvious lack of consolidation and an absence of integration of infrastructure and land uses. Furthermore, the lack of climate-appropriate design for buildings in order to save on the cost of building materials and modern technologies has led to an exponential waste of energy caused by air conditioning.

While in the past the Gulf region was populated by a small number of nomads and oasis farmers, the immigration of millions of guest workers and the rapid growth of the local population has led to large metropolises with the highest waste of energy per capita in the world. In addition to the energy waste of cars and air conditioning, the increasing need for water, which is produced by desalination plants, is the third major contributor to the exponential need for energy causing a high dependence on local oil production. Furthermore, export revenues from oil production are still needed by governments for major public investment in the development and maintenance of technical and social infrastructure. With oil resources dwindling, the introduction of more and more taxes is expected to be inevitable, which will lead to Gulf cities becoming less attractive as a global business location. This would result in less investment, fewer companies relocating to the Gulf and present companies moving. Moreover, taxes and higher living costs would make it difficult to attract experienced and educated guest workers, who have become the driving force of the new economies in the Gulf. Thus, it can be stated that the current form of post-oil urbanism is to a large extent still dependent on oil as the driving development factor and without major changes and efforts on the part of political decision-makers there will be major problems with achieving the goal of establishing oil-independent service centres in the Gulf.

---

1 Davis 2007, p. 61
2 http://www.emaar.com/index.aspx?page=emaar-international, 16.07.2009
3 http://www.spiegel.de/reise/aktuell/0,1518,621666,00.html, 16.07.2009.
4 http://www.telegraph.co.uk/travel/travelnews/4380051/Dubais-polluted-beaches-closed-to-public.html, 27.07.2009.
5 http://www.dse.vic.gov.au, 27.07.2009.

8 The Future Challenges of Post-oil Urbanism

8.1 The Implementation of a Holistic Development Strategy

8.1.1 The Formulation of a Comprehensive Development Vision

In consideration of all negative development tendencies, mainly caused by the speculation-driven real estate market, the general development goal of the current urbanisation has to be questioned. Today, all GCC-countries are ambitious to diversify their economies in a rather short period of time, which has led to a widespread liberalisation of the regional markets. In spite the initial goal of economic cooperation within the GCC each government has more or less exclusively focused on the development of its own country, which has led to a widespread competition regarding all post-oil economic sectors including tourism, trade or knowledge-based industries. Consequently, current projects targeting these economic sectors have become rather similar and exchangeable from country to country. This can be observed in the case of tourism due to the development of a large number of hotels and leisure facilities in addition to the construction and expansion of airports. Today, several airports and airport cities are under construction serving beside the growing number of tourists an expanding cargo industry. The transformation of the Gulf to one of the biggest logistic centres and global trading hubs due to its geopolitical location between Europe and Asia has become the biggest hope of most rulers to establish an economic basis during the post-oil era. Hence, in many places airports, ports and logistic centres have been built, expanded or their construction has commenced in order to provide large capacities to overtake this prospering sector. The result has been an exponentially growing capacity in the whole Gulf region, which is expected to exceed the actual regional and global demand. While few places, which offer the best amenities, will succeed within this competition, most remaining locations will not be able to develop to global trading hubs. In addition to tourism and trade the increasing competition concerns almost every economic field, including knowledge-based industries. Today, a large number of universities, technology parks and health care institutions are developed in form of small cities within the city. Their organisation as free economic zones and their development with modern infrastructure have led to a high attractivity of investments and thus the relocation of a large number of international companies in the Gulf. Due to the growing number of alternative locations many companies decide to relocate from country to country searching the most profitable environment regarding low costs, advanced infrastructure and educated labour. The big attractivity of tax-free zones has led to the major problem that any future introduction of taxes would threaten the local economies due to the move of companies and investors. But based on decreasing oil resources many countries are facing the inevitable need to introduce taxes in future in order to finance public expenses. While trade, tourism and knowledge-based industries are in this regard to a certain extent still subsidized, the growth of these post-oil economies has become the most important reason for the speculation-driven real estate market. Beside a large number of real estate projects, this sector has led to the immigration of thousands of guest workers, who are engaged in the construction business. Along with this construction boom the financial sector has been rapidly expanded, particularly due to the activities of investment banks. The result has been an enormous amount of large-scale real estate projects of enormous scale based on speculative interests. Due to rather low restrictions and the lack of planning the real estate developments led to an overloaded infrastructural network and a generally decreasing liveability of the built environment. Thus, the development goal to grow as fast as possible by providing low restrictions has led to the increasing problem of attracting shortterm investments and the tendency to move investments from place to place if the speculative interests decrease. The major problem of this development is that in many places the consolidation of the built environment is decelerated.

Due to these negative outcomes of the current development a more sustainable urbanisation has to be introduced by replacing the isolated goal of each country to become the future centre of a post-oil Gulf by a more cooperative development goal to lead the region itself into a future without oil. Cooperation, communication and coordination regarding all development concerns will be essential in order to prevent social and environmental problems in many places. Thus, the old tribal thinking of focusing on the own survival and economic benefits has become a basic obstacle to create one development vision for the Gulf and to develop a solide structure of centres with different specialisations regarding economic sectors. The decreasing oil resources and the fragile desert environment are realities, which have to be respected regarding the future growth of settlements. Hence, smart growth has to replace the currently favoured principle of exponential growth in order to ensure the balance with the natural environment. But only if the GCC-countries agree to a common development strategy, which involves the introduction of several restrictions regarding speculation-driven developments, the ecological balance of this region can be achieved. Single countries will face major difficulties to change towards a more balanced growth due to the growing economic pressure. Cooperation instead of competion has to be the major focus of a regional development strategy leading to the coordinated elaboration of regional plans and projects.

Except a few examples, such as the idea to develop a train network connecting all centres of the GCC, there are currently no comprehensive regional plans. Even within countries, such as the UAE, the competition between emirates and cities has led to a difficult basis for any kind of cooperation. In order to establish the roots for a new phase of post-oil urbanism these obstacles have to be resolved by recognizing common interests, such as a healthy environment and social peace. In this regard the GCC will have to become that, what it was originally envisioned, a platform of regional cooperation.

8.1.2 The Reconfiguration of Urban Governance

In addition to the reconfiguration of the general development goals of the entire region, the local urban governance itself has to be adjusted to a new approach of the current urbanisation in the Gulf. In contrast to the previous laisser-faire attitude towards the private sector and an ongoing decentralisation of decision-making processes the public sector would have to be generally strengthened to become an effective and equal partner within the development. In this regard the development model of Dubai, in which the town itself is managed as a private company, has to be questioned. A purely profit-oriented development has led to huge problems of consolidation due to a growing lack of sufficient public infrastructure and services. Furthermore, the missing restrictions and the missing central control have caused the development of a built environment, which is relying on a high waste of energy due to climate inapropriate architecture and a general lack of integration regarding land uses and new technologies. This waste leads to a threatened ecological balance and environmental damage, which in turn threatens the economic sustainability. While the fast decentralisation and deregulation caused rapide urban growth due to the high attractivity of short term investments, the resulting lack of consolidation and the decreasing quality of the built environment are currently turning into a major development problem.

The very specific circumstance of urban governance in the Gulf has been the alliance between rulers and private sector, which have created joint ventures to develop large-scale projects. The ruler himself has on the one side become a private entrepreneur, while on the other side he remained in charge of the political decision making. Consequently, many political decisions followed entrepreneurial interests instead of general public concerns. Thus, many restrictions were removed leading to open and deregulated markets, particularly regarding the trade with real estate. This initiation of speculation however has to be seen in two ways: On the one hand rulers had the hope to attract fast growth in order to intensify the economic diversification, while on the other hand rulers benefited directly as partners of speculators due to their land ownership. The result of this development has been a debilitated public sector, which has become more and more a passive supplier of services instead of being an equal partner of urban governance. This new role of the public sector has been furthermore enforced by its outdated administrative structure, which is a relict of the previous oil urbanisation. In addition to the large number of employees working in the administration, it has been the rather low education and qualification of many of those employees, which has caused a sluggish buraeucracy. Thus, the current organisation of the public sector is hardly able to deal with the new circumstances of a competitive and flexible private sector as driving force of the development. Subsequently, the physical planning of the future urban structure and the survey of the current urban development became outdated.

The current reaction of many governments has been the installation of new public agencies and authorities, which are structured similar to private companies, to overtake the responsibilities of ministries and municipalities. Due to a rather slow disorganisation of the old administration, there has been the overlapping of certain decision making fields, particularly regarding urban planning. Thus, while the general physical planning often remains at the old ministries or municipalities, the new agencies are in charge of many decisions regarding future developments or the initiation of new strategic plans. This decentralisation of decision-making has led to an increasingly weak position of the public sector, which is passively reacting on new circumstances rather than setting up the framework of the future development. The consequence has been an unconsolidated built environment due to missing infrastructure, which has been caused by the lack of regulations forcing the private sector to participate within the infrastructural development and insufficient planning.

This missing participation of the private sector has led to the increasing threat of an infrastructural collapse of many cities. This threat however can be seen as the biggest chance to reconsider and reconfigure the organisation of the current urban governance. Most importantly it has to be recognized that centralised planning and coordination regarding the general physical planning of the urban structure cannot be replaced by a decentralised decision making process, in which developers gain more and more responsibilities. In order to consolidate the urban structure regulations have to be centrally introduced and their compliance has to be centrally observed by a strengthened public sector. However, the outdated administrative structures have to be finally replaced by a more competitive structure of qualified authorities. The most important aspect of the new structure has to be the integration of all decision making fields interfering with each other in order to establish a reliable and effective organisation. The new public sector has to be in charge of creating comprehensive plans and executing their implementation on the basis of a clear regulatory system.

This strengthened public sector has to guide the development by providing the best planning and coordination in order to enforce consolidation, which will in turn benefit the private sector regarding longterm perspectives. In this regard it can be expected that a number of speculation-driven investments will be replaced by middleterm and longterm investments due to a growing trust into the future development of cities. This particularly concerns investments of the private sector into infrastructure projects, which will inevitably become a more and more interesting investment alternative. Hence, the private sector would turn into a partner of the general urban development sharing the major concerns regarding the development of the overall urban structure. This new public private partnerships would be based on the mutual interest to develop and establish a longterm and sustainable urbanisation, of which both parties will benefit. Major obstacle of reaching this new reorganisation of urban governance, which is based on a strong public sector as guiding force, is the regional and global economic pressure to open up for shortterm investments. But in consideration of the already occuring threats for economic, social and environmental stability there will be no alternative for reinforcing the general public interests by strengthening the public sector, which has to set up the guidelines of the future urbanisation.

In addition to the reconfiguration of the public sector a further important aspect of future urban governance in the Gulf has to be the widespread promotion of public participation. The contribution of urban communities within the planning process can help to prevent infrastructural deficits and to enhance the quality of the built environment within neighbourhoods. In this regard communities for instance can be engaged to be partly self-responsible for sustaining public leisure areas. The feedback from inhabitants has to be generally seen as an important indicator, which public investments have priority and which investments can follow in future. In the cases of most Gulf countries the installation of educational programs will be needed in order to spread the awareness of being responsible part of the urbanisation. Biggest obstacle in this regard is the large percentage of foreign guest workers, who traditionally have a rather low identification with the location. Thus, widespread participation and bottom-up movements will only be possible if there is an political integration of this large part of the population. After all, the future urban governance is responsible to integrate the interests of the population by actively introducing methods and strategies to encourage participation.

8.1.3 The Introduction of Integrative Planning
In addition to a new organisation of urban governance, in which the role of the public sector is getting strengthened, urban planning itself has to be adjusted in order to cope with the new development challenges. While during the oil urbanisation master plans were mainly reduced on the design of the future land use distribution and the expansion of the road grid, new strategic plans are currently needed in order to integrate the economic, social and environmental development goals. The old master plans have usually become outdated due to the widespread development of projects, which do not follow the previous planning. The result has been a patchwork-like urban structure with many "cities" within one city leading to a general lack of cohesiveness and consolidation, particularly regarding the development of a sufficient infrastructural network. While each master developer has taken care to prepare his own master plan integrating all aspects of his project, there has been little coordination between master developers to work on concerns of the overall urban development. Beside the ongoing competition between developers the missing guidelines from the public sector contributed to a more and more decentralised planning process and thus to a development of large areas without major coordination.

Today, this lack of planning has developed to a major threat, which is enhanced by the fact that the scale of the developments, which have been commenced within the last ten years, was initially not expected. Due to widespread optimism and the growing hope to develop in high speed to a global economic centre the preparation of new plans and the introduction of regulations were in most cases ignored and postponed. The results have been major problems of consolidation, which have especially become more and more noticeable by growing traffic congestions. Consequently, the preparation of new comprehensive plans has become essential in order to successfully centralise and control the unregulated urban development. Only the integration of all aspects of the development within one strategic plan, which is designed to be directly implemented, would promise a more consolidated urbanisation. In addition to the general physical planning in form of a land use map, which includes all current and future developments of the private and public sector, there has to be the approach to link the plan with general development programs and projects, such as socio-economic development strategies. Furthermore, there has to be the attempt to guide future urban growth in order to prevent urban sprawl and in order to develop an urban structure with defined centres and subcentres.

Most challenging aspect of the new planning is to ensure the flexibility of adjusting the plan on new developments. While old master plans did not expect and therefore did not integrate the phenomenon of large scale projects, new strategic plans have to ensure the development of an urban structure, which allows certain changes by an investor-driven development. This integration of possible future developments within the current planning, particularly regarding a more capable infrastructural network, is the only way to reduce the very high public expenses. This in turn has to come along with clear guidelines and restrictions on the new developments regarding land uses and urban densities. In this way future developments will have to adjust themselves by integrating the development aspects of the overall urban structure in order to participate in improving the built environment, which has to become the mutual interest of public and private sector.

In order to prepare future-oriented strategic plans, there has to be a strong participation of all main stakeholders of the urban development and their agreement on all major decisions. While old master plans have often been designed solely by the public sector with hardly any participation of the private sector, the new strategic plans will have to include the interests of investors. Most challenging aspect of preparing these new plans is to find the biggest common denominator between public and private interests. In order to find compromises the main argument has to be the comprehensive goal to improve the economic, social and environmental conditions. The usual tradition to commission an international consultant to prepare more or less isolated a new master plan has to be replaced by a new approach of a transparent planning process by initiating discussion forums and workshops. Furthermore, the consultant has to be involved in the implementation process by developing the capacity of the administration instead of handing in prepared documents. In order to enable consultants to prepare strategic plans there has to be a reliable data base, which is continuously updated, and an intense moderation of the planning process by the public sector. Moreover, international consultants have to join work and task groups with local planners in order to integrate the knowledge of previous developments. One of the most important aspects of preparing the new plan will be the cooperation of all ministries and municipalities under the umbrella of one central authority in order to prevent parallel planning processes and in order to prepare the future implementation of commonly agreed strategies. All in all, planning has to be understood as essential basis to reach key development goals and thus all stakeholders and partners of the development have to participate and cooperate in preparing one comprehensive plan as guideline.

8.2 The Integration of Urban Qualities

8.2.1 The Development of Healthy Urban Cells

The recent urban development led to a rapidly growing number of large-scale projects transforming the urban landscape into a patchwork-like structure. Large undeveloped areas between construction sites and built-up areas are causing an increasing urban sprawl. While the overall urban density is decreasing due to large distances between new and old urban areas, the urban density has increased in certain districts due to the widespread construction of high rises. Thus, the recent urbanisation has led to an enormous growth turning the former oil settlements into large-scale construction sites. The speed and scale of this development was enabled by a more and more decentralised form of urban governance. The resulting lack of central planning has led to a rather uncoordinated urban development, which is currently neither integrating the construction of sufficient technical infrastructure nor services, such as schools or public leisure areas. Subsequently, the overall quality of the built environment is currently decreasing in spite the undoubtfully existing exemplary developments in certain areas.

One of the main occuring problems is the lack of new urban centres, which in addition to their function as business districts would act as important growth poles directing the future urban development. In most places the former oil city structure of one main CBD has remained due to the continuous expansion of the former business districts, which has been mainly caused by the interest of the private sector to establish their businesses at a central location. Furthermore, there has been often no plan to promote the development of new urban centres at strategically located areas. In addition to missing initiatives of the public sector to define those centres, the speculation-driven development in form of large-scale masterplanned projects has prevented the establishment of new business districts in certain areas due to the widespread favouratism of residential projects. Although there has been a tendency to integrate commercial districts within these large-scale developments, there has been apart from the development of small shopping malls and entertainment facilities rather little interest of businesses to relocate due to a general lack of centrality and accessibility. In addition to the lack of coordination regarding the development of new urban centres and subcentres there has been a widespread lack of integrating services within residential areas. While exclusive masterplanned developments often include entertainment and leisure facilities for their residents, many urban areas, particularly at the fringes of the old city centres, are not integrating sufficient open spaces.

In order to enhance the quality of the built environment there has to be at first a clear definition of main centres and sub centres, which are connected by a strong transportation network including public transport. This network of centres will be capable to direct the urban growth in a more coordinated manner and integrate commercial uses and services at the new urban areas. The aim has to be the establishment of self-contained urban cells, which complement one another to shape an overall effective structure. In this regard the centrally coordinated development of the infrastructural backbone has to dictate the size

271

Fig. 1: The Metro project in Dubai.

and main land use of each district, which has to get an active part of the entire urban structure. In this context each district should ideally integrate all main services for its residents in order to reduce daily travels. Furthermore, the integration of open spaces, such as parks, would help to enhance the quality of the built environment and to improve the ventilation in densely built areas. In this regard, the construction of high rise buildings has to be restricted regarding height, distances between buildings and the provision of sufficient parking spaces in order to prevent badly ventilated and congested areas.

All in all, most urban quality diminishing factors are currently originating from the extensive use of cars, which are responsible for traffic congestions, air pollution and a growing parking space problem in densely built areas. The development of "healthy urban cells" with a balanced density and mixed land uses will contribute to shorter ways and thus less traffic. Furthermore, a net of defined centres, connected by a strong infrastructural backbone, will help to reduce the growing problem of traffic during rush hours. In order to reduce the general dependency on cars the establishment of a structure of several centres instead of one main centre has to come along with the integration of public transport systems. Metro systems, for instance, have proven in many places to effectively reduce the traffic of commuters and to attract new developments. Hence, the construction of a metro system would for instance allow the development of dormitory towns for low income groups in the outskirts, where better living conditions are provided then at the crowded urban centres. Subsequently, in addition to investments into the infrastructure public and private sector has to be involved in improving the living conditions of lower income groups by providing sufficient housing programs.

One of the biggest future challenges will be the development of an apropriate balance of urban densities. On the one hand high urban densities provide the possibility to establish short ways between residences, working places and services. On the other hand the missing open space leads to a diminished urban quality. In this regard both current development tendencies in form of dense high rise agglomerations and exclusive suburbian settlements with integrated golf courses have to be adressed by introducing either a mix of different typologies or a medium-high typology in order to reduce or to increase the urban densities accordingly. All in all, the future development of the built environment has to be based on balanced growth integrating all aspects, which are necessary to establish an urban structure consisting of self contained urban cells merging to cities. The current separation and isolation of developments can only be consolidated if plans are implemented, which define the general structure in form of centres and main infrastructure networks. A further important aspect will be the introduction of clear urban growth boundaries in order to prevent urban sprawl and to increase the construction on undeveloped lots within cities. Moreover, the ongoing growth of high rise agglomerations along the coast has to be limited in order to sustain the necessary ventilation of inland settlements. After all, the urban growth has to be generally directed in a manner that reduces the negative impacts on the environment. Hence, further land reclamation has to be restricted in order to preserve the coastal flora and fauna. Without a strict direction of future urban growth, Gulf cities will face major problems to develop a consolidated built environment, which will help to prevent environmental and social inbalance.

While it will be essential to introduce comprehensive planning of land use and infrastructure, further innovative strategies have to be found how to improve the current built environment and thus to enhance consolidation. One possible strategy would be the attraction of international sport events, such as the Olympics, or cultural events, such as Expos. These events could support the process of consolidation if economic benefits are for instance invested into the construction of public infrastructure. Furthermore, events can be used to revitalise urban areas, which have been neglected in the last years. But in order to use events as consolidation strategy they have to take place at strategic locations within the urban area itself and not along highways in the desert. Recent examples, such as the Formula One in Bahrain or the Asian Games in Qatar, had little impact on consolidating the overall urban structure. Apart from new highways there has been no comprehensive strategy how to use these events for improving the general urban structure.

All in all, the general improvement of the built environment will need the integration and coordination of many actions within a sufficient timeframe. The first planning actions will have to focus on defining a structure of centres and subcentres and how to implement public transport. Moreover, clearly defined urban growth boundaries and land use plans in combination with an effective regulatory system have to direct the future urban development.

8.2.2 The Application of Modern Technologies
One of the most urgent problems of the current built environment is the high energy consumption per capita, which has led to the fact that Gulf cities have currently the biggest ecological footprint worldwide. While general physical planning is necessary

Fig. 2: Bahrain World Trade Centre.

to shorten ways between residential areas, business districts and services, there has to be the introduction of innovative technologies helping to reduce the currently growing energy waste. In addition to petrol for cars most energy is needed to desalinate water and to cool buildings. Today, solely fossil fuels are used to produce energy due to their still comparitively cheap price. Subsequently, the waste of the last oil and gas resources is not only accelerated by increasing exports, but also by the rapidly growing local consumption. In order to use renewable energy resources, such as wind and sun, new technolgies have to be introduced. Although the Gulf region provides plenty of sun and wind, there has been one major obstacle preventing the use of common technologies, which have been developed in Europe or North America. The constantly flying sand leads to thin layers on solar panels or mirrors, as well as it can damage wind turbines. Subsequently, the effectiveness of these technologies is limited and due to their high acquisition and maintenance costs the construction of new fossil fuel power plants is still favoured. Thus, these technologies have to be refined and adjusted to the regional circumstances. Furthermore, innovative new concepts, such as solar updraft towers, might be further alternatives to diversify the energy production. In addition to make usage of renewable energy sources new technologies should be introduced to improve current fossil fuel power plants. Hydrogen, which can be produced by partial oxidation of fossil fuels, might be an alternative fuel for future energy plants in the Gulf if the technologies are further progressing.

While technologies can be used to produce energy more effectively, technologies can also be applied to reduce the energy consumption itself. This can be achieved by introducing new transport systems, such as monorails or metros, which would reduce the daily traffic of commuters. Furthermore, technologies, such as district cooling plants, can be used to reduce the energy waste of air conditioning, by serving many buildings with chilled water. In order to reduce the high demand for air conditioning itself new building insulation technologies have to be introduced. Last but not least, technologies to safe and to reuse water would lower the total energy consumption due to the rather energy costly production of desalinated water. Thus, the recycling of greywater by using mechanical or biological systems can contribute to the development of more ecological Gulf cities.

All in all, reducing energy consumption by investing into new technologies faces one major obstacle, which is the still remaining wealth of fossil fuels. Beside the comparatively high investments new technologies have proven to be not very effective due to missing adjustments on regional circumstances. The biggest chance for integrating new technologies is investing into local research and local production in order to enter the market of energy technologies, which is expected to grow globally within the next years. In this way the Gulf can remain one of the worldwide biggest energy providers, but instead of fossil fuels it will export technologies how to use renewable energy sources and how to reduce energy waste. Initiatives, such as the Masdar development in Abu Dhabi, can be regarded as examples illustrating the future possibilities. Although new technologies, such as wind turbines at high rises in Bahrain and Dubai, are currently often used to attract media attention, they have undouptfully a certain impact on the general awareness regarding the existence of innovative solutions to reduce energy waste.

8.2.3 The Introduction of New Building Standards
Beside the rather un-ecological structure of Gulf cities and missing technologies to reduce energy waste the built environment has suffered from an architecture, which in most cases has neither any relation to the local climate nor to the local culture. Subsequently, there has been a rather diverse mix of global architecture following worldwide trends with rather minor adjustments to the regional circumstances. While reputable international architects have been engaged to design the main landmarks, most of the architecture originates from catalogues of construction companies, which are generally interested to satisfy the current market. Thus, most residential buildings, particularly villas, have been designed with a wide range of ornaments from oriental to neo-classical designs. However, the typology behind the various facades is in most cases standardised and multiplied leading to large monotone suburbian areas. Beside the typical two-storey house surrounded by gardens and walls, there has been recently the trend to develop multi-story apartment blocks in a similar pattern. Most critical aspects of this development are the rather monotone looking urban landscape and that buildings have been designed without responding to local climatic conditions. Instead of constructive methods to reduce the heating-up and thus to reduce the use of air conditioning, most buildings are built with large windows and thin walls.

Since the first imports of modern housing typolgies in the middle of the 20th century rather little efforts have been undertaken to adjust the architecture on the desert climate or the local culture. While cheap energy made air conditioning the easiest solution, tall walls surrounding the lots preserved the privacy of the traditional Islamic families. The result of the first phase of the oil urbanisation has been one of the most unecological types of town

Fig. 3: Masdar in Abu Dhabi.

Fig. 4: Xeritown in Dubai.

with large suburban areas hidden behind walls. Today, the widespread construction of high rise buildings is a new phenomenon transforming urban areas into high rise clusters. Their steel and glas architecture is currently the biggest expression of disregarding the environmental circumstances. Moreover, the import of global architectural forms has led to a replaceable urban image leading to a rather serious problem of a missing expression of identity. More than ever Gulf cities have become intruding space ships landing on a foreign planet, which is not understood but chosen because of its resources. Thus, current post-oil urbanism has generally not led to a significant change regarding the tradition of importing architecture instead of generating a new local architectural language. Current examples, such as the construction of fake wind towers on rooftops, cannot deceive of the general lack of innovative new designs creating the future genius loci of Gulf cities.

In order to establish an own architectural language local universities and research institutions have to be more involved within the current development. In addition to their role of observers and critics they have to become trendsetters delivering new approaches of architecture and urban design. Recent years and development tendencies have proven that some investors are searching for new innovations in order to differ from the mainstream projects and to market themselves as sustainable. This can be regarded as realistic chance to develop benchmark projects, which help to establish a new understanding of architecture. Most importantly there has to be a solution for a residential typology, which allows a balanced urban density. A mix of two to three-storey townhouses and medium-scale apartment blocks might be a possible alternative for the common typologies of low rise villas and residential high rises. Furthermore, innovative constructive methods can be applied to reduce the heating up of buildings. Thus, facade elements can be used to spend shade on windows and the building direction, in which the long side of buildings faces south can help to reduce the heating up time of walls during mornings and afternoons. In addition, systems of cross ventilation and tall room heights can improve the indoor climate, particularly during nights. Last but not least, walls have to be designed with a sufficient thickness and with apropriate building materials, such as local building materials, which furthermore would help to improve the adjustment to both the climate and to the surrounding landscape.

In order to introduce new building standards new regulations have to be introduced, which force developers to use certain techniques in order to reduce energy waste. The introduction of green building rating systems, such as LEED (Leadership in Energy and Environmental Design), which has been introduced in the UAE, can be considered as first step in the right direction, but without implementing an obligatory standard many buildings will be still built without regarding ecological concerns.

8.3 Future Scenarios of Post-oil Urbanism

8.3.1 The Scenario of Centralisation
Although post-oil urbanism still looks back on a very short history and can be considered still at its beginning, recent development tendencies have already drawn a picture what the future of Gulf cities might look like. The need of economic diversification has led to the widespread idea to liberalise markets and to attract private investments. No case illustrates the rapide privatisation of the urban development better than Dubai, which has become a model of current post-oil urbanism. Dubai showed how to reach fast growth by opening the markets for speculations and thus it has caused a large regional competition. Subsquently most Gulf cities entered the race of attracting as many investors as possible in order to build up a future post-oil economy. This, often as "Dubaification" described phenomenon has indeed led to rapide urban growth all over the region. Most countries and emirates started to imitate successful developments in neighbouring countries. Subsequently, airports and habours are either expanded or newly built in order to increase the capacities. Similar to the trading industry almost every place has discovered tourism as main future economic sector, which has led to a vast number of hotel complexes, shopping malls and leisure facilities. Tourism has also been used to market huge real estate projects, whose trade became the centrepiece for a booming financial sector. Consequently, almost every GCC-country has introduced freehold property laws, allowing an easy trade of real estate, which has become the most favourite regional commodity after oil. Last but not least, each country has begun to provide free economic zones in order to attract industries including knowledge-based industries.

Today, the speculation to become a global service centre is leading to the development of a regional oversupply of airports, harbours, tourism projects, real estate or free zones. Subsequently, the open competition has led to a rather fortunate environment for speculators searching for the best conditions of investments. These conditions in turn depend on many factors, of which a low-tax environment can be considered as the most decisive. In order to sustain low taxes in future remaining oil and gas resources will still play a major role, which automatically leads to the assumption that oil-rich places, such as Abu Dhabi, possesses a fortunate development perspective. However, places, such as Dubai, who have already established a service driven industry can already provide the needed infrastructure and services and thus are currently in development advance. At the end, a combination of factors will most likely define which place develops to the biggest regional trading hub with potentials to enter the big global business. Due to the establishment of

277

the largest airport and harbour many further industries will gradually concentrate at this main trading centre, which can be expected to offer the best opportunities for business. While this centre will continue to grow, other centres gradually will lose their importance and shrink. Subsequently, many of these centres will economically depend on the main centre. In the best case they will become the location of outsourced economic sectors, such as wellness tourism or heavy industries. Hence, the future map of this scenario shows one dominant urban agglomeration, while the remaining centres of the Arabian Peninsula will gradually lose importance. After all, the future settlement structure is most likely to be dominated by a centralist character with one main post-oil metropolis, which has reached in size and economic importance the status of a "Global City".

Due to the continuation of the current urbanism the typical development tendencies regarding the built environment cannot be expected to change significantly. Thus, the construction of high rises will still be favoured in many central and coastal areas due to growing land prices. Low rise residential typologies in turn will be built at the outskirts. Furthermore, the construction of superlatives, such as man-made islands or the worldwide tallest buildings, can be expected to be continued in certain ways due to their catalyst function to attract speculators. In addition to the development of future landmarks many buildings will be produced in masses with a standardised repertoire of designs leading to a large monotonious urban landscape. Hence, the future Gulf metropolis will be most likely divided into mixed-use high rise agglomerations with the highest urban densities and low rise suburban fringes. While more and more people will live and work at the CBD, there will be a growing need to develop dormitory settlements in form of large-scale apartment blocks outside the business districts and connected by public transit. Although public transportation will inevitably become the most important transport system for most commuters, the urban landscape will still express the result of the long term focus on the car as the only means of transportation. Big highways and flyovers will occupie a large percentage of the urban area and turn into barriers between districts. All in all, the scenario of a "Mega Gulf City", which is based on current development tendencies, would inevitably include large social segregation. This will be expressed in urban areas with the highest urban densities in form of apartment blocks and luxurious villas with private beaches and golf clubs.

While the future mega city will reach a certain point of consolidation, in which its inhabitants have adjusted their way of living to the circumstances, other cities and urban agglomerations will have failed in their attempt to consolidate due to missing investments. The end of the oil production and the relocation of most post-oil economies to the main urban agglomeration will make it hardly possible to sustain most of the previously built developments. This would mean the end of many high rises and man-made islands, whose maintenance costs will be too high to be subsidized. After all, several ruins will remind on the previous real estate boom and large former urban areas will be taken back by the surrounding desert. While most guest workers will leave these places, many local inhabitants will move to the old centres due to missing infrastructural supply in the outskirts or they will relocate to the remaining cities. Thus, in the cases of most places the current urbanisation would have to be seen as a temporary event creating "Instant Cities" for a short period of time attracted by the last oil resources.

Although Dubai has grown at the fastest and largest scale the realisation of most current projects has become doubtful since the break-out of the international financial crisis. Thus, other places can be considered as serious competitors, which have the potential to succeed the on-going competition and to overshadow Dubai within the next decades. While Abu Dhabi has still large oil resources, Qatar possesses the worldwide biggest gas resources after Russia and Iran. Bahrain in turn benefits from the fortunate location close to the oil production in Saudi Arabia. In comparison to other places both Bahrain and Dubai can look back on decades of economic diversification, which again means a certain advance of experience. In addition to the wealth of fossil fuels or the provision of the best infrastructure the race will be decided by many outer circumstances, particularly the development of the global economy. The ambitious target to transform into a global service centre and trading hub will inevitably lead to a growing dependency on the global economic situation. Thus, the Gulf will face global competitors, particularly India, which is currently expected to expand its economic importance within the global network. Subsequently, when the oil production is going to an end a Gulf mega city will undergo a large verification of its inner potentials to keep up with growing international markets. At the end, it might come down to the flexibility of the future population of this Gulf city in order to react and adjust on ever-changing circumstances. Otherwise, post-oil urbanism has truely been the final chapter of the urbanisation in the Gulf leaving behind the memories of a construction firework, metaphorically describeable as a bloom in the desert.

8.3.2 The Scenario of Centralised Decentralisation

While the first scenario is based on the expectation that there will be no major changes regarding the development goal and model, the second scenario assumes that there will be modifications and adjustments of the current post-oil urbanism. Biggest catalyst of this development direction might be the current international financial crisis, which has led to a slow-down of the construction boom. This break can be considered as chance to rethink the recent development and to question the general preference of fast growth before smart growth. Thus, the growing awareness of the evident risks, caused by a speculation-driven development, could lead to a new phase of cooperation between the GCC-countries. In contrast to an on-going competition a strong regional alliance would enable the Gulf cities to introduce more restrictions and to focus on longterm goals. These goals differ from the current development strategy to enhance economic wealth by short term investments due to the fact that they include development strategies regarding environment and society. In order to attract middle and long term investments from the private sector there have to be integrated development strategies and a trustworthy legal framework. After all, the GCC-countries would remain serious competitors within the globalisation, but instead of one large mega urban agglomeration a number of cities would grow on the basis of a strong network of cooperation.

Consequently, the speculation-driven development would slow down and investments would focus on the consolidation of the already built urban areas. Along with the expansion of the general technical infrastructure there will be the introduction of public transport systems, which are in turn used to connect centres and sub-centres. In addition to a network of public transport within single urban agglomerations there might be a chance to establish a new regional railway system linking all urban centres in the Gulf. Furthermore, main stations and stations of this regional and local network of transportation can define new centres and sub-centres and thus transform the previous centralist structure to a polycentral structure, which is based on an effective infrastructural network. Every urban agglomeration would be specialised on certain post-oil economic fields and thus be able to sustain their populations. In contrast to the current situation of rapide growth of new economic sectors, a slower development will provide the local populations with the chance to adjust to the new economic circumstances and thus to become more integrated within the new development. Moreover, the new emphasis on longterm goals might attract guest workers to settle with their families in the Gulf for their lifetimes, what might lead to cosmopolitan societies instead of the current phenomenon of airport societies. Last but not least, the environment would benefit from balanced growth, which allows the introduction of new standards preventing the fast exploitation of its resources.

This modification of current urban development tendencies, which means the replacement of regional competition with regional cooperation, would inevitably lead to another transformation of the built environment. Instead of construction superlatives in form of the tallest high rises and man-made islands in order to attract speculators many future projects would have a moderate scale due to less pressure to stand out and less growing land prices. Subsequently many undeveloped lots, particularly close to the business districts, would be built and many old buildings replaced leading to higher urban densities. Furthermore, regulations would enhance the consolidation by establish new public-private partnerships in order to build up technical and social infrastructure. In contrast to the current development direction that the infrastructural supply follows the construction of buildings the future infrastructural supply would become a tool to restrict urban sprawl and to enhance the development in certain areas. Subsequently, compact self contained districts would evolve, which are connected by a strong infrastructural network. Beside the transformation to a more effective urban structure the built environment would be changed by an adjusted architecture of buildings, which integrates new technological and constructional methods in order to react on climatic conditions and decreasing energy resources. Last but not least, less focus on short term investments would allow the integration of more public areas in form of parks and promenades, which would improve the future liveability of urban districts.

This future scenario of the balanced growth of many urban centres within one network would be rather flexible and robust regarding regional and global threats. A strong network of political and economic interaction would be able to react on regional conflicts or global crisis by relocating certain economic sectors within the network. Furthermore, in many places a network of centres and sub-centres has proven to be more effective regarding the realisation of environmental and social development goals. Subsequently, the built environment would express the new circumstances in form of a more regionally apropriate architecture and compact self-contained urban districts instead of constructive superlatives and a pattern of isolated themed projects of large real estate developers. However, this further transformation of post-oil urbanism towards consolidation requires a strong cooperation between all GCC-countries. In this regard the global financial crisis and its impact on the Gulf can be seen as chance to realise the common threats and develop a new level of cooperation to achieve economic, environmental and social sustainability. Urban planners and architects will have to play their role to contribute to the development of a future built environment, which is based on a comprehensive development strategy integrating all aspects of the post-oil urbanisation. Consequently, current constructive superlatives would become built manifestations of the first phase of post-oil urbanism representing a short period of exponential growth. After all, the second phase, which is marked by a consolidated development, will transform the Gulf into a net of highly specialised urban centres using state-of-the-art technologies and methods in order to adjust to the desert environment.

8.3.3 Outlook
Both previously described scenarios illustrate that post-oil urbanism has reached a point, where political decision makers have to react on new upcoming challenges. Due to the unlikelihood that all GCC-countries would agree on one development strategy enforcing balanced growth, it might be more realistic that there will be single alliances between neighbouring countries or emirates. Thus, the future causeway between Bahrain and Qatar in addition to the already existing causeway from Bahrain to Saudi Arabia might enhance the economic interaction between these places and therefore the cooperation regarding urban development strategies. Similarly, it can be expected that the emirates within the UAE and the northern region of Oman will be more and more economically connected. After all, it will remain decisive if growth within these networks will focus on the most competitive place or if there is a more equal distribution of growth on all centres and subcentres. In this regard the biggest political decision will always remain between continuing an open competition by deregulating markets instead of initiating new alliances by agreeing on common development goals.
In addition to old rivalries between the GCC countries and even single emirates within the UAE, the biggest obstacle of introducing a common development strategy is the growing global competition, which is putting pressure on the region to keep up with growing economic centres, particularly in South Asia. While the Gulf has benefited from

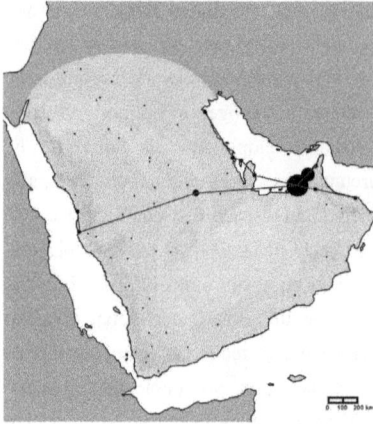

Fig. 5: The scenario of one dominant centre.   *Fig. 6: The scenario of a polycentric region.*

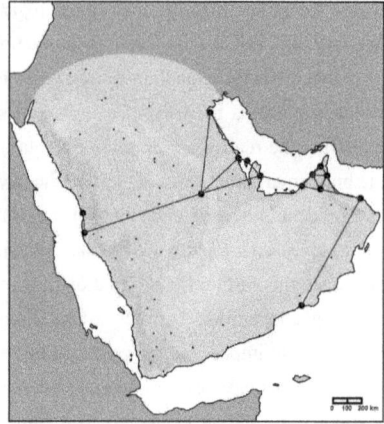

its geopolitical location between global markets, it cannot be denied that its wealth of fossil fuels has directly or indirectly remained an important attracting point for most new economic sectors. Thus, the oil wealth enabled and still enables the development of modern infrastructure and it sustains a low tax environment, which both have led to a development advance in comparison to other competitors in Asia. But shrinking oil resources and the high dependency on foreign labour currently question the economic competitiveness regarding longterm perspectives. Last but not least, future domestic and external conflicts might threaten the successful economic diversification, which is still to be considered at its starting point.

The recent attempt to liberalise the markets in order to establish the Gulf as global service hub is currently leading to a transformation of the built environment. In recent years the Gulf has become a centre of constructive superlatives in form of a large number of man-made islands shaped in various designs in addition to the construction of large high rise conglomerates. While large areas, particularly along the coasts and close by city centres, are more and more occupied by these high rise buildings, the outskirts are affected by rapide urban sprawl caused by low rise developments and speculation. The recent decentralisation of the physical planning currently prevents a coordinated development of centres and sub-centres. Thus, main business centres often remain at central locations and most new urban areas turn into dormitory settlements. In general it can be stated that the speculation-driven development focuses on the construction of residential projects, while the development of social infrastructure, such as schools, and the integration of public leisure areas are often neglected. Therefore, the new settlement areas will in most cases remain dependent on the infrastructure and facilities of the city centres and thus they will not become self-contained entities. Due to the isolated process of planning most projects do not regard adjacent developments. This de-central urban development has led to a patchwork-like urban pattern with a lack of integrated land use and infrastructure. Main consequence of this development is the missing structure of centres and sub-centres, which integrate identification points within the cities. Each development area is a closed entity, which is dependent on the overall structure, but does hardly add any conjunctive elements. The facts that the developments themselves are neither self-contained nor do they merge to one consolidated urban area lead to the main problems of current post-oil urbanism, which is apparently unable to coordinate urban growth.

In spite the financial opportunities to develop ecological desert towns or to establish new standards of an integrated urban development it can be stated that current post-oil urbanism is mainly dominated by speculative interests. Thus, the doctrine to develop short term commercially successful projects currently prevents the transformation into a worldwide unique place of experimenting for sustainable urban and architectural designs. Rare examples, such as the Masdar-development in Abu Dhabi, are exceptional cases illustrating the possibilities to develop new standards of ecological design in the region. Although there is a growing awareness regarding the necessity of new building standards, their fast introduction will be rather unlikely. In most cases it will still remain the choice of the private investors themselves to invest into new building techniques and technologies in order to market their projects as sustainable or to reduce maintenance costs.

In addition to a lack of building regulations the de-centralised decision making process within the urban governance is responsible for an uncoordinated urban growth without any general strategic plan as binding guideline of the developments. Thus, one of the biggest challenges of today is the initiation and implementation of new plans and strategies, which will be capable to enhance consolidation and thus future sustainability and liveability. These future plans have to integrate all strategies in a comprehensive plan in order to be successfully implemented. Examples, such as the National Strategic Plan of Bahrain, illustrate the current attempt to develop these plans, which differ from previous master plans due to their approach to address the new realities of a privatised urbanisation. The enhanced transparency of future goals is intended to enhance the interest of the private sector for longterm investments. This particularly concerns the establishment of a new kind of public private partnerships in order to develop for instance public infrastructure or social housing.

In spite the current financial crisis, in which the urbanisation in the Gulf is majorly effected, post-oil urbanism cannot be considered to have reached its final chapter yet. In contrast post-oil urbanism is still at the beginning to configure the possibilities of establishing a consolidated urban growth. The big development potentials to become global service centres due to the geopolitical location and the already existing infrastructure are still potentials, which undeniably could sustain future Gulf cities. However, the model of post-oil urbanism, which has been introduced by the Emirate of Dubai at the end of the 90s, can already be seen as an outdated strategy. A new model has to be elaborated, which is not based on the idea of exponential growth resulting in mega-cities in the desert. Future Gulf cities will have to face major challenges to become resource-effective and to provide liveable surroundings for their populations. Thus, the recent transformation of the built environment might be the result of post-oil urbanism's first chapter, which was formed by a speculation-driven development disregarding environmental and socio-economic realities.

Today post-oil urbanism is still struggling to finds its goals and methods to develop a sustainable form of cities in the Gulf. First step would have to be the abolishement of a non-productive competition within the GCC by agreeing on common development goals in order to face the future economic, social and environmental challenges. The second step would have to be the elaboration of integrated strategies by preparing one comprehensive plan for the entire region. After all, this plan can be implemented as basis, on which local strategies will be prepared, and thus become the first step of an improving cooperation within the GCC. This merging between Gulf cities has to replace the old rivalries between tribes and countries. The consequence could be a competitive region consisting of a network of towns instead of one or few dominating centres facing major environmental and social problems. Subsequently, the built environment will have to face a new transformation, in which the urban structure is formed by merging self-contained urban cells instead of isolated urban patchworks.

List of Figures

Figure 6   By author 2006.
Figure 7   Ministry of Municipalities and Agricultural Affairs (Bahrain) 2008.
Figure 8   By author 2008.
Figure 9   By author 2008.
Figure 10   By author 2008.
Figure 11   Nouri, 2008.
Figure 12   By author 2008.
Figure 13   http://www.constructionweekonline.com/article-6666-atkins-bags-durrat-al-bahrain-contract/, 31.01.2010.
Figure 14   By author 2008.
Figure 15   By author 2008.
Figure 16   By author 2008.

Chapter 4
Figure 1   Ministry of Municipalities and Agricultural Affairs (Bahrain) 2008.
Figure 2   Survey Directorate (Bahrain) 2008.
Figure 3   http://www.skyscrapercity.com/showthread.php?t=511665, 31.01.2010.
Figure 4   Survey Directorate (Bahrain) 2008.
Figure 5   http://www.skyscrapercity.com/showthread.php?t=511665, 31.01.2010.
Figure 6   http://www.skyscrapercity.com/showthread.php?t=511665, 31.01.2010.
Figure 7   Ministry of Municipalities and Agricultural Affairs (Bahrain) 2008.
Figure 8   By author 2010.
Figure 9   http://www.bahrainbay.com/image-gallery.html, 31.01.2010.
Figure 10   http://www.bahrainbay.com/image-gallery.html, 31.01.2010.
Figure 11   http://www.arabianbusiness.com/512978-al-hafeera-wins-contract-to-build-reef-island-bridge, 31.01.2010.
Figure 12   By author 2010.
Figure 13   By author 2010.
Figure 14   Survey Directorate (Bahrain) 2008.
Figure 15   By author 2010.
Figure 16   Ministry of Municipalities and Agricultural Affairs (Bahrain) 2008.
Figure 17   By author 2010.
Figure 18   By author 2010.
Figure 19   Ministry of Municipalities and Agricultural Affairs (Bahrain) 2008.
Figure 20   Ministry of Municipalities and Agricultural Affairs (Bahrain) 2008.
Figure 21   By author 2010.
Figure 22   Ministry of Municipalities and Agricultural Affairs (Bahrain) 2008.
Figure 23   By author 2010.
Figure 24   By author 2010.
Figure 25   By author 2010.
Figure 26   By author 2010.
Figure 27   By author 2010.
Figure 28   By author 2010.
Figure 29   By author 2010 (Survey Directorate)
Figure 30   By author 2010.

Chapter 5
Figure 1   EDB (Bahrain) 2008.
Figure 2   By author 2010.
Chapter 6
Figure 1   Google Earth 2010.
Figure 2   By author 2010.
Figure 3   Urban Planning Council (Abu Dhabi) 2007.
Figure 4   By author 2010.
Figure 5   By author 2010.
Figure 6   By author 2010.
Figure 7   http://www.skyscrapercity.com/showthread.php?t=332488&page=6, 31.01.2010.
Figure 8   http://www.e-architect.co.uk/saudi_arabia/king_abdullah_economic_city.htm, 31.01.2010.

Chapter 7
Figure 1   By author 2010.
Figure 2   By author 2010.

Chapter 8
Figure 1   http://dubaimetro.com, 09.07.2012.
Figure 2   By author 2010.
Figure 3   http://aswinworld.wordpress.com/2009/04/02/si-habra-masdar-ciudad-de-foster-para-el-2016/, 31.01.2010.
Figure 4   http://derstandard.at/1265852018998/Architektur-Die-Zukunft-der-Vergangenheit, 31.01.2010.
Figure 5   By author 2010.
Figure 6   By author 2010.

Bibliography

Internet:
ABC. http://www.abc.net.au/4corners/special_eds/20060710/, 19.10.2008.
Abrajallulu. http://www.abrajallulu.com/www/index.php, 22.03.2008.
Abu Dhabi Airport. http://www.abudhabiairport.ae/theairport/index.asp, 11.05.2009.
Al Areen. http://www.alareenresort.com/alareenoverview.htm, 03.03.2009.
Al Areen. http://www.alareenresort.com/history.htm, 03.03.2009.
Al Areen. http://www.alareenresort.com/masterplan.htm, 03.03.2009.
Al Areen. http://www.alareenresort.com/progress.htm, 03.03.2009.
Al Furjan. http://www.alfurjan.com/en/media-centre/press-release/nakheel-launches-al-furjan.html, 04.01.2009.
Al Hathloul, Saleh. 2003: Riyadh Architecture in one hundred years. www.csbe.org/e_publications/Riyadh_architecture/index.htm, 05.10.2008.
Al Raha Beach. http://www.alrahabeach.com/main.php, 11.05.2009.
AME info. http://www.ameinfo.com/16342.html, 13.10.2008.
AME info. http://www.ameinfo.com/man_made_islands_article/, 28.10.2008.

AME info. http://www.ameinfo.com/67119.html, 28.10.2008.
AME info. http://www.ameinfo.com/140234.html, 30.10.2008.
AME info. http://www.ameinfo.com/113976.html, 30.10.2008.
AME info. http://www.ameinfo.com/143945.html, 31.10.2008.
AME info. http://www.ameinfo.com/76184.html, 04.01.2009.
AME info. http://www.ameinfo.com/166451.html, 08.01.2009.
AME info. http://www.ameinfo.com/31019.html, 15.01.2009.
AME info. http://www.ameinfo.com/161240.html, 28.01.2009.
AME info. http://www.ameinfo.com/102262.html, 18.01.2009.
AME info. http://www.ameinfo.com/134782.html, 29.12.2009.
AME info. http://www.ameinfo.com/193154.html, 11.05.2009.
Arab Decision. http://www.arabdecision.net/show_func_3_12_5_0_3_1345.htm, 02.11.2008.
Arabian business. http://www.arabianbusiness.com/518541-dubai-tourist-numbers-set-to-hit-10mn?ln=en, 21.10.2008.
Arabian business. http://www.arabianbusiness.com/property/unit/523-jumeirah-park, 04.01.2009.
Arabian business. http://www.arabianbusiness.com/500339-putting-the-culture-back-into-the-village, 06.01.2009.
Arabian business. http://www.arabianbusiness.com/542751-dubai-maritime-city-on-schedule-for-2012-completion, 07.01.09.
Arabian business. http://www.arabianbusiness.com/index.php?option=com_companylist&view=list&companyid=5805, 16.01.2009.
Arabian business. http://www.arabianbusiness.com/542962-qatar-bahrain-causeway-unlikely-to-start-in-january, 11.05.2009.
Arabian business. http://www.arabianbusiness.com/543142-first-flight-out-of-al-maktoum-air-port-by-end-of-2009, 14.01.2009.
Arabian business. http://www.arabianbusiness.com/530704-bahrain-opens-door-to-kingdom, 14.02.2009.
Arabian business. http://www.arabianbusiness.com/537297-emaar-denies-problems-on-dubai-tram-project, 28.01.2009.
Arabian business. http://www.arabianbusiness.com/539418-bus-lanes-cycle-zones-part-of-12bn-traffic-master-plan, 29.01.2009.
Arab News. http://www.arabnews.com/?page=6&section=0&article=56605&d=26&m=12&y=2004, 14.06.2009.
Ar Riyadh Development Authority. http://www.arriyadh.com/En/Ab-Arriyad/index.asp, 14.10.2008.
Ar Riyadh Development Authority 2005: Investment Climate 2005. www.arriaydh.com/En/Economy/LeftBar/ADA_Report_Final.pdf, 22.10.06.
Asteco. http://web.asteco.com/resources/pdf/Q4/200811_astqrt008_q3_bahrain_qatar.pdf, 16.04.2009.
BAHIW. http://www.bahiw.com/biwprofile.htm, 03.03.2009.
BAHIW. http://www.bahiw.com/masterplan.htm, 03.03.2009.
Bahrain Bay. http://www.bahrainbay.com/resources/pdf/BB_Zone1.pdf, 01.04.2009.
Bahrain Bay. http://www.bahrainbay.com/resources/pdf/BB_Zone2.pdf, 01.04.2009.
Bahrain Bay. http://www.bahrainbay.com/resources/pdf/BB_Zone3.pdf, 01.04.2009.
Bahrain Bay. http://www.bahrainbay.com/resources/pdf/BB_Zone4.pdf, 01.04.2009.
Bahrain Bay. http://www.bahrainbay.com/resources/pdf/BB_Zone5.pdf, 01.04.2009.
Bahrain Bay. http://www.bahrainbay.com/resources/pdf/BB_Zone6.pdf, 01.04.2009.
Bahrain Bay. http://www.bahrainbay.com/resources/pdf/BB_Zone7.pdf, 01.04.2009.

Bahrain EDB. http://www.bahrainedb.com/AboutEDBIntro.aspx, 14.02.2009.

Bahrain EDB. http://www.bahrainedb.com/uploadedFiles/BahrainEDB/Media_Center/Economic%20Vision%202030%20(English).pdf, 03.05.2009.

Bahrain Freehold Properties. http://www.bahrainfreeholdproperties.com/freeholdproperty_aboutbahrain.aspx, 14.02.2009.

Bahrain Rights. http://www.bahrainrights.org/en/node/2431, 08.05.2009.

BARWA. http://www.barwa.com.qa/english/LocalProjects.aspx, 11.05.2009.

BBC. http://news.bbc.co.uk/2/hi/middle_east/7446923.stm, 24.08.2009.

BFH. http://www.bfharbour.com/html/aboutbahrain/milestones.htm, 14.02.2009.

BFH. http://www.bfharbour.com/html/faq/faq_development.htm, 26.02.2009.

Blue City. http://www.bluecity.me/HOME/tabid/36/Default.aspx, 23.08.2009.

Burj Al Alam. http://theburjalalam.com/?paged=8, 04.01.2009.

Burj Dubai. http://www.burjdubaiskyscraper.com/, 28.10.2008.

Central Market. http://www.centralmarket.ae/main.html, 09.05.2009.

CIA Factbook. http://www.cia.gov/library/publications/the-world-factbook/geos/ba.html, 13.10.2008.

CIA Factbook. https://www.cia.gov/library/publications/the-world-factbook/geos/sa.html#People, 24.08.2009.

City of Arabia. http://www.cityofarabiame.com/about-us/about-city-of-arabia.html, 11.01.2009.

City of Arabia. http://www.cityofarabiame.com/our-projects/mall-of-arabia/overview-mall-of-arabia.html, 11.01.2009.

Civic Arts. http://www.civicarts.com/mohammed-bin-rashid-gardens.php, 28.01.2009.

Construction Digital. http://www.constructiondigital.com/MarketSector/Civil-Construction-and-Engineering/Construction-of-Qatar-Bahrain-Friendship-Bridge-to-start_38445.aspx, 28.12.2009.

Construction Week Online. http://www.constructionweekonline.com/article-4391-bahrain_construction_looking_to_uptown/, 05.05.2009

Construction Week Online. http://www.constructionweekonline.com/projects-276-mohammed_bin_zayed_city_development/, 11.05.2009.

Cowshed Properties. http://www.cowshedproperties.com/about-dubai/the-waterfront/, 12.01.2009.

DAH Property. http://www.dahproperty.com/ThePalmJabelAli.asp, 11.01.2009.

Dayaratne, Ranjith. http://203.77.197.231:81/isvs/isvs-4-1/paper-dump/full-papers/1.pdf, 05.10.2008.

Demographia. http://www.demographia.com/db-worldua2015.pdf, 30.10.2008.

DHCC. http://www.dhcc.ae/EN/AboutDHCC/Pages/Location.aspx, 06.01.2009.

DIFC. http://www.difc.ae/district/index.html, 06.01.2009.

DIPark. http://www.dipark.com/index.php?lang=en, 13.01.2009.

Diyar. http://www.diyar.bh/en-faq.html, 28.02.2009.

Doha Traveller. http://dohatraveller.info/, 11.05.2009.

Downtown JA. http://www.downtownjebelali.com/#, 18.01.2009.

DSC. http://www.dsc.gov.ae/DSC/Pages/Statistics%20Data.aspx?Category_Id=0226, 28.10.2008.

DSE. http://www.dse.vic.gov.au/DSE/nrencm.nsf/LinkView/EB6C4024FDA76297CA256EEB-0021C55F9AC29B2310AEF5F1CA256EF9000B230F, 27.07.2009.

DSO. http://www.dso.ae/en/about-dsoa/vision.html, 08.01.2009.

DSO. http://www.dso.ae/en/about-dsoa/master-plan.html, 08.01.2009.

Dubai. http://www.dubai.de/artikel/230-Palm-Deira-waechst-kontinuierlich-vor-der-Kueste-Dubais.html, 07.01.2009.

Dubai Airport. http://www.dubaiairport.com/DIA/English/TopMenu/About+DIA/ DIA+and+History/, 25.10.2008.

Dubai City Guide. http://www.dubaicityguide.com/geninfo/news_dtls.asp?newsid=11198, 04.01.2009.

Dubai City Guide. http://www.dubaicityguide.com/GENINFO/news_dtls.asp?newsid=20243, 08.01.2009.

Dubai FAQS. http://www.dubaifaqs.com/al-sufouh-tram.php, 28.01.2009.

Dubai FAQS. http://www.dubaifaqs.com/dubai-academic-city.php, 08.01.2009.

Dubai Holding. http://dubaiholding.com/en/our-companies/tecom-investments/, 04.01.2009.

Dubai Holding. http://dubaiholding.com/en/media-centre/news/2008/October/mizin-to-unveil-dubais-first-boutique-villa-community-at-cityscape-08/, 13.01.2009.

Dubai Holding. http://dubaiholding.com/en/media-centre/news/2008/October/dubai-industrial-city-commences-leasing-2-8-million-sq-ft-of-open-storage-yards/, 17.01.2009.

Dubai Industrial City. http://www.dubaiindustrialcity.ae/Pages/home/faq.aspx, 13.01.2009.

Dubai Internet City. http://www.dubaiinternetcity.com/, 04.01.2009.

Dubailand. http://www.dubailand.ae/project_details.html, 08.01.2009.

Dubailand. http://www.dubailand.ae/facts_figures.html, 11.01.2009.

Dubai Marina. http://www.dubai-marina.com/, 04.01.2009.

Dubai Marina. http://www.dubai-marina.com/skyscrapers.html, 04.01.2009.

Dubai Marina Realty. http://www.dubaimarinarealty.com/CityProjects.html, 04.01.2009.

Dubai Maritime City. http://www.dubaimaritimecity.com/, 07.01.2009.

Dubai Media City. http://www.dubaimediacity.com/, 21.08.2008.

Dubai Online. http://www.dubai-online.com/news/plans-for-dubai-tram-system-unveiled/, 28.01.2009.

Dubai Pearl. http://www.dubaipearl.com/About.aspx, 02.01.2009.

Dubai Promenade. http://www.dubaipromenade.ae/en/masterplan/an-inspired-masterplan.html, 11.01.2009.

Dubai Promenade. http://www.dubaipromenade.ae/en/facts-and-figures/facts-and-figures.html, 11.01.2009.

Dubai Properties. http://www.dubai-properties.ae/en/Projects/JumeirahBeach/Index.html, 04.01.2009.

Dubai Properties. http://www.dubai-properties.ae/pdf/Business-Bay-bro.pdf, 05.01.2009.

Dubai Properties. http://www.dubai-properties.ae/en/Projects/BusinessBay/ExecutiveTowers/ Index.html#, 05.01.2009.

Dubai Properties. http://dubai-properties.ae/en/Flexible_Images/pdf/mudon_factsheet_en.pdf, 11.01.2009.

Dubai Properties. http://dubai-properties.ae/en/Projects/Tijara_Town/master_plan.html, 18.01.2009.

Dubai Properties. http://www.dubai-properties.ae/the-villa-overview.html, 11.01.2009.

Dubai Property 4U. http://www.dubaiproperty4u.co.uk/impz.php, 15.01.2009.

Dubai Search. http://www.dubai-search-and-find.com/healthcare-city.html, 06.01.2007.

Dubai Towers. http://www.dubaitowersdubai.com/, 05.01.2009.

Durrat Bahrain. http://www.durratbahrain.com/en/explore/durrat-marina/durrat.html, 01.03.2009.

Durrat Bahrain. http://www.durratbahrain.com/en/explore/golf-course/golf.html, 01.03.2009.

http://www.durratbahrain.com/en/explore/crescent/zones/zones.html, 01.03.2009.

Durrat Bahrain. http://www.durratbahrain.com/en/investors-ar/investors.html, 01.03.2009.

Durrat Bahrain. http://www.durratbahrain.com/en/project/location.html, 01.03.2009.

DWC. http://www.dwc.ae/dwc.html, 14.01.2009.

DWC. http://www.dwc.ae/dwc_dubai_logistics_city.html, 14.01.2009.
DWC. http://www.dwc.ae/images/pdf/28%20nov%2007%20DUBAI%20WORLD%20CEN-TRAL.pdf, 14.01.2009.
DWC. http://www.dwc.ae/dwc_real_estate_division.html, 14.01.2009.
DWC. http://www.dwc.ae/dwc_real_estate_division_commercial_city.html, 14.01.2009.
E-Architect. http://www.e-architect.co.uk/bahrain/nomas_towers.htm, 27.04.2009.
Economist. http://www.economist.com/finance/displaystory.cfm?story_id=13186145, 11.05.2009.
Education Emirates Network. http://education.theemiratesnetwork.com/zones/qatar_education_city.php, 11.05.2009.
EI City. http://www.ei-city.net/, 11.05.2009.
Emaar. http://www.emaar.com/index.aspx?page=emaaruae-downtownburj-burjdubaiboulevard, 05.01.2009.
Emaar. http://www.emaar.com/index.aspx?page=emaar-international, 16.07.2009.
Emaar. http://www.emaar.com/index.aspx?page=umm-al-quwain-marina, 11.05.2009.
Emporis. http://www.emporis.com/en/wm/bu/?id=210141, 16.01.2010.
Energy Information Admininistration. http://www.eia.doe.gov/emeu/cabs/Bahrain/Oil.html, 19.10.2008.
Estates Dubai. http://www.estatesdubai.com/labels/Palm%20Deira.html, 07.01.2009.
Estimate. http://www.theestimate.com/public/122900.html, 14.02.2009.
Falcon City. http://www.falconcity.com/faq.asp, 10.01.2009.
FDI Magazine. http://www.fdimagazine.com/news/fullstory.php/aid/2585/Qatar:_The_cost_of_growth.html, 11.05.2009.
Foster + Partner. http://www.fosterandpartners.com/Projects/1461/Default.aspx, 11.05.2009.
Fujairah Paradise. http://www.fujairahparadise.com/inde.htm, 11.05.2009.
Futtaim. http://www.futtaim.com/content/companyprofile.asp?profileid=1325, 05.01.2009.
Gafoor, Abdul. 1995: Islamic Banking. http://users.bart.nl/~abdul/chap4.html, 19.10.2008.
Gardens. http://www.thegardens.ae/thegardens.html, 04.01.2009.
GIS Development. http://www.gisdevelopment.net/proceedings/mest/2007/Papers/day3/P51.pdf, 19.02.2009.
Global Village. http://www.globalvillage.ae/AboutUs/Facts.aspx, 10.01.2009.
Global Security. http://www.globalsecurity.org/military/facility/manama.htm, 22.02.2009.
Go Wealthy. http://www.gowealthy.com/gowealthy/wcms/en/home/real-estate/uae/dubai/dubai-land/al-barari/index.html, 18.01.2009.
Guggenheim. http://www.guggenheim.org/press_releases/release_159.html, 21.10.2008.
Gulf Base. http://www.gulfbase.com/site/interface/NewsArchiveDetails.aspx?n=60483, 31.10.2008.
Gulf Daily News. http://www.gulf-daily-news.com/NewsDetails.aspx?storyid=211302, 30.04.2009.
Gulf Daily News. http://www.gulf-daily-news.com/NewsDetails.aspx?storyid=230865, 05.05.2009.
Gulf News. http://www.gulfnews.com/articles/04/09/25/133296.html, 04.01.2009.
Gulf News. http://www.gulfnews.com/business/General/10256900.html, 04.01.2009.
Gulf News. http://archive.gulfnews.com/gnfocus/gibtm_exhibition2007/more_stories/10109162.html, 25.03.2009.
Gulf News. http://archive.gulfnews.com/uae/ajman/more_stories/32581.html, 11.05.2009.
Gulf News. http://archive.gulfnews.com/articles/06/10/06/10072630.html, 22.01.2009.
Gulf News. http://archive.gulfnews.com/indepth/creekextension/more_stories/10123188.html, 05.01.2009.

Gulf Weekly. http://www.gulfweekly.com/article.asp?Sn=6446&Article=21894, 22.04.2009.

Halcrow. http://www.halcrow.com/html/our_projects/projects/yas_island.htm, 11.05.2009.

Independent. http://www.independent.co.uk/news/media/aljazeera-the-new-power-on-the-small-screen-512562.html, 21.10.2008.

Index Mundi. http://indexmundi.com/g/g.aspx?v=115&c=ba&l=en, 30.04.2009.

Index Mundi. http://www.indexmundi.com/energy.aspx?country=om&product=oil&graph=prod uction+consumption, 19.10.2008.

Inhabitat. http://www.inhabitat.com/2008/03/10/ras-al-khaimah%E2%80%99s-gateway-city-to-rival-masdar/, 11.05.2009.

International City. http://www.internationalcity.ae/en/the-city/facts-and-figures.html, 08.01.2009.

International Freight Week. http://www.internationalfreightweek.com/upl_images/news/GCClo-gisticsmarkettodoublegrowthonbackofenergyboom26May08UAE.pdf, 14.01.2009.

IRIS. http://www.iriswll.com/bahrain-real-estate-projects.php, 16.04.2009.

IUE. http://www.iue.it/RSCAS/RestrictedPapers/conmed2003free/200303Khalaf05.pdf, 16.02.2009.

Islamic-banking. http://www.islamic-banking.com/shariah/index.php, 19.10.2008.

Jumeirah Golf Estates. http://www.jumeirahgolfestates.com/en/section/selling-now/faq/general, 13.01.2009.

Jumeirah Heights. http://www.jumeirahheights.com/en/master-plan/, 03.01.2009.

Jumeirah Islands. http://www.jumeirahislands.com/en/overview/jumeirah-islands-fact-sheet.html, 04.01.2009.

Jumeirah Lake Towers. http://www.jumeirahlaketowers.ae/jlt_launch/about_jlt.aspx, 04.01.2009.

Jumeirah Park. http://www.jumeirahpark.com/, 04.01.2009.

Jumeirah Village. http://www.jumeirahvillage.com/facts.php, 04.01.2009.

Khaleej Times. http://www.khaleejtimes.com/DisplayArticle.asp?xfile=data/business/2005/De-cember/business_December176.xml&section=business&col=, 05.01.2009.

Kieferle + Partner. http://www.kieferle-partner.com/news_5423_20080808489c0f1ab285e-Jew-el-of-the-Creek.html, 06.01.2009.

King Abdullah City. http://www.kingabdullahcity.com/en/AboutKAEC/Visionaries.html, 24.08.2009.

King Abdullah City. http://www.kingabdullahcity.com/en/CityInProgress/CityPhases.html, 24.08.2009.

Kish Trade Promotion Centre. http://www.kishtpc.com/Freetrade%20ZONES.htm, 19.10.2008.

Kuwait Past. http://www.kuwaitpast.com/era.html, 11.05.2009.

Liskw. http://liskw.wordpress.com/2008/07/21/kuwait-university-furute-plans/, 14.06.2009.

Madinat Jumeirah. http://www.madinatjumeirah.com/, 28.10.2008.

Masdar. http://www.masdar.ae/en/Menu/index.aspx?MenuID=48&CatID=27&mnu=Cat, 11.05.2009.

Masdar. http://www.masdar.ae/en/Menu/DescriptionPage.aspx?MenuID=48&Catid=27&cat_id=27&SubcatID=25&mnu=SubCat, 11.05.2009.

Masdar. http://www.masdar.ae/en/mediaCenter/newsDesc.aspx?News_ID=65&fst=mc&nws=1, 11.05.2009.

MEW. http://www.mew.gov.bh/default.asp?action=category&id=34, 19.02.2009.

MIC. http://www.mic.com.qa/mic/web.nsf/web/mic_masterp#, 11.05.2009.

Middle East Electricity. http://www.middleeastelectricity.com/upl_images/news/Jumeirah%20Lake%20Towers%20becomes%20free%20zone%20%20July%2019%202006%20Gulf%20News.pdf, 05.01.2009.

Migration Information. http://www.migrationinformation.org/dataHub/GCMM/Dubaidatasheet. pdf, 30.10.2008.

Mizin. http://www.mizin.ae/arjan.html, 13.01.2009.

Municipality Bahrain. http://websrv.municipality.gov.bh/ppd/doc/rule_buidingregulations_ en.pdf, 26.02.2009.

Municipality Bahrain. http://websrv.municipality.gov.bh/ppd/doc/rule_buidingregulations_ en.pdf, 26.02.2009.

Mubarek, Faisal A. http://faculty.ksu.edu.sa/3177/Documents/Suburbanization%20Draft%204. pdf, 14.10.2008.

Nakheel. http://www.nakheel.com/en/developments, 07.01.2009.

Nakheel Harbour. http://www.nakheelharbour.com/#/faq, 13.01.2009.

Nations Encyclopedia. http://www.nationsencyclopedia.com/economies/Asia-and-the-Pacific/ Bahrain-AGRICULTURE.html, 22.02.2009.

Nations Encyclopedia. http://www.nationsencyclopedia.com/Asia-and-Oceania/Qatar-HISTO-RY.html, 11.05.2009.

NY Times. http://www.nytimes.com/global/dubai/four.html, 08.01.2009.

Oil Drum. http://www.theoildrum.com/story/2006/10/5/215316/408, 19.10.2008.

OMA. http://www.oma.eu/index.php?option=com_projects&view=project&id=443&Itemid=10, 11.05.2009.

Oxford Business Group. http://www.oxfordbusinessgroup.com/country.asp?country=63, 11.05.2009.

Palm Dubai. http://thepalmdubai.com/palm-island-dubai/about-the-palm-islands/news-updates/ palm-jumeirah/06.05.09-Palm-Jumeirah-work-on-schedule.html, 04.01.2009.

Palm Dubai. http://thepalmdubai.com/palm-island-dubai/about-the-palm-islands/news-updates/ palm-jebel-ali/03.05.05-Remarkable-features-unveiled-on-Palm-Jebel-Ali.html, 11.01.2009.

Palm Jumeirah. http://www.palmjumeirah.ae/the-palm-story.php, 02.01.2009.

Pearl Qatar. http://www.thepearlqatar.com/SubTemplate1.aspx?ID=165&MID=115, 11.05.2009.

Pentominium. http://www.pentominium.com/, 04.01.2009.

Population Statistics. http://www.populstat.info/Asia/saudiarc.htm, 14.10.2008.

Property Portal. http://www.propertyportal.ae/alkaheel.php, 11.01.2009.

Purple Journal. http://thepurplejournal.wordpress.com/2008/08/03/dubai-internet-city-the-middle-easts-biggest-it-infrastructure/, 02.01.2009.

QP. http://www.qp.com.qa/qp.nsf/web/bc_mic, 11.05.2009.

Railway Technology. http://www.railway-technology.com/projects/dubai-metro/, 22.01.2009.

Realestate Dubai Property. http://www.realestate-dubai-property.com/property/dubai/impz/development/51.cntns, 13.01.2009.

Realestate Emirates Network. http://realestate.theemiratesnetwork.com/developments/dubai/ dubailand.php, 08.01.2009.

Realestate Emirates Network. http://realestate.theemiratesnetwork.com/developments/abu_dhabi/al_reem_island.php, 09.05.2009.

Realestate Emirates Network. http://realestate.theemiratesnetwork.com/developments/abu_dhabi/shams_abu_dhabi.php, 09.05.2009.

Realestate Emirates Network. http://realestate.theemiratesnetwork.com/developments/abu_dhabi/najmat_abu_dhabi.php, 09.05.2009.

Realestate Emirates Network. http://realestate.theemiratesnetwork.com/developments/umm_al_ quwain/white_bay.php, 11.05.2009.

Realestate Emirates Network. http://realestate.theemiratesnetwork.com/developments/umm_al_ quwain/al_salam_city.php, 11.05.2009.

Realestate Emirates Network. http://realestate.theemiratesnetwork.com/developments/umm_al_

quwain/emirates_modern_industrial_area.php, 11.05.2009.
Realestate Emirates Network. http://realestate.theemiratesnetwork.com/developments/ras_al_khaimah/al_hamra_village.php, 11.05.2009.
Realestate Emirates Network. http://realestate.theemiratesnetwork.com/developments/ras_al_khaimah/the_cove.php, 11.05.2009.
Realestate Emirates Network. http://realestate.theemiratesnetwork.com/articles/freehold_property.php, 11.05.2009.
Realestate Emirates Network. http://realestate.theemiratesnetwork.com/developments/oman/the_wave.php, 23.08.2009.
Realestate Emirates Network. http://realestate.theemiratesnetwork.com/developments/qatar/lusail_city.php, 11.05.2009.
Realestate Emirates Network. http://realestate.theemiratesnetwork.com/developments/ras_al_khaimah/mina_al_arab.php, 11.05.2009.
Realestate Emirates Network. http://realestate.theemiratesnetwork.com/developments/sharjah/nujoom_islands.php, 11.05.2009.
Realestate Emirates Network. http://realestate.theemiratesnetwork.com/articles/freehold_property.php, 11.05.2009.
Realestate Emirates Network. http://realestate.theemiratesnetwork.com/developments/ajman/, 11.05.2009.
Realtyna. http://www.realtyna.com/dubai_real_estate/dubai-investment-park.html, 13.01.2009.
Reef Island. http://www.reef-island.com/main.asp, 27.02.2009.
Riffa Views. http://riffaviews.com/about/, 03.03.2009.
Riffa Views. http://riffaviews.com/signature-estates/lagoons-estate/, 03.03.2009.
Sarayarak. http://www.sarayarak.com/#7, 11.05.2009.
Sheikhhmohammed. http://www.sheikhhmohammed.co.ae, 28.10.2008.
SIC. http://www.sic.ae/SIC_Brocure.pdf, 11.05.2009.
SLRB. http://www.slrb.gov.bh/AboutSLRB/default.aspx?PageId=95&Lnk=Link1, 19.02.2009.
Spiegel. http://www.spiegel.de/reise/aktuell/0,1518,621666,00.html, 16.07.2009.
Statoids. http://www.statoids.com/ubh.html, 22.02.2009.
Sukoon Tower. http://www.sukoontower.com/SukoonTower-Jan16.pdf, 22.04.2009.
Tabreed. http://www.tabreed.org/MediaCenter.aspx?NewsType=News&ID=34, 26.04.2009.
Tameer. http://www.tameer.com/qmr/luxuries-residential.pdf, 17.04.2009.
Tatweerdubai. http://www.tatweerdubai.com/En/cd-2-1, 08.01.2009.
Tatweerdubai. http://www.tatweerdubai.com/En/cd-4-6#, 13.01.2009.
Telegraph. http://www.telegraph.co.uk/property/3344183/Palm-before-a-storm.html, 02.01.2009.
Telegraph. http://www.telegraph.co.uk/travel/travelnews/4380051/Dubais-polluted-beaches-closed-to-public.html, 27.07.2009.
Trade Arabia. http://www.tradearabia.com/NEWS/CONS_135696.html, 19.02.2009.
Transportation. http://www.transportation.gov.bh/en/modules.php?name=Content&pa=showpage&pid=16, 19.02.2009.
Trump Dubai. http://www.trumpdubai.com/, 02.01.2009.
UAE government. http://www.uae.gov.ae/Government/tourism.htm, 21.10.2008.
UAE Interact. http://www.uaeinteract.com/uaeint_misc/pdf/perspectives/11.pdf, 19.10.2008.
UAE Property. http://www.uaepropertytrends.com/ptrends/mvnforum/viewthread?thread=2135, 13.01.2009.
UP. http://www.up.org.qa/upeng/modules.php?name=News&file=article&sid=7, 11.05.2009.
Waterfront. http://www.waterfront.ae/, 12.01.2009.
Water Garden City. http://www.watergardencity.com, 05.05.2009.

Wave Muscat. http://thewavemuscat.com/Development/Developers.aspx, 23.08.2009.

Wave Muscat. http://thewavemuscat.com/Overview/Masterplan.aspx, 23.08.2009.

Weavingideas. http://www.weavingideas.com/en/applications-products/architecture/projects/singleview/seiten/0/mode/1/hsref/aspire-tower.html, 11.05.2009.

Wikipedia. http://de.wikipedia.org/wiki/Islamic_Banking, 19.10.2008.

Wikipedia. http://en.wikipedia.org/wiki/Sana%27a,05.10.2008.

Wikipedia. http://en.wikipedia.org/wiki/Wadi, 05.10.2008.

Wikipedia. http://de.wikipedia.org/wiki/Emirates_Hills, 04.01.2009.

Wikipedia. http://en.wikipedia.org/wiki/Dubai_Media_City, 04.01.2009.

Wikipedia. http://en.wikipedia.org/wiki/Jumeirah_Lake_Towers, 04.01.2009.

Wikisource. http://en.wikisource.org/wiki/Constitution_of_the_Kingdom_of_Bahrain_ (2002)#Section_3_The_Legislative_Authority_National_Assembly, 16.02.2009.

Works Ministry. http://www.works.gov.bh/default.asp?action=article&id=160, 19.02.2009.

Works Ministry. http://www.works.gov.bh/default.asp?action=category&id=14, 19.02.2009.

World. http://www.theworld.ae/mp_infrastructure.html, 07.01.2009.

World. http://www.theworld.ae/mp_islands.html, 07.01.2009.

World Architecture News. http://www.worldarchitecturenews.com/index. php?fuseaction=wanappln.projectview&upload_id=2133, 01.05.2009.

Yas Island. http://yasisland.ae/Default_en_gb.html?1, 11.05.2009.

Zawya. http://www.zawya.com/story.cfm/sidZAWYA20080307082009, 18.01.2009.

Zawya. https://www.zawya.com/story.cfm/sidZAWYA20071212050038/Bahrain:%20King%20 orders%20setting%20up%20of%20Electricity,%20Water%20Authority, 15.02.2009.

Literature:

Adham, Khaled. 2008: Rediscovering the Island: Doha`s Urbanity from Pearls to Spectacle; in Elsheshtawy 2008, 218 – 258.

Al Hathloul, Saleh. 1996: The Arab- Muslim City. Tradition Continuity and Change in the Physical Environment; Riyadh: Dar Al Sahan.

Al Hathloul, Saleh; Mughal, Muhammed Aslam. 2004: Urban growth management the Saudi experience; Habitat International 28 (2004), 609-623.

Al Kalali, Nayef. 2005: Overview of Coastal Reclamation Projects & Plans in the Kingdom of Bahrain; Manama: Ministry of Works and Housing.

Al Mosaind, Musaad. 1998: Freeway traffic congestion in Riyadh, Saudi Arabia. Attitudes and policy implications; in Journal of Transport Geography Vol. 6, No. 4 (1998), 263-272.

Al Naim, Mashari A.. 2008: Identity in transitional context: Open-ended local architecture in Saudi Arabia; in: Archnet-iJAR, Volume 2 – Issue 2, July 2008, 125- 146.

Benton- Short, Lisa; Price, Marie; Friedmann, Samantha. 2005: Globalisation from Below. The Ranking of Global Immigrant Cities; in: International Journal of Urban and Regional Research; 2005, Volume 29.4, 945- 959.

Blum, Elisabeth; Neitzke, Peter. 2009: Dubai – Stadt aus dem Nichts: Ein Zwischenbericht über die derzeit größte Baustelle der Welt; Basel: Birkhäuser Verlag.

Cross Border Legal Publishing FZ LLC. 2007: Property Investment Guide 2007-2008; UAE: Emirates Printing Press.

Davidson, Christopher M.. 2008: Dubai – The vulnerability of success; New York: Columbia University Press.

Davis, M. 2007:Dreamworld of Neoliberalism – Evil Paradises; New York: The New Press.

De Montequin, François-Auguste. 1983: The essence of urban existence in the world of Islam; in: Islamic Architecture and Urbanism: Selected papers from a Symposium by the college of Architecture and Planning, 5-10 January 1980, King Faisal University Dammam.

Diener, C.; Gangler, A..; Fein A.. 2003: Transformationsprozesse in Oasensiedlungen Omans; in: Trialog; 76, 2003, 15- 21.

Elsheshtawy, Yasser. 2008: The Evolving Arab City – Tradition, Modernity & Urban Development; New York: Routledge.

Fadan, Yousef. 1983: Traditional houses of Makka; in: Islamic Architecture and Urbanism: Selected papers from a Symposium by the college of Architecture and Planning, 5-10 January 1980, King Faisal University Dammam.

Garba, Shaibu. 2003: Managing urban growth and development in the Riyadh metropolitan area, Saudi Arabia; in Habitat International 28 (2004), 593-608.

Golf Construction. April 2008: Dubai Review; Manama: Al Hilal Group.

Golf Construction. August 2007: Abu Dhabi; Manama: Al Hilal Group.

Golf Construction. December 2007:Kuwait Review; Manama: Al Hilal Group.

Golf Construction. January 2006: Bahrain Review; Manama: Al Hilal Group.

Golf Construction. January 2007: Bahrain Review; Manama: Al Hilal Group.

Golf Construction. January 2008: Bahrain Review; Manama: Al Hilal Group.

Golf Construction. January 2009: Bahrain Review; Manama: Al Hilal Group.

Golf Construction. March 2009: Qatar Review; Manama: Al Hilal Group.

Hakim, Besim. 2007: Revitalizing traditional towns and heritage districts; in: Archnet-iJAR, Volume 1 – Issue 3, November 2007, 153- 166.

Hamouche, Mustapha B.. 2008: Manama: The Metamorphosis of an Arab Gulf City; in Elsheshtawy 2008, 184 - 217.

Heck, Gerhard. 2004: Dubai. Köln: DuMont Reiseverlag.

Heck, Gerhard; Wöbcke, Manfred. 2005: Arabische Halbinsel. Ostfildern: DuMont Reiseverlag, 5. Auflage.

Konzelmann, Gerhard. 2005: Die Emirate. Das Paradies im Nahen Osten; München: Herbig Verlag.

Korn, Lorenz. 2003: Transformationsprozesse in Oasensiedlungen Omans; in: Trialog; 76, 2003, 27- 30.

Lauber, Wolfgang. 2003: Angepasstes Bauen in trockenheißen Klima; in: Trialog; 76, 2003, 36- 39.

Mahgoub, Yasser. 2008: Kuwait: Learning from a Globalized City; in Elsheshtawy 2008, 152 – 284.

Meinel, Ute Devika. 2003: Die Intifada im Ölscheichtum Bahrain. Münster: Lit Verlag.

Melamid, Alexander. 1980: Urban Planning in Eastern Arabia; in: Geographical Review; Oct., 1980, 473-477.

Melamid, Alexander. 1989: Dubai City; in: Geographical Review; 1989, 3, Volume 79, 345- 347.

Ministry of Housing. 1996: General Report on Housing and Urban Development in Bahrain. Istanbul: The United Nations Conference on Human Settlements – Habitat II, City Summit.

Moutamerat. 2007: Al Manakh – International Design Forum Dubai; Amsterdam: Idea Books.

Mubarek, Faisal. 2004: Urban growth boundary policy and residential suburbanisation. Riyadh, Saudi Arabia; in Habitat International 28 (2004), 567-591.

Nabi, Mohammed Nurun. 2000. Rapid Urban Expansion and Physical Planning in Bahrain. Manama: Ministry of Housing.

OMS Advertising L.L.C. 2007: Property Guide a to z. Dubai: Raidy Emirates Printing Group.

Pacione, Michael. 2005: City Profile Dubai; in: Cities; 3, 2005, Volume 22, 255- 265.

Pape, Heinz. 1977: Er Riad. Stadtgeografie und Stadtkartografie der Hauptstadt Saudi-Arabiens; Paderborn: Schöningh Verlag.

Putz, Ulrike; Stieber, Benno 2006: Dubai. VAE und Oman; in: Merian; 2006, 5.

Reichert, Horst. 1978: Die Verstädterung der Eastern Provinz von Saudi Arabien, Dissertation am Institut für Städtebau, Universität Stuttgart.

Sayrafi, Yousef. 1981: Islam versus Planung? Situation der staatlichen und örtlichen Planung in Saudi-Arabien, Dissertation an der Fakultät für Bauwesen, Technische Hochschule Aachen.

Sassen, Saskia. 1997: Metropolen des Weltmarkts. Frankfurt/Main: Campus Verlag; 2. Auflage.

Schmid, Heiko. 2008: Ökonomie der Faszination – Dubai und Las Vegas als Beispiele inszenierter Stadtlandschaften; Habilitation an der Fakultät für Chemie und Geowissenschaften, Ruprecht-Karls-Universität Heidelberg.

Scholz, Fred. 1999: Die kleinen Golfstaaten. Gotha: Justus Perthes Verlag Gotha GmbH, 2. Auflage.

Schrammel, J. 1993: Orientalisch-Islamische Stadtstruktur; in Reader SIAAL; WS 92/93, 17-23.

Skidmore, Owings and Merril. 2006: Bahrain National Plan – Draft National Framework Plan; Manama: Economic Development Board.

Storch, Christian. 2009. Bahrain National Planning Development Strategy; in: Dubai Magazin; 2/09.

Wirth, Eugen. 1988: Dubai: Ein modernes städtisches Handels- und Dienstleistungszentrum am Arabisch-Persischen Golf. Erlangen: Selbstverlag der Fränkischen Geographischen Gesellschaft.

Interviews:

Ahmad, Khaled Galal. Assistant Professor, Architecture Department, United Arab Emirates University. Emirate of Abu Dhabi, UAE.

Al Asfoor, Saba. Technical Services Director, Northern Area Municipality. Kingdom of Bahrain.

Al Ansari, Fuad. Teaching Assistant, University of Bahrain. Kingdom of Bahrain.

Al Bardooli, Adham. Engineer, Al Areen Holding Company. Kingdom of Bahrain.

Al Gaoud, Shaheera. Public Relations Coordinator, Durrat Al Bahrain. Kingdom of Bahrain.

Al Ghatam, Muneera Khalifa. Head of Mapping, Topographic Survey Directorate. Kingdom of Bahrain.

Al Ghatam, Wafa Rahman. Teaching Assistant, University of Bahrain. Kingdom of Bahrain.

Al Ghazal, Ali. Urban planner, Urban Planning Affairs, Ministry of Municipalities & Agriculture. Kingdom of Bahrain.

Al Jowder, Ahmed Abdul rahman. Head of Planning Studies Section, Ministry of Municipalities & Agriculture. Kingdom of Bahrain.

Al Kooheji, Hamad. Sales Executive, Durrat Al Bahrain. Kingdom of Bahrain.

Al Kubaisy, Falah. Research & Development Advisor, Ministry of Municipalities & Agriculture. Kingdom of Bahrain.

Al Moayyed, Maamoon. Chief Executive Officer, Edameh. Kingdom of Bahrain.

Al Pachachi, Nouf Saleh. Marketing Executive, Ossis Property developers BSC. Kingdom of Bahrain.

Al Saed, Majeed. Diector, City Engineering. Kingdom of Bahrain.

Al Sawad, Khairya Isa. Architect, Northern Area Municipality. Kingdom of Bahrain.

Al Watani, Abbas. Advisor, Housing Projects Directorate, Ministry of Housing. Kingdom of Bahrain.

Alameh, Rabih. Sales Director, United Development Company. Doha, Qatar.

Aliar, Thajudeen. Journalist, Dubai Municipality, Public Relation & Organization Department. Emirate of Dubai, UAE .

Alraouf, Ali. Assistant Professor, University of Bahrain. Kingdom of Bahrain.

Alwadi, Aziz Mohamed. Senior General Engineer, Northern Area Municipality. Kingdom of Bahrain.

Ameen, Feras Abbas. Head of Regional Planning Department, Ministry of Municipalities & Agriculture. Kingdom of Bahrain.

Awad, Jihad. Assistant Professor, Faculty of Engineering, Ajman University of Science & Technology Network. Emirate of Ajman, UAE.

Chojnacki, Wieslaw. Senior Project Manager, Housing and Urban Development Committee. Kingdom of Bahrain.

Dayaratne, Ranjith. Assistant Professor, University of Bahrain. Kingdom of Bahrain.

De, Prasanta. Urban planner, Surpreme Committee for Town Planning. Muscat, Sultanate of Oman.

Doughty, Brett. Project Director, Hyder. Kingdom of Bahrain.

El Aramouni, Anis. Communication Coordinator, United Development Company. Doha, Qatar.

El Gritly, Nagy Mahmud. Urban Planner, Urban Planning & Development Authority. Doha, Qatar.

El Masri, Soheil. Director, Architectural Design, Gulf House Engineering. Kingdom of Bahrain.

Hadi, Waheed Ahmed. Director, Topographic Survey Directorate. Kingdom of Bahrain.

Hamouche, Mustafa. Assoc. Professor, University of Bahrain. Kingdom of Bahrain.

Husnein, Adnan. Assistant Professor, Urban Planning Department, Al Hosn University. Emirate of Abu Dhabi, UAE.

Isa, Mohamed Abdulla. Executive Director, Tameer, Al Khaleej Development Company. Kingdom of Bahrain.

Karaan, Elias. General Manager, Lulu Tourism Company BSC. Kingdom of Bahrain.

Lunt, Robert. Director Special Projects, Blue City Oman. Sultanate of Oman.

Matar, Ali Hussein. Senior Civil Engineer, Housing and Urban Development Committee. Kingdom of Bahrain.

MacDonald, Robert. Senior Design Manager, Blue City Oman. Sultanate of Oman.

Madibo, Ali Mousa. Urban planner, Arab Urban Development Institute. Riyadh, Saudi Arabia.

McPolin, Dominic. Executive Coordinator, Central Planning Unit. Ministry of Works. Kingdom of Bahrain.

Montague, Simon. Planning Director, Cracknell, Emirate of Dubai. UAE.

Müller, Rudi. Vice President, Design & TIO, Bahrain Bay Development BSC. Kingdom of Bahrain.

Nabi, Mohammed Nurun. Advisor, Urban Planning Affairs, Ministry of Municipalities & Agriculture. Kingdom of Bahrain.

Nouri, Mohammed. Project Director, Ossis Property developers BSC. Kingdom of Bahrain.

Schmid, Heiko. Lecturer, Ruprecht-Karls-University of Heidelberg. Heidelberg, Germany.

Shamma, Kareem. Chief Executive Officer, Manara Developments. Kingdom of Bahrain.

Tan, Tomas. Senior Project Manager, Housing and Urban Development Committee. Kingdom of Bahrain.

www.ingramcontent.com/pod-product-compliance
Lightning Source LLC
Chambersburg PA
CBHW021031210326
41598CB00016B/978